A First Course in Group Theory

Bijan Davvaz

A First Course in Group Theory

 Springer

Bijan Davvaz
Department of Mathematics
Yazd University
Yazd, Iran

ISBN 978-981-16-6367-3   ISBN 978-981-16-6365-9   (eBook)
https://doi.org/10.1007/978-981-16-6365-9

This Springer imprint is published by the registered company Springer Nature Singapore Pte Ltd.
The registered company address is: 152 Beach Road, #21-01/04 Gateway East, Singapore 189721,
Singapore

Preface

The aim of this book is to provide a readable account of the examples and fundamental results of groups from a theoretical point of view and a geometrical point of view. The concept of a group is one of the most fundamental in modern mathematics. Groups are systems consisting of a set of elements and a binary operation that can be applied to two elements of the set, which together satisfy certain axioms. These require that the group is closed under the operation (the combination of any two elements produces another element of the group), that it obey the associative law, that it contains an identity element (which, combined with any other element, leaves the latter unchanged), and that each element have an inverse (which combines with an element to produce the identity element).

Since the book does not make any assumptions about the reader's background, it is suitable for newcomers to group theory and even those who never studied algebra. To explain many subjects we used figures and images to help the readers.

The book is organized into eleven chapters. To get to any depth in group theory requires set theory, combinatorics, number theory, matrix theory, and geometry. Thus, not only we have devoted Chaps. 1–2 to the introductory concepts of set theory, combinations, number theory, and symmetry, but we have also devoted some parts of Chap. 7 to matrix theory. Although the main subject of the book is groups but in many parts we will need some information about other algebraic structures, like rings, fields and vector spaces. In Chap. 3, we give the definitions of group and subgroup, examples and some elementary properties. Several excellent and important examples of groups like cyclic groups, permutation groups, group of arithmetical functions, matrix groups and linear groups are investigated in Chaps. 4–7. In Chap. 8, Lagrange's theorem as the most important theorem in finite groups is discussed. Chapters 9–11 address the normal subgroups, factor groups, derived subgroup, homomorphism, isomorphism and automorphism of groups. A consequence is Cayley's theorem in which every group could be realized as a permutation group.

Some chapters or sections are labelled as optional; this means that the readers can ignore them for the first study. Each section ends in a collection of exercises. The purpose of these exercises is to allow students to test their assimilation of the material, to challenge their knowledge and ability. Moreover, at the end of each chapter we have two special sections: (1) Worked-Out Problems; (2) Supplementary Exercises.

In each worked-out problems section, I decided to try teach by examples, by writing out to solutions to problems. In choosing problems, three major criteria have been considered, to be challenging, interesting and educational. Moreover, there are many exercises at the end of each chapter as supplementary exercises. They are harder than before and serve to present the interesting concepts and theorems which are not discussed in the text.

The list of references at the end of the book is confined to works actually used in the text.

Yazd, Iran   Bijan Davvaz

Contents

About the Author

Bijan Davvaz is Professor at the Department of Mathematics, Yazd University, Iran. Earlier, he served as the Head of the Department of Mathematics (1998–2002), Chairman of the Faculty of Science (2004–2006), and Vice-President for Research (2006–2008) at Yazd University, Iran. He earned his Ph.D. in Mathematics with a thesis on "Topics in Algebraic Hyperstructures" from Tarbiat Modarres University, Iran, and completed his M.Sc. in Mathematics from the University of Tehran, Iran. His areas of interest include algebra, algebraic hyperstructures, rough sets and fuzzy logic. On the editorial boards for 25 mathematical journals, Prof. Davvaz has authored 6 books and over 600 research papers, especially on algebra, fuzzy logic, algebraic hyperstructures and their applications.

Notation

$G, H, \ldots$	sets, groups, $\ldots$
$\in, a \in A$	membership
$\emptyset$	empty set
$\{x \mid p(x)\}$	set of all x that $p(x)$
$A \subseteq B$	A is a subset of B
$A \subset B$	A is a proper subset of B
$A_1 \cap A_2 \cap \cdots \cap A_n$ and $\bigcap_{i \in I} A_i$	intersection of sets
$A_1 \cup A_2 \cup \cdots \cup A_n$ and $\bigcup_{i \in I} A_i$	union of sets
$A - B$	difference set of A and B
$\mathbb{N}$	natural numbers
$\mathbb{Z}$	integers, additive group and ring of integers
$\mathbb{Q}$	field of rational numbers
$\mathbb{R}$	field of real numbers
$\mathbb{C}$	field of complex numbers
$\mathbb{F}$	field
$\mathbb{F}^*$	if $\mathbb{F}$ is a field, $\mathbb{F}^* = \mathbb{F} - \{0\}$
$R[x]$	ring of polynomials over a ring R
$\mathbb{Z}_n$	ring and group of integers modulo n
U_n	group of units modulo n
$a|b$	a divides b
$a \nmid b$	a does not divide b
$a \equiv b \,(\mathrm{mod}\, n)$	congruence modulo n
(a, b)	greatest common divisor of integers a and b
$f : A \to B$	a function of A to B
$f(a)$	image of a under f
$f(A)$	image of A under f
Imf	image of f
$f^{-1}(B)$	inverse image of B under f
id_X	the identity function $X \to X$
φ	Euler function
μ	Möbius function

f^{-1}	the inverse of the function f
$f \circ g$	composite function
$\binom{n}{k}$	binomial coefficient $n!/k!(n-k)!$
$|A|$	number of elements in a finite set A
$|G|$	order of the group G
$\circ(a)$	order of element a
$H \leq G$	H is a subgroup of G
$H \trianglelefteq G$	H is a normal subgroup of G
a^{-1}	the inverse of a
$Z(G)$	center of G
$C_G(a)$	centralizer of a in G
$C_G(X)$	centralizer of the set X in G
$N_G(X)$	normalizer of the X in G
$\langle a \rangle$	cyclic group generated by a
$\langle X \rangle$	subgroup generated by a set X
AB	$\{ab \mid a \in A \text{ and } b \in B\}$
$A = (a_{ij})$	matrix whose entry in row i and column j is a_{ij}
$\det(A)$	determinant of matrix A
I_n	the identity $n \times n$ matrix
A^t	transpose of the matrix A
$tr(A)$	trace of the matrix A
$Mat_{m \times n}(\mathbb{F})$	the set of all $m \times n$ matrices over $\mathbb{F}$
aH	left coset
Ha	right coset
G/H	factor group
$[G : H]$	index of the subgroup H in the group G
$G \cong H$	G is isomorphic to H
C_n	cyclic group of order n
S_X	symmetric group on X
S_n	symmetric group of degree n
A_n	alternating group
D_n	dihedral group of order $2n$
Q_8	quaternion group
$GL_n(\mathbb{F})$	general linear group
$SL_n(\mathbb{F})$	special linear group
$O_n(\mathbb{F})$	orthogonal group
$SO_n(\mathbb{F})$	special orthogonal group
$UT_n(\mathbb{F})$	the set of upper triangular matrices such that all entries on the diagonal are non-zero
$LT_n(\mathbb{F})$	the set of lower triangular matrices such that all entries on the diagonal are non-zero
$\langle X, Y \rangle$	inner product of two vectors
$[x, y]$	$x^{-1}y^{-1}xy$

G'	derived subgroup of a group G
$Ker f$	Kernel of the homomorphism f
$End(G)$	the set of all endomorphism from G to itself
$Aut(G)$	automorphism group
$Inn(G)$	inner automorphism group

Chapter 1
Preliminaries Notions

The concept of a set is used in various disciplines, particularly in group theory. In this chapter we review elementary set theory. The concepts of relations, partitions, functions, and ordered sets are presented after a discussion of the algebra of sets. Many problems in group theory require that we count the number of ways that a particular event can occur, so we study the topics of permutations and combinations. Moreover, much of group theory involves properties of integers. For this reason the last section deals with the properties of integers. We collect the ones we need for future reference such as divisibility, prime numbers, and congruences.

1.1 Sets and Equivalence Relations

The notion of set is so simple that it is usually introduced informally and regarded as self-evident. By a *set* we mean a collection of objects characterized by some defining property that allows us to think of the objects as a whole entity. The defining property has to be such that it must be clear whether the given objects belong to the set or not. The objects possessing the property are called *elements* or *members* of the set. We denote sets by common capital letters A, B, C, etc. and elements of the sets by lower case letters a, b, c, etc. If S is a set (whose elements maybe numbers or any other objects) and x is an element of S, then we write $x \in S$. If it so happens that x is not an element of S, then we write $x \notin S$. For a property p and an element x of a set S, we write $p(x)$ to indicate that x has the property p. A set may be defined by a property. The notation $A = \{x \in S \mid p(x)\}$ indicates that the set A consists of all elements x of S having the property p. The *empty set* is the set having no elements which will be denoted by the symbol $\emptyset$. In set theory there are several fundamental operations for constructing new sets from given sets.

© The Author(s), under exclusive license to Springer Nature Singapore Pte Ltd. 2021
B. Davvaz, *A First Course in Group Theory*,
https://doi.org/10.1007/978-981-16-6365-9_1

Definition 1.1 Suppose that A and B are two sets. We say A is a *subset* of B and we write $A \subseteq B$ if every element of A is an element of B.

If $A \subseteq B$, then we also say A is included in B or that B includes A. Thus, $A = B$ if and only if $A \subseteq B$ and $B \subseteq A$. Note that any set is a subset of itself. At the other extreme, $\emptyset$ is a subset of every set.

Definition 1.2 Let X be a set and A be a subset of X.

(1) The *union* of A and B, denoted by $A \cup B$, is the set of all elements x such that x belongs to at least one of the two sets A or B.
(2) The *intersection* of A and B, denoted by $A \cap B$, is the set of all elements x which belongs to both A and B.
(3) The *difference set* of A and B, denoted by $A \setminus B$, is the set of all elements x which belongs to A and not in B.
(4) The *complement* of A respect to X refers to elements of X not in A and denoted by A^c.

Two sets are said to be *disjoint sets* if they have no elements in common. Equivalently, two disjoint sets are sets whose intersection is the empty set.

Definition 1.3 A set which contains only finitely many elements is called a *finite set*. A set is *infinite* if it is not finite.

The number of elements in a finite set A is denoted by $|A|$. The study of operations of union, intersection and difference, together with inclusion relation, goes with the name of *algebra of sets*. The following identities, which holds for any sets, are some of the elementary facts of the algebra of sets. It is routine to check that the following identities, for any three sets A, B, and C:

(1) $A \cup B = B \cup A$,
(2) $A \cap B = B \cap A$,
(3) $A \cup (B \cup C) = (A \cup B) \cup C$,
(4) $A \cap (B \cap C) = (A \cap B) \cap C$,
(5) $A \cup (B \cap C) = (A \cup B) \cap (A \cup C)$,
(6) $A \cap (B \cup C) = (A \cap B) \cup (A \cap C)$,
(7) $(A \cup B)^c = A^c \cap B^c$,
(8) $(A \cap B)^c = A^c \cup B^c$.

The equalities (7) and (8) are called *De Morgan's Laws*. Let A and B be two sets. The set of all *ordered pair* (a, b) with $a \in A$ and $b \in B$ is called the *Cartesian product* of A and B and is denoted by $A \times B$. In symbols $A \times B = \{(a, b) \mid a \in A, \ b \in B\}$. For the ordered pair (a, b), a is called the *first coordinate* and b is called the *second coordinate*. For any three sets A, B, and C, we have

(1) $A \times (B \cap C) = (A \times B) \cap (A \times C)$,   (2) $A \times (B \cup C) = (A \times B) \cup (A \times C)$.

The simplest numbers are the positive numbers, 1, 2, 3, and so on, used for counting. These are called *natural numbers*. The set of natural numbers is denoted by $\mathbb{N}$. The *whole numbers* are the natural numbers together with 0. The *integer set* includes

Fig. 1.1 A positive length
whose square is 2

Fig. 1.2 Venn diagram for
well-known number sets

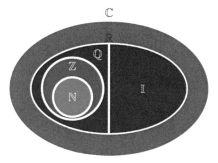

whole numbers and negative whole numbers. Integers can be positive, negative, or zero. The set of integers is denoted by $\mathbb{Z}$. A rational number is any number that can be expressed as the quotient or fraction a/b of two integers, a numerator a and a non-zero denominator b. The set of all rational numbers is denoted by $\mathbb{Q}$. Now, consider a square with sides having length one, see Fig. 1.1. If d represents the length of the diagonal, then from the geometry we know that $1^2 + 1^2 = d^2$, i.e., $d^2 = 2$. This means that there is a positive length whose square is 2, which we write as $\sqrt{2}$. But $\sqrt{2}$ cannot be a rational number. The fact that $\sqrt{2}$ is not rational is proved indirectly, by showing that the assumption that $\sqrt{2}$ is rational leads to a contradiction. It is evident that there are a lot of rational numbers, yet there are "gaps" in $\mathbb{Q}$. Numbers that are not rational are called *irrational numbers*.

The set of all irrational numbers is denoted by $\mathbb{I}$. These are the numbers that we normally use and apply in real-world applications. The set of all real numbers is denoted by $\mathbb{R}$. A *complex number* is formed by adding a real number to a real multiple of i, where $i = \sqrt{-1}$. Thus, we could write $\mathbb{C} = \{a + bi \mid a, b \in \mathbb{R}\}$, the set of complex numbers. Clearly, we have $\mathbb{N} \subseteq \mathbb{Z} \subseteq \mathbb{Q} \subseteq \mathbb{R} \subseteq \mathbb{C}$ and $\mathbb{I} \subseteq \mathbb{R}$. Pictorially we can illustrate these inclusions by a Venn diagram (see Fig. 1.2).

It is quite possible that elements of a set may themselves be sets. For example, the set of all subsets of a given set A has sets as its elements. This set is called the *power set* of A and is denoted by $\mathcal{P}(A)$. When dealing with sets whose elements are themselves sets, it is fairly a common practice to refer them as *family of sets*; however, this is not a definition. Let I be a set and assume that with each element $i \in I$ there is associated a set A_i. The family of all such sets A_i is called an *indexed family of sets* indexed by the set I and is denoted by $\{A_i \mid i \in I\}$. We can extend the notion of union and intersection to arbitrary family of sets.

Definition 1.4 Let $\mathcal{F}$ be a family of sets.

(1) The *union* in $\mathcal{F}$ denoted by $\bigcup\limits_{A \in \mathcal{F}} A$ is the set of all elements that are in A for some $A \in \mathcal{F}$. That is $\bigcup\limits_{A \in \mathcal{F}} A = \{x \in U \mid x \in A \text{ for some } A \in \mathcal{F}\}$.

(2) The *intersection* in $\mathcal{F}$ denoted by $\bigcap\limits_{A \in \mathcal{F}} A$ is the set of all elements that are in A for all $A \in \mathcal{F}$. That is $\bigcap\limits_{A \in \mathcal{F}} A = \{x \in U \mid x \in A \text{ for all } A \in \mathcal{F}\}$.

Let X and Y be sets, not necessary distinct, and let R be a relation from X to Y. Then, the *inverse* R^{-1} of the relation R is the relation from Y to X such that $R^{-1} = \{(y, x) \mid (x, y) \in R\}$. The *domain* of R denoted by $Dom(R)$ is the set $Dom(R) = \{x \in X \mid (x, y) \in R \text{ for some } y \in Y\}$. The *image* of R, denoted by $Im(R)$, is the set $Im(R) = \{y \in Y \mid (x, y) \in R \text{ for some } x \in X\}$.

Definition 1.5 A relation R on a set X is called an *equivalence relation* if it is *reflexive*, *symmetric*, and *transitive*, i.e., for all $x, y, z \in X$

(1) $x R x$ (reflexive),
(2) $x R y$ implies $y R x$ (symmetric),
(3) $x R y$ and $y R z$ imply $x R z$ (transitive).

Example 1.6 (1) Let $X = \mathbb{R}$ and define the *square relation* $R = \{(x, y) \mid x^2 = y^2\}$. The square relation is an equivalence relation.

(2) Let R be the relation on $\mathbb{Z} \times \mathbb{Z}$ such that $(x, y) R(z, t)$ if and only if $x + t = y + z$. The relation R is an equivalence relation.

Example 1.7 (1) Let P denote the set of people. Let R be the relation on P such that for all $x, y \in P$ is defined as $x R y$ if and only if x is the sister of y. Then R is not an equivalence relation.

(2) The relation "is less than or equal to", denoted "$\leq$", is not an equivalence relation on the set of real numbers. For any $x, y, z \in \mathbb{R}$, "$\leq$" is reflexive and transitive but not necessary symmetric.

Given a non-empty set X, there always exist at least two equivalence relation on X. One of these relations is $R = X \times X$ and another one is the *diagonal relation* Δ_X (also called the *identity relation*) defined by $\Delta_X = \{(x, x) \mid x \in X\}$ which relates every element with itself. Pictorially, if X is represented as a line interval, then $X \times X$ is a square and Δ_X is the main diagonal of the square (see Fig. 1.3). The relation Δ_X is the smallest equivalence relation on X, whereas $X \times X$ is the largest.

Fig. 1.3 Diagonal relation

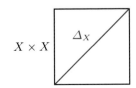

Fig. 1.4 Equivalence classes
of the relation R

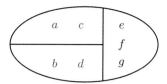

If R is an equivalence relation on X, we define the *equivalence class* $x \in X$ to be the set $[x] = \{y \in X \mid xRy\}$. We write X/R to show the set of all equivalence classes. The procedural version of this definition is

$$\forall y \in X, \ y \in [x] \iff xRy.$$

Example 1.8 Let $X = \{a, \ b, \ c, \ d, \ e, \ f, \ g\}$ and define the relation R on X as follows:

$$R = \{(a,a), \ (b,b), \ (c,c), \ (d,d), \ (e,e), \ (f,f), \ (g,g), \ (h,h), \ (a,c),$$
$$(c,a), \ (b,d), \ (d,b), \ (e,f), \ (f,e), \ (e,g), \ (g,e), \ (f,g), \ (g,f)\}.$$

Then, R is an equivalence relation. We can easily draw a picture illustrating how the equivalence classes divided X into three pieces. In Fig. 1.4, we see that $X/R = \left\{ \{a,c\}, \ \{b,d\}, \ \{e,f,g\} \right\}$.

Theorem 1.9 *Let R be an equivalence relation on a set X. Then the following are equivalent.*

(1) xRy *(the elements are related);*
(2) $[x] = [y]$ *(the classes intersect non-trivially);*
(3) $[x] \cap [y] \neq \emptyset$ *(the classes are equal).*

Proof $(1 \Rightarrow 2)$: Assume that xRy. Let $c \in [x]$ be an arbitrary element. Then, we have xRc. Since xRy and R is symmetric, it follows that yRx. Moreover, since yRx and xRc, it follows that yRc. This yields that $c \in [y]$. So, we conclude that $[x] \subseteq [y]$. Similarly, we can show that $[y] \subseteq [x]$. Consequently, we have $[x] = [y]$.
$(2 \Rightarrow 3)$: Since $x \in [x]$, it follows that $[x]$ is non-empty, so the result is obvious.
$(3 \Rightarrow 1)$: Suppose that $[x] \cap [y] \neq \emptyset$. Then, there exists $c \in [x] \cap [y]$. This implies that xRc and yRc. By the symmetry and transitivity properties, we obtain xRy. ∎

Corollary 1.10 *If X is a set, R is an equivalence relation on a set X, and x and y are elements of X, then either $[x] \cap [y] = \emptyset$ or $[x] = [y]$.*

A partition of a set X is a finite or infinite family of non-empty, mutually disjoint subsets whose union is X.

Fig. 1.5 A partition of the
set X

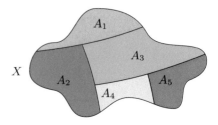

Definition 1.11 A *partition* of a set X is a family of disjoint non-empty subsets of X that have X as their union. In other words, the family of subsets $A_i, i \in I$ (where I is an index set) forms a partition of X if and only if

(1) $A_i \neq \emptyset$, for each $i \in I$;
(2) $A_i \cap A_j = \emptyset$ when $i \neq j$;
(3) $\bigcup_{i \in I} A_i = X$.

Example 1.12 Figure 1.5 shows that the family A_1, A_2, A_3, A_4, and A_5 of subsets of X forms a partition of X.

Theorem 1.13 *Let R be an equivalence relation on a non-empty set X. Then, the family of equivalence classes, X/R, forms a partition of X.*

Proof Clearly, $X/R = \{[x] \mid x \in X\}$ is a family of non-empty subsets of X. Next, if $[x] \neq [y]$, then by Corollary 1.10, we have $[x] \cap [y] = \emptyset$. Finally, we have to show that $\bigcup_{x \in X} [x] = X$. This is also trivial, because each $x \in X$ belongs to $[x]$. This completes the proof.   ■

Theorem 1.14 *Given a partition $\{A_i \mid i \in I\}$ of the subsets of X, then there is an equivalence relation R that has the sets A_i ($i \in I$) as its equivalence classes.*

Proof We define the relation R on X by xRy if and only if there exists $i \in I$ such that $x, y \in A_i$. Since each $x \in X$ is contained in A_i, for some $i \in I$, it follows that xRx, i.e., R is reflexive. The symmetry of R is a result of the definition. In order to prove the transitivity, assume that x, y, and z are three arbitrary elements of X satisfying xRy and yRz. Then, by the above definition, there exist $i, j \in I$ such that $x, y \in A_i$ and $y, z \in A_j$. This implies that $y \in A_i \cap A_j$. So, we conclude that $A_i = A_j$. Consequently, $x, z \in A_i$, and hence xRz. Therefore, R is an equivalence relation on X.

Now, let $x \in X$ be an arbitrary element. Then, there is one and only one $i \in I$ such that $x \in A_i$. Consequently, we have $[x] = A_i$. So, we showed that each equivalence class in X/R is a set in the family $\{A_i \mid i \in I\}$. Conversely, suppose that A_i is any set in the partition. Since A_i is non-empty, it follows that there is $x \in X$ such that $x \in A_i$. This yields that $A_i = [x]$.   ■

Exercises

1. Prove that the De Morgan rules for $\{A_i \mid i \in I\}_{\gamma \in \Gamma}$, a family of subsets of U, namely

   (a) $\left(\bigcup_{i \in I} A_i \right)^c = \bigcap_{i \in I} A_i^c$;

   (b) $\left(\bigcap_{i \in I} A_i \right)^c = \bigcup_{i \in I} A_i^c$.

2. Let A, B, and C be sets. Prove that

   (a) $A \setminus (B \cup C) = (A \setminus B) \cap (A \setminus C)$;

   (b) $A \setminus (B \cap C) = (A \setminus B) \cup (A \setminus C)$.

3. Let $A = \{1, 2, 3, \ldots, 10\}$. Prove that if we select any 20 ordered pairs from $A \times A$, then we can always find two of the chosen pairs that give the same sum when the numbers within the pair are added together.

4. How many relations are there on a set with n elements?

5. Let R be a relation from X to Y and let A be any subset of X. Define $R(A)$, called R-image of A by $R(A) = \{y \in Y \mid aRy, \text{ for some } a \in A\}$. If A and B are subsets of X, prove that

   (a) $R(A \cup B) = R(A) \cup R(B)$;

   (b) $R(A \cap B) \subseteq R(A) \cap R(B)$.

6. Let X be the set of all lines. For any $x, y \in X$ define xRy if and only if x is perpendicular to y. Show that R is symmetric but neither reflexive nor transitive.

7. Let $X = \{a, b, c\}$.

   (a) Find all relations on X;

   (b) Find all equivalence relations on X;

   (c) Find all those relations on X which are reflexive but neither symmetric nor transitive;

   (d) Find all those relations on X which are symmetric but neither reflexive nor transitive;

   (e) Find all those relations on X which are transitive but neither reflexive nor symmetric;

   (f) Find all those relations on X which are neither reflexive, nor symmetric, nor transitive.

8. For each real numbers a and b, define aRb if $a - b$ is an integer. Show that R is an equivalence relation on the set of real numbers.

9. For each integers a and b, define aRb if $ab \geq 0$. Is R an equivalence relation on $\mathbb{Z}$?

10. Let $X = \mathbb{R} \times \mathbb{R}$. Define a relation R on X by $(x, y)R(z, t)$ if and only if the distance from (x, y) to (z, t) is an irrational number. Determine whether or not R is an equivalence relation.

11. Show that given any family of equivalence relations on a set X, their intersection is an equivalence relation on X.

12. Let $X = \{1, 2, \ldots, 10\}$. For every $k \in X$, define

$$A_k = \{B \in \mathcal{P}(X) \mid k \text{ is the least element of } B\}.$$

Let $P = \{\{\emptyset\}, A_1, \ldots, A_{10}\}$.

(a) Prove that P is a partition of $\mathcal{P}(X)$;
(b) Let R be the equivalence relation on $\mathcal{P}(X)$ corresponding to P. How many equivalence classes does R have?

1.2 Functions

Let X and Y be two sets. A *function* f from X into Y is a subset of $X \times Y$ (and hence is a relation from X into Y) with the property that each $x \in X$ belongs to precisely one pair (x, y). In other words no two distinct ordered pairs have the same first coordinate. Instead of $(x, y) \in f$ we usually write $y = f(x)$. Then y is called the *image of x under f* . The set X is called the *domain* of f. The *range* of f is the set $\{y \in Y \mid y = f(x) \text{ for some } x \in X\}$. That is, the range of f is the subset of Y consisting of all images of elements of X. If f is a function from X into Y, we write $f : X \to Y$. A *constant function* is a function whose range consists of a single element. We can define some operations on functions. New functions are formed from given functions by adding, subtracting, multiplying, and dividing function values. These functions are known as the *sum*, *difference*, *product*, and *quotient* of the original functions. More concisely, let f and g be two functions and D the largest set on which f and g are both defined. Then, we define

$$(f + g)(x) = f(x) + g(x), \qquad (fg)(x) = f(x) \cdot g(x),$$
$$(f - g)(x) = f(x) - g(x), \qquad \left(\frac{f}{g}\right)(x) = \frac{f(x)}{g(x)} \quad (g(x) \neq 0),$$

for all $x \in D$. We suppose that the readers are familiar to the elementary functions, which include powers of x, roots of x, trigonometric functions, logarithms and exponential functions, and hyperbolic functions.

Definition 1.15 Let $f : X \to Y$ and $g : Y \to Z$ be two functions. Then, we define a new function $g \circ f$, the *composition* of f and g by $g \circ f : X \to Z$ whose value at x is given by $(g \circ f)(x) = g(f(x))$, for all $x \in X$. In another notation, we have

$$g \circ f = \{(x, z) \in X \times Z \mid \exists y \in Y \text{ such that } (x, y) \in f \text{ and } (y, z) \in g\}.$$

Definition 1.16 A function $f : X \to Y$ is said to be

(1) *injective* or *one to one* if $f(x_1) = f(x_2)$ implies $x_1 = x_2$,
(2) *surjective* or *onto* provided that $y \in Y$, then there exists at least $x \in X$ such that $y = f(x)$,

Fig. 1.6 Image of A
under f

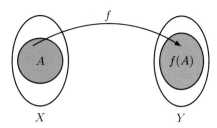

Fig. 1.7 Inverse image of B
under f

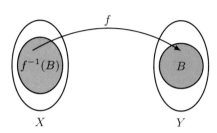

(3) *bijective* if it is one to one and onto. A bijective is also called a *one to one correspondence*.

Definition 1.17 Two sets are said to have the *same cardinality* if there exists a one to one correspondence between them. An infinite set is said to be *countably infinite* (*enumerable*) if it has the same cardinality as the set of positive integers. An infinite set is *uncountably infinite* (*non-denumerable*) if it is infinite but not countably infinite.

Definition 1.18 Let $f : X \to Y$ be a function and let $A \subseteq X$ and $B \subseteq Y$. The *image* of A under f, which we denote by $f(A)$, is

$$f(A) = \{f(x) \mid x \in A\}.$$

The *inverse image* of B under f, which we denote by $f^{-1}(B)$, is

$$f^{-1}(B) = \{x \in X \mid f(x) \in B\}.$$

The reader may picture this as illustrated in Figs. 1.6 and 1.7.

Lemma 1.19 *Let $f : X \to Y$ be a function.*

(1) *If $A \subseteq B \subseteq X$, then $f(A) \subseteq f(B)$;*
(2) *If $C \subseteq D \subseteq Y$, then $f^{-1}(C) \subseteq f^{-1}(D)$.*

Proof The proof is left as an exercise. ∎

Theorem 1.20 *Let $f : X \to Y$ be a function and let $\{A_i \mid i \in I\}$ be a family of subsets of X. Then*

(1) $f\left(\bigcup_{i \in I} A_i\right) = \bigcup_{i \in I} f(A_i);$ (2) $f\left(\bigcap_{i \in I} A_i\right) \subseteq \bigcap_{i \in I} f(A_i).$

Proof The proof is left as an exercise. ∎

The inclusion symbol in Theorem 1.20 (2) may not be replaced by an equal sign, as the next example shows.

Example 1.21 Suppose that $X = \{a, b, c\}$, $Y = \{d, e\}$, $I = \{1, 2\}$, $A_1 = \{a, b\}$, $A_2 = \{b, c\}$, and let $f : X \to Y$ is defined by $f(a) = f(c) = d$ and $f(b) = e$. Then, we have $f(A_1 \cap A_2) = f(\{b\}) = \{e\}$, whereas $f(A_1) \cap f(A_2) = \{d, e\}$. This yields that $f(A_1 \cap A_2) = f(\{b\}) \neq f(A_1) \cap f(A_2)$.

Theorem 1.22 *Let* $f : X \to Y$ *be a function and let* $\{B_i \mid i \in I\}$ *be a family of subsets of* Y. *Then*

(1) $f^{-1}\left(\bigcup_{i \in I} B_i\right) = \bigcup_{i \in I} f^{-1}(B_i);$ (2) $f^{-1}\left(\bigcap_{i \in I} B_i\right) = \bigcap_{i \in I} f^{-1}(B_i).$

Proof The proof is left as an exercise. ∎

Exercises

1. Let $f : X \to Y$ be a function, and let $B \subseteq Y$. Prove that $f^{-1}(Y \setminus B) = X \setminus f^{-1}(B)$.
2. Let $f : X \to Y$ be a function, and let A and B be subsets of X. Give an example which shows that it is not true that $f(A \setminus B) = f(A) \setminus f(B)$.
3. Let $f : X \to Y$ be a function, and let $A \subseteq X$ and $B \subseteq Y$. Prove that

   (a) $A \subseteq f^{-1}(f(A));$ (b) $f(f^{-1}(B)) \subseteq B.$

4. Let $f : X \to Y$ be a function, and let $A \subseteq X$ and $B \subseteq Y$. Give examples which show that the following statements are not true.

   (a) $f^{-1}(f(A)) = A;$ (b) $f(f^{-1}(B)) = B.$

5. Let $f : X \to Y$ be a function, and let $A \subseteq X$ and $B \subseteq Y$. Prove that

   (a) $f(A \cap f^{-1}(B)) = f(A) \cap B;$ (b) $f(f^{-1}(B)) = f(X) \cap B.$

6. Prove that there is a one to one correspondence between the set $\mathbb{N}$ of natural numbers and the set of all even natural numbers.
7. Prove that there is a one to one correspondence between the set $\mathbb{Z}$ of integers and the set of all odd integers.

1.3 Ordered Sets

In this section, we look at certain kinds of ordered sets and then we introduce the notion of Hasse diagram.

Definition 1.23 Let R be a relation on a set X. We say that R is *antisymmetric* if for all $x, y \in X$, xRy and yRx imply $x = y$.

Definition 1.24 A relation R on a set X is called a *partial order relation* if it is reflexive, antisymmetric, and transitive.

Example 1.25 Two fundamental partial order relations are the relation "$\leq$" on a set of real numbers and the relation "$\subseteq$" on a family of sets.

Because of special role played by the relation $\leq$ in the study of partial order relations, the symbol $\preceq$ is often used to refer to a partial order relation.

Example 1.26 Given $n \in \mathbb{N}$ and let X be the family of partitions of $\{1, 2, \ldots, n\}$. For $A, B \in X$ define $A \preceq B$ if and only if each class of A is contained in a class of B. Then, $(X, \preceq)$ is a partially ordered set.

Suppose that $\preceq$ is a partial order relation on a set X. Elements x and y of X are said to be *compatible* if either $x \preceq y$ or $y \preceq x$. Otherwise, x and y are called *non-compatible*.

Definition 1.27 A *total order relation* $\preceq$ on a set X is a partial order relation such that all elements of X are compatible.

Definition 1.28 A *partially ordered set* is a pair $(X, \preceq)$, where X is a set and $\preceq$ is a partial order relation on X.

A *totally ordered set* is a pair $(X, \preceq)$, where X is a set and $\preceq$ is a total order relation on X.

Definition 1.29 Let X be a set that is partially ordered with respect to a relation $\preceq$. A subset C of X is called a *chain* if the elements in each pair of elements in C are compatible.

The *length of a chain* is one less than the number of elements in the chain. If C is a chain in X, then C is a totally ordered set with respect to the restriction of $\preceq$ to C. A chain C in a partial order set is *maximal* if no more elements can be added on it.

If $(X, \preceq)$ is a partially ordered set, then $x \prec y$ means $x \preceq y$ and $x \neq y$. Given $x, y \in X$, we say that y cover x if $x \prec y$ and there does not exist $z \in X$ such that $x \prec z \prec y$.

Definition 1.30 The *Hasse diagram* of a partially ordered set $(X, \preceq)$ is the graph with vertices x in X and

(1) If $x \prec y$, then y is drawn above x in the diagram;
(2) If y covers x, then there is an edge between x and y in the diagram.

Example 1.31 Consider Example 1.26 and suppose that X is the family of all partitions of $\{1, 2, 3\}$. Let

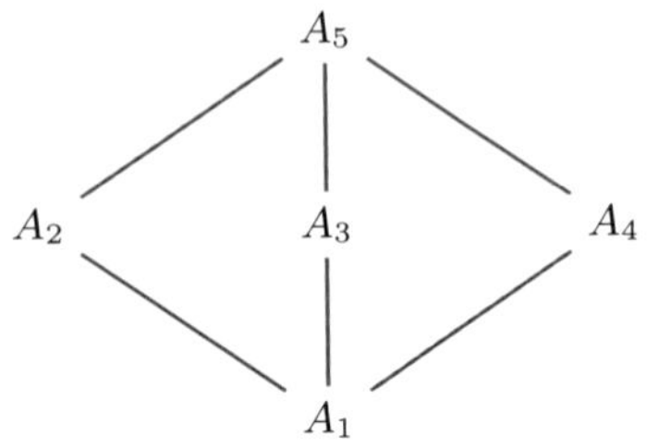

Fig. 1.8 Hasse diagram for
the family of all partitions of
$\{1, 2, 3\}$ in Example 1.31

Fig. 1.9 Hasse diagram for
$(\mathcal{P}(\{a, b, c\}, \subseteq)$

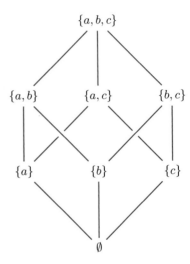

$A_1 = \big\{\{1, 2, 3\}\big\}$,
$A_2 = \big\{\{1\}, \{2, 3\}\big\}$,
$A_3 = \big\{\{2\}, \{1, 3\}\big\}$,

$A_4 = \big\{\{3\}, \{1, 2\}\big\}$,

$A_5 = \big\{\{1\}, \{2\}, \{3\}\big\}$.

Then, $X = \{A_1, A_2, A_3, A_4, A_5\}$, and we have $A_5 \prec A_1, A_2, A_3, A_4$ and $A_2, A_3, A_4 \prec A_1$. The Hasse diagram for this partially ordered set is given in Fig. 1.8.

Example 1.32 Let $A = \{a, b, c\}$ and $X = \mathcal{P}(A)$. By Example 1.25, we know that $(X, \subseteq)$ is a partial ordered set. The Hasse diagram for this partially ordered set is given in Fig. 1.9.

Definition 1.33 Let $(X, \preceq)$ be a partially ordered set and $A \subseteq X$.

(1) An element $u \in X$ is said to be an *upper bound* for A if $a \preceq u$, for all $a \in A$;
(2) An upper bound u_0 for A is the *least upper bound* (or *supremum*) for A if $u_0 \preceq u$, for each upper bound u of A;
(3) An element $m \in X$ is said to be *maximal* if for all $x \in X$, $m \preceq x$ implies $m = x$.

Definition 1.33 has a dual form as follows.

Definition 1.34 Let $(X, \preceq)$ be a partially ordered set and $A \subseteq X$.

(1) An element $l \in X$ is said to be a *lower bound* for A if $l \preceq a$, for all $a \in A$;
(2) A lower bound l_0 for A is the *greatest lower bound* (or *infimum*) for A if $l \preceq l_0$, for each lower bound l of A;
(3) An element $m' \in X$ is said to be *minimal* if for all $x \in X$, $x \preceq m'$ implies $m' = x$.

Example 1.35 Let A be a non-empty set and $\mathcal{B}$ be a subset of the partially ordered set $(\mathcal{P}(A), \subseteq)$. Then the least upper bound of $\mathcal{B}$ is $\bigcup_{B \in \mathcal{B}} B$, and the greatest lower bound of $\mathcal{B}$ is $\bigcap_{B \in \mathcal{B}} B$.

Zorn's lemma, also known as Kuratowski-Zorn lemma originally called *maximum principle*, statement in the language of set theory is often used to prove the existence of a mathematical object when it cannot be explicitly produced.

Theorem 1.36 (Zorn's Lemma) *Let $(X, \preceq)$ be a partially ordered set in which every totally ordered subset has an upper bound. Then, X has a maximal element.*

Exercises

1. Let F be the set of all functions $f : \mathbb{R} \to \mathbb{R}$ and let

$$R = \{(f, g) \in F \times F \mid f(x) \le g(x), \text{ for all } x \in \mathbb{R}\}.$$

   Prove that (F, R) is a partially ordered set.
2. Let $A = \{a, b, c, d\}$ and consider the partially ordered set $(\mathcal{P}(A), \subseteq)$. Find all upper bounds, all lower bounds, the least upper bound, and the greatest lower bound for the set $\{\{a, b\}, \{a, b, d\}\}$.
3. Let $X = \{a_{00}, a_{10}, a_{11}, a_{12}, a_{13}, a_{20}, a_{21}, a_{22}\}$ be a partial ordered set with Hasse diagram shown in Fig. 1.10. Let $A = \{a_{00}, a_{10}, a_{12}\}$. Determine

   (1) The upper and lower bounds of A;
   (2) The least upper bound and greatest lower bound of A if they exist;
   (3) The maximal and minimal elements of X.

4. Let $(X, \preceq)$ be a partially ordered set and $A \subseteq X$. If the least upper bound (or greatest lower bound) of A exists, prove that it is unique.
5. Let $(X_1, \preceq_1)$ and $(X_2, \preceq_2)$ be two partially ordered sets. The *lexicographical order* $\preceq$ on $X_1 \times X_2$ is given by specifying that

$$(a, b) \preceq (c, d) \iff a \prec c \text{ or } (a = c \text{ and } b \preceq d).$$

   Prove that $(X_1 \times X_2, \preceq)$ is a partially ordered set.
6. If $(X, \preceq)$ is a partially ordered set and x, y are two non-compatible elements of X, prove that there is a partially order $\preceq'$ extending $\preceq$ such that $x \preceq' y$.

Fig. 1.10 Hasse diagram for
Exercise 3

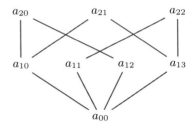

1.4   Combinatorial Analysis

Many problems in group theory can be solved by simply counting the number of different ways that a certain choice can occur. The mathematical theory of counting is formally known as *combinatorial analysis*. The following principle of counting is basic.

Theorem 1.37 (Basic Principle of Counting) *Suppose that two experiments are to be performed. Then if the first experiment can result in any one of m possible outcomes and if for each outcome of the first experiment, there exist n possible outcomes of the second experiment, then together there exist mn possible outcomes of the two experiments.*

Proof The basic principle of counting may be proved by enumerating all the possible outcomes of the two experiments as follows:

$$(1, 1), \ (1, 2), \ \dots, \ (1, n)$$
$$(2, 1), \ (2, 2), \ \dots, \ (2, n)$$
$$\vdots$$
$$(m, 1), \ (m, 2), \ \dots, \ (m, n)$$

where we say that the outcome is (i, j) if the first experiment results in its ith possible outcome and the second experiment then results in the jth of its possible outcome. Therefore, the set of possible outcomes consists of m rows, each row containing n elements, which proves the result.   ■

A generalization version of the basic principle states that if r experiments are such that the first may result in any of n_1 possible outcomes, and if for each of the possible outcomes of the first $i - 1$ experiments there are n_i $(i = 1, \dots, r)$ possible outcomes of the ith experiment, then there are a total of $n_1 n_2 \dots n_r$ possible outcomes of the r experiments.

Example 1.38 How many functions defined on n points are possible if each functional value is either 0 or 1? Assume that the points are $1, 2, \dots, n$. Since $f(i)$ must be either 0 or 1 for each $i = 1, 2, \dots, n$, it follows that there are 2^n possible functions.

How many different ordered arrangements of the letters a, b, and c are possible? By direct enumeration we see that there are 6, namely

$$abc, \quad acb, \quad bac, \quad bca, \quad cab, \quad cba.$$

Each arrangement is known as a permutation. So, a *permutation* of the set $\{a_1, a_2, \ldots, a_n\}$ is an arrangement of these elements in some order without omissions or repetition. Permutations have important role in the group theory. We will study them deeply in Sects. 5.1 and 5.2.

Sometimes we are interested in determining the number of different collections of r objects that could be formed from total of n objects.

We define $\binom{n}{r}$, for $r \leq n$ by

$$\binom{n}{r} = \frac{n!}{(n-r)!r!},$$

where $n!$ denotes the product $n(n-1) \ldots 3 \cdot 2 \cdot 1$ of the first n consecutive natural numbers if $n > 0$ and $0! = 1$ by convention.

Theorem 1.39 $\binom{n}{r}$ *represents the number of possible combinations of n objects taken r at a time.*

Proof As $n(n-1) \ldots (n-r+1)$ represents the number of different ways that a collection of r objects could be selected from n objects when the order of selection is relevant, and as each collection of r objects will be counted $r!$ times in this count, it follows that the number of different collection of r objects that could be formed from a set of n objects is

$$\frac{n(n-1) \ldots (n-r-1)}{r!} = \frac{n!}{(n-r)!r!}.$$

This completes the proof. ∎

Thus, $\binom{n}{r}$ represents the number of different subsets of size r that could be selected from a set of n objects when the order of selection is not considered relevant.

Theorem 1.40 (Binomial Theorem) *If x and y are two variables and n is a positive integer, then*

$$(x+y)^n = \sum_{k=0}^{n} \binom{n}{k} x^k y^{n-k}. \tag{1.1}$$

Proof Consider the product $(x_1 + y_1)(x_2 + y_2) \ldots (x_n + y_n)$. Its expansion consists of the sum of 2^n terms and each term is the product of n factors. Moreover, each of the 2^n terms in the sum contains as a factor either x_i or y_i for each $i = 1, 2, \ldots, n$. Now, how many of the 2^n terms in the sum have a factor k of the x_is and $n - k$ of the y_is? Since each term consisting of k of the x_is and $n - k$ of the y_is corresponds to a choice of a collection of k from the n values $x_1, x_2, \ldots, x_n$, it follows that there exist $\binom{n}{k}$ such terms. Finally, if we take $x_i = x$ and $y_i = y$ $(i = 1, \ldots, n)$, then we obtain Eq. (1.1). ∎

If $n_1 + n_2 + \cdots + n_r = n$, we define $\binom{n}{n_1, n_2, \ldots, n_r}$ by

$$\binom{n}{n_1, n_2, \ldots, n_r} = \frac{n!}{n_1! n_2! \ldots n_r!}.$$

Theorem 1.41 *If $n_1 + n_2 + \cdots + n_r = n$, then $\binom{n}{n_1, n_2, \ldots, n_r}$ represents the number of possible divisions of n distinct objects into r distinct collections of respective sizes $n_1, n_2, \ldots, n_r$.*

Proof The proof is left as an exercise. ∎

Theorem 1.42 (Multinomial Theorem) *If $x_1, x_2, \ldots, x_r$ are variables and n is a positive integer, then*

$$(x_1 + x_2 + \cdots + x_r)^n = \sum_{\substack{(n_1, \ldots, n_r): \\ n_1 + \cdots + n_r = n}} \binom{n}{n_1, n_2, \ldots, n_r} x_1^{n_1} x_2^{n_2} \ldots x_r^{n_r}.$$

Proof The proof is left as an exercise. ∎

Theorem 1.43 *There are $\binom{n + r - 1}{n}$ distinct non-negative integer-valued sequences $(x_1, x_2, \ldots, x_r)$ satisfying*

$$x_1 + x_2 + \cdots + x_n = n. \qquad (1.2)$$

Proof Consider a sequence of n ones and $r - 1$ zeros. To each permutation of this sequence we correspond a solution of Equation (1.2), namely the solution where

$x_1 = $ the number of ones to the left of the first zero,

$x_2 = $ the number of ones between the first and second zeros,

$x_3 = $ the number of ones between the second and third zeros,

$\vdots$

$x_n = $ the number of ones to the right of the last zero.

For instance, if $n = 8$ and $r = 5$, then the sequence $(1, 1, 0, 0, 1, 1, 1, 0, 1, 1, 1)$ corresponds to the solution $x_1 = 2$, $x_2 = 0$, $x_3 = 3$ and $x_4 = 3$. It is not difficult to see that this correspondence between permutations of a sequence of n ones and $r - 1$ zeros and solutions of Eq. (1.2) is a one to one correspondence. Now, the result follows because there are $\frac{(n+r-1)!}{n!(r-1)!}$ permutations of n ones and $r - 1$ zeros. ∎

Exercises

1. Prove that
$$\binom{2n}{n} = \sum_{k=0}^{n} \binom{n}{k}^2.$$

2. Prove that there are $\binom{n-1}{n-r}$ positive integer-valued solutions of

$$x_1 + x_2 + \cdots + x_r = n.$$

3. Prove that there are exactly $\binom{r}{k}\binom{n-1}{n-r+k}$ solutions of

$$x_1 + x_2 + \cdots + x_r = n.$$

for which exactly k of the xs equal to 0.

4. In how many ways can n indistinguishable balls be distributed into r urns in such a way that the ith urn contains at least m_i balls. Assume that

$$n \geq \sum_{i=1}^{r} m_i.$$

Answer: $\binom{n - \sum m_i + r - 1}{n - \sum m_i}$.

5. Show that
$$\sum_{i=0}^{n} (-1)^{n-1} \binom{n}{i} = 0.$$

6. Suppose a street grid starts at position $(0, 0)$ and extends up and to the right, see Fig. 1.11

Fig. 1.11 A street grid

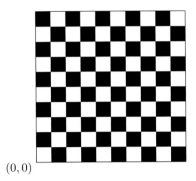

$(0, 0)$

A shortest route along streets from $(0, 0)$ to (i, j) is $i + j$ blocks long, going i blocks east and j blocks north. How many such routes are there? Suppose that the block between (k, l) and $(k + 1, l)$ is closed, where $k < i$ and $l < j$. How many shortest routes are there from $(0, 0)$ to (i, j)?

1.5 Divisibility and Prime Numbers

In this section, all numbers are integers, unless specified otherwise.

We introduce basic concepts of elementary number theory such as divisibility and greatest common divisor. The principal results are Division algorithm, which provides a quotient and a reminder when we divide two numbers and Bezout's Lemma, which establish the existence of the greatest common divisor of any two integers as a linear combination of those two.

Definition 1.44 We say d divides n and we write $d \,|\, n$ whenever $n = cd$ for some c. If d divides n, we also say that d is a *divisor* or *factor* of n. The symbol $d \nmid n$ means that d does not divide n.

The Principle of Induction. If S is a subset of $\mathbb{Z}$ such that

(1) $1 \in S$,
(2) $n \in S$ implies $n + 1 \in S$,

then all integers greater than 1 belong to S.

The basic assumption we make about the set of integers is the well-ordering principle.

The Well-Ordering Principle. Any non-empty set of non-negative integers has a smallest number.

The well-ordering principle is logically equivalent to the principle of induction. We assume that the reader is familiar with this principle and its use in proving theorems by induction.

Theorem 1.45 (Mathematical Induction) *Let $P(n)$ be a statement about the positive integers such that*

(1) $P(1)$ is true;
(2) If $P(k)$ happens to be true for some integer $k \geq 1$, then $P(k + 1)$ is also true.

Then, $P(n)$ is true for each $n \geq 1$.

Proof Suppose that the theorem is false. Then, by the well-ordering principle, there exists a smallest integer $m \geq 1$ for which $P(m)$ is not true. Since $P(1)$ is true, it follows that $m \neq 1$. So, $m > 1$. Now, we have $1 \leq m - 1 < m$, so by the choice of m, $P(m - 1)$ must be true. But in this case, by the inductive hypothesis, we must have $P(m)$ is true. This is a contradiction. Therefore, we conclude that there can be no integer for which P is not true, so the theorem is proved. ∎

The following theorem is the foundation of much of what we will do during the book.

Theorem 1.46 (Division Algorithm) *If a and b are integers with $b > 0$, then there exist unique integers q and r such that*

(1) $a = bq + r$,
(2) $0 \leq r < b$.

Proof Let a and b be fixed integers with $b > 0$, and let $S = \{a - bx \mid x \in \mathbb{Z} \text{ and } b - ax \geq 0\}$. We claim that S is non-empty. There exist two possibilities:

(1) If $a \geq 0$, then $a - bx$ is non-negative for $x = 0$.
(2) If $a < 0$, then $-a > 0$. Since b is a positive integer, it follows that $b \geq 1$. So, we get $-ab \geq -a$, or equivalently $a - ab \geq 0$. Hence, $a - bx$ is non-negative for $x = a$.

Therefore, we conclude that S is non-empty. Since S is a non-empty subset of $\mathbb{N}$, by the well-ordering principle we know that S contains a smallest element, call it r. Since $r \in S$, it follows that r is of the form $r = a - bq$ for some q, or equivalently $a = bq + r$. We claim that $r < b$. Assume that on the contrary $r \geq b$. Then, $r - b \geq 0$. So, we have

$$r - b = (a - bq) - b = a - b(q + 1) < r. \tag{1.3}$$

But r is the smallest element of S, while Eq. (1.3) seems to say there is another element of S smaller than r. This is a contradiction. Consequently, $r < b$. In order to complete the proof, we must show that the integers q and r are unique. To do this, also, assume that there exist integers q' and r' such that $a = bq + r = bq' + r'$ and $0 \leq r, r' < b$. Without loss of generality, suppose that $r' \geq r$. Then, we have $0 = bq + r - bq' - r' = b(q - q') - (r' - r)$, or equivalently $b(q - q') = r' - r$. Since $b > 0$ and $r' - r > 0$, it follows that $q - q' > 0$. But, if $q - q' > 0$, then $r' - r > b$. This is a contradiction, because $r' - r < b$. Hence, we conclude that $q - q' = 0$, or equivalently $q = q'$, so $r = r'$. ∎

We say that q is the *quotient* and r is the *remainder* obtained when b is divided into a. If d divides two integers a and b, then d is called a *common divisor* of a and b. So, 1 is a common divisor of each pair of integers a and b.

Definition 1.47 The positive integer d is said to be the *greatest common divisor* of a and b if

(1) $d|a$ and $d|b$;
(2) Any divisor of a and b is a divisor of d, i.e., $c|a$ and $c|b$ implies $c|d$.

We shall use the notation (a, b) for greatest common divisor of a and b. We can extend the definition of greatest common divisor in a straightforward way. Given n integers $a_1, a_2, \ldots, a_n$ not all zero, we define their greatest common divisor $(a_1, a_2, \ldots, a_n)$ to be the greatest integer which divides all the given numbers.

Theorem 1.48 (Bezout's Lemma) *If both a and b are not zero, then (a, b) exists, is unique, and in addition, we can find integers x_0 and y_0 such that $(a, b) = ax_0 + by_0$.*

Proof Suppose that $S = \{ax + by \mid x, y \in \mathbb{Z}\}$. Since one of the a or b is not zero, it follows that there exist non-zero integers in S. If $ax + by$ is in S, it follows that $a(-x) + b(-y)$ is also in S. Thus, S always contains some positive integers. Hence, by the well-ordering principle, there is a smallest positive number $d \in S$. Since $d \in S$, by the form of the elements of S we know that $d = ax_0 + by_0$, for some integers x_0 and y_0. We claim that d is our required greatest common divisor. First, note that if $c|a$ and $c|b$, then $c|ax_0 + by_0$, or equivalently $c|d$. Now, we must show that $d|a$ and $d|b$. By the Division algorithm, $a = qd + r$, where $0 \leq r < d$, i.e., $a = q(ax_0 + by_0) + r$. Therefore, we have $r = a(1 - qx_0) + b(-qy_0)$, so r must be in S. Since $r < d$ and $r \in S$, so by the choice of d, we conclude that r cannot be positive. Hence, $r = 0$. Consequently, we have $a = qd$ and so $d|a$. Analogously, we obtain $d|b$.

For the uniqueness of d, if $d' > 0$ also satisfied $d'|a$, $d'|b$ and $d|d'$ for each d such that $d|a$ and $d|b$, then we would have $d'|d$ and $d|d'$. This yields that $d = d'$. ∎

Theorem 1.49 (Euclidean Algorithm) *Given positive integers a and b, where $b \nmid a$. Let $r_0 = a$, $r_1 = b$ and apply the division algorithm repeatedly to obtain a set of remainders $r_2, r_3, \ldots, r_n, r_{n+1}$ defined by*

$$\begin{aligned}
r_0 &= r_1 q_1 + r_2, & 0 < r_2 &< r_1, \\
r_1 &= r_2 q_2 + r_3, & 0 < r_3 &< r_2, \\
&\;\;\vdots \\
r_{n-2} &= r_{n-1} q_{n-1} + r_n, & 0 < r_n &< r_{n-1}, \\
r_{n-1} &= r_n q_n, & r_{n+1} &= 0.
\end{aligned}$$

Then, r_n, the last non-zero remainder in this process is the greatest common divisor of a and b.

Proof In view of the last equality $r_{n-1} = r_n q_n$ we have $r_n | r_{n-1}$. The next to last shows that $r_n | r_{n-2}$. By induction, we observe that $r_n | r_i$, for each integer i. In particular, $r_n | r_1 = b$ and $r_n | r_0 = a$. So, r_n is a common divisor of a and b. Now, suppose that c be any common divisor of a and b. Then, respectively, we obtain $c | r_2, c | r_3, \ldots, c | r_n$. Therefore, we conclude that $r_n = (a, b)$.   ∎

Theorem 1.50 *The greatest common divisor has the following properties:*

(1) $(a, b) = (b, a)$;
(2) $(a, (b, c)) = ((a, b), c)$;
(3) $(ac, bc) = |c|(a, b)$;
(4) $(a, 1) = (1, a) = 1$ *and* $(a, 0) = (0, a) = |a|$.

Proof We prove only (3). The proofs of the other statements are straightforward.

Suppose that $d = (a, b)$ and $e = (ac, bc)$. We wish to prove that $e = |c|d$. Since $d = (a, b)$, it follows that $d = ax + by$, for some integers x and y. Then, $cd = acx + bcy$, which implies that $e | cd$. Moreover, since $cd | ac$ and $cd | bc$, it follows that $cd | e$. Therefore, we conclude that $|e| = |cd|$, or equivalently $e = |c|d$.   ∎

The integers a and b are *relatively prime* if $(a, b) = 1$. More generally, the integers $a_1, a_2, \ldots, a_n$ are called relatively prime if $(a_1, a_2, \ldots, a_n) = 1$, and they are called *pairwise relatively prime* if any two of them are relatively prime.

Theorem 1.51 *Let a and b be two integers. Then, a and b are relatively prime if and only if there exist integers x and y such that $1 = ax + by$.*

Proof If $d = (a, b)$, then by Theorem 1.48, $d = ax + by$ for some integers x and y.

Conversely, let $1 = ax + by$ for some integers x and y. If $d = (a, b)$, then $d | a$ and $d | b$. So, $d | ax + by$, or equivalently $d | 1$. This implies that $d = 1$.   ∎

Theorem 1.52 (Euclid's Lemma) *If $a | bc$ and $(a, b) = 1$, then $a | c$.*

Proof Since $(a, b) = 1$, it follows that $1 = ax + by$, for some integers x and y. Then, $c = acx + bcy$. Now, since $a | acx$ and $a | bcy$, it follows that $a | acx + bcy$, or equivalently $a | c$.   ∎

Theorem 1.53 *If $(a, b) = d$, then $(a/d, b/d) = 1$.*

Proof Suppose that $(a/d, b/d) = c$. Then, $c | (a/d)$ and $c | (b/d)$. So, there exist integers u and v such that $a = dcu$ and $b = dcv$. Hence, dc is a common divisor of a and b. If $c > 1$, then this is a contradiction, because $cd > d$ but d is the greatest common divisor of a and b. Therefore, we conclude that $c = 1$.   ∎

We recall the notions of prime and composite numbers. The principle result is Fundamental Theorem of Arithmetic, which shows that each integer greater than 1 can be represented as a product of prime numbers in only one way apart from the order of factors.

Definition 1.54 An integer $n > 1$ is *prime* if the only positive divisors of n are 1 and n. An integer $n > 1$ which is not prime is called *composite*.

Prime numbers are usually denoted by p, p', p_i, q, q', q_i.

Theorem 1.55 *Let p be a prime number. If $p|ab$, then $p|a$ or $p|b$.*

Proof Suppose that $p|ab$ and $p \nmid a$. Since p has only trivial divisors, it follows that $(p, a) = 1$. Now, by Euclid's lemma, we conclude that $p|b$. ■

Theorem 1.55 is easily extended to:

Corollary 1.56 *Let p be a prime number. If $p|a_1 a_2 \ldots a_n$, then $p|a_i$, for some $1 \le i \le n$.*

Proof The result follows by induction. ■

Theorem 1.57 *Each integer n greater than 1 is either a prime number or a product of prime numbers.*

Proof The existence of such a factorization is proved by induction. Assume that each integer less than n can be written as a product of primes. If n is a prime, then we have a factorization of n consisting of one prime factor. If n is composite, then $n = n_1 n_2$ with $1 < n_1, n_2 < n$ and it follows from the induction hypothesis that each of n_1 and n_2 is a product of primes. Therefore, we conclude that n is a product of primes too. ■

There is a certain uniqueness about the composition of an integer into primes. We make this precise now.

Theorem 1.58 (Fundamental Theorem of Arithmetic) *Each integer n greater than 1 can be expressed as a product of primes in a unique way apart from the order of the prime factors.*

Proof We use induction on n. The theorem is true for $n = 2$. Suppose that it is true for all integers greater than 1 and less than n. We shall prove it is also true for n. If n is prime, then there is nothing more to prove. Suppose that the theorem is false for n. So, n is composite with two different factorizations. Let

$$n = p_1 p_2 \ldots p_k = q_1 q_2 \ldots q_l, \tag{1.4}$$

where each p_i and q_j is prime and where the two factorizations are different. Since $p_1|q_1 q_2 \ldots q_l$, by Corollary 1.56, it follows that p_1 divides one of the prime numbers $q_1, q_2, \ldots, q_l$. Renumbering these numbers, we may assume that $p_1|q_1$, which of course means that $p_1 = q_1$. Dividing n by p_1 we obtain a smaller number

$$\frac{n}{p_1} = p_2 p_3 \ldots p_k = q_2 q_3 \ldots q_l.$$

If $k > 1$ or $l > 1$, then $1 < n/p_1 < n$. By induction hypothesis, n/p_1 has the unique factorization property. So, the two factorization of n/p_1 must be identical apart from the order of the factors. Consequently, $k = l$ and the factorizations in (1.56) are also identical apart from the order. Therefore, we see that the primes arising for the factorization of n are unique. This contradicts the lack of such uniqueness for n. This completes the proof.   ∎

In the factorization of an integer n, a particular prime p may occur more than once. If the distinct prime factors of n are $p_1, p_2, \ldots, p_k$ and if p_i occurs as a factor α_i times, we can write $n = p_1^{\alpha_1} p_2^{\alpha_2} \ldots p_k^{\alpha_k}$.

Theorem 1.59 (Euclid's Theorem) *There is an infinite number of primes.*

Proof Suppose that there are only a finite number of primes, say $p_1, p_2, \ldots, p_k$. Now, consider the integer $N = 1 + p_1 p_2 \ldots p_k$. Since $N > 1$, it follows that either N is prime or N is a product of primes. Since $N > p_i$, for each $1 \leq i \leq k$, it follows that N cannot be prime. Moreover, no p_i divides N (if $p_i | N$, then p_i divides the difference $N - p_1 p_2 \ldots p_k = 1$). Thus, N is not prime nor is divisible by any prime. This violates Theorem 1.57. Therefore, we deduce that the assumption that there are only finitely many primes is not true.   ∎

Theorem 1.60 *If $n > 1$ is composite, then n has a prime divisor $p \leq \sqrt{n}$.*

Proof Let $n = ab$, where $1 < a, b < n$. We claim that at least one of the a or b is less than or equal to $\sqrt{n}$. For if not, then $a > \sqrt{n}$ and $b > \sqrt{n}$. Thus, $n = ab > \sqrt{n}\sqrt{n} = n$, which is impossible. Without loss of generality, we assume that $a \leq \sqrt{n}$. Since $a > 1$, it follows that there exists a prime number p such that $p | a$. Since $p | a$ and $a | n$, it follows that $p | n$. Consequently, we have $p \leq a \leq \sqrt{n}$, and this completes the proof.   ∎

Gauss introduced a remarkable notation which simplifies many problems concerning divisibility of integers.

Definition 1.61 Let $n > 0$ be a fixed integer and $a, b \in \mathbb{Z}$. We define $a \equiv b \pmod{n}$ if $n | a - b$. This relation is referred to as *congruence modulo n*, and n is called the *modulus* of the relation.

Lemma 1.62 *The relation congruence modulo n defines an equivalence relation on the set of integers.*

Proof For each $a \in \mathbb{Z}$, since $n | a - a$, it follows that $a \equiv a \pmod{n}$, i.e., the relation $\equiv$ is reflexive. Further, for each $a, b \in \mathbb{Z}$, if $a \equiv b \pmod{n}$, then $n | a - b$ and so $n | b - a$, which implies that $b \equiv a \pmod{n}$, i.e., the relation $\equiv$ is symmetric. Finally, for each $a, b, c \in \mathbb{Z}$, if $a \equiv b \pmod{n}$ and $b \equiv c \pmod{n}$, then $n | a - b$ and $n | b - c$. Consequently, $n | a - b + b - c$ or $n | a - c$ and hence $a \equiv c \pmod{n}$, i.e., the relation $\equiv$ is transitive.   ∎

Theorem 1.63 *If $a \equiv b \pmod{n}$ and $c \equiv d \pmod{n}$, then we have*

(1) $ax + cy \equiv bx + dy \pmod{n}$, *for all integers x and y;*
(2) $ac \equiv bd \pmod{n}$;
(3) $a^m \equiv b^m \pmod{n}$, *for all positive integer m;*
(4) $f(a) \equiv f(b) \pmod{n}$, *for each polynomial f with integer coefficients.*

Proof The proof is left as an exercise.   ■

Theorem 1.63 tells us that two congruences with the same modulus can be added, subtracted, or multiplied, member by member, as though they were equations.

Theorem 1.64 *If $c > 0$, then $a \equiv b \pmod{n}$ if and only if $ac \equiv bc \pmod{nc}$.*

Proof We have $n \mid b - a$ if and only if $cn \mid c(b - a)$.   ■

Theorem 1.65 *If $ac \equiv bc \pmod{n}$ and $d = (n, c)$, then*

$$a \equiv b \left(\operatorname{mod} \frac{n}{d} \right).$$

Proof Since $ac \equiv bc \pmod{n}$, it follows that $n \mid c(a - b)$. This implies that $(n/d) \mid (c/d)(a - b)$. Since $(n/d, c/d) = 1$, it follows that $n/d \mid a - b$.   ■

Theorem 1.66 $a \equiv b \pmod{n}$ *if and only if a and b leave the same remainder when divided by n.*

Proof Suppose that $a \equiv b \pmod{n}$. Then, $b = a + nq$, for some integer q. If a leaves the remainder r when divided by n, then we have $a = nq' + r$ with $0 \leq r < n$. Thus, we get $b = a + nq = nq' + r + nq = n(q' + q) + r$, so b leaves the same remainder when divided by n.

The converse is straightforward and we omit the proof.   ■

Theorem 1.67 *If $(a, n) = 1$, then the congruence $ax \equiv b \pmod{n}$ has a solution $x = c$. In this case, the general solution of the congruence is given by $x \equiv c \pmod{n}$.*

Proof Since $(a, n) = 1$, it follows that $ar + ns = 1$, for some integers r and s. This implies that $arb + nsb = b$, or equivalently $abr \equiv b \pmod{n}$. This yields that $c = br$ is a solution of the congruence $ax \equiv b \pmod{n}$. In general, if $x \equiv c \pmod{n}$, then $ax \equiv ac \equiv b \pmod{n}$. Now, we claim that any solution of $ax \equiv b \pmod{n}$ is necessary congruent to c modulo n. We already know that $ac \equiv b \pmod{n}$. Then, $ax - ac \equiv 0 \pmod{n}$. Consequently, $n \mid a(x - c)$. Since $(a, n) = 1$, we conclude that $n \mid x - c$ and so $x \equiv c \pmod{n}$.   ■

An important special case occurs when n is a prime p.

Corollary 1.68 *If p is a prime number, then the congruence $ax \equiv b \pmod{p}$ has a unique solution x modulo p provided $a \not\equiv 0 \pmod{p}$*

Exercises

1. If $(a, b) = (a, c) = 1$, prove that $(a, bc) = 1$.
2. If $(a, b) = 1$ and $d | a + b$, prove that $(a, d) = (b, d) = 1$.
3. Find a positive integer n such that $3^2 | n$, $4^2 | n + 1$ and $5^2 | n + 2$.
4. The *Fibonacci sequence* 1, 1, 2, 3, 5, 8, 13, 21, 34, ... is defined by the recursion formula $a_{n+1} = a_n + a_{n-1}$ with $a_1 = a_2 = 1$. Prove that $(a_n, a_{n+1}) = 1$, for each positive integer n.
5. Let n be a composite positive integer and let p be the smallest prime divisor of n. Show that if $p > \sqrt[3]{n}$, then n/p is prime.
6. If p is a prime, prove that one cannot find non-zero integers a and b such that $a^2 = pb^2$.
7. Let n be a composite positive integer and let p be the smallest prime divisor of n. Show that if $p > \sqrt[3]{n}$, then n/p is prime.
8. If $2^n - 1$ is prime, prove that n is prime.
9. If $2^n + 1$ is prime, prove that n is a power of 2.
10. Prove that $5n^3 + 7n^5 \equiv 0 \pmod{12}$ for all integer n.
11. If $(m, n) = 1$, given integers a and b, prove that there exists an integer x such that $x \equiv a \pmod{m}$ and $x \equiv b \pmod{n}$.
12. Prove that each number of the set of $n - 1$ consecutive integers

$$n! + 2, \ n! + 3, \ \ldots, \ n! + n$$

is divisible by a prime which does not divisible any other integer of the set.

1.6 Worked-Out Problems

Problem 1.69 Show that

$$\sum_{k=1}^{n} k = \binom{n+1}{2}.$$

Solution We arrange $1 + 2 + \cdots + n$ red circles in a triangle, mirror it, and we observe that this fills all positions of a square of size $(n + 1) \times (n + 1)$ except for the diagonal. Figure 1.12 shows this for $n = 9$. Consequently, we obtain

$$2 \sum_{k=1}^{n} k = (n + 1)^2 - (n + 1) = n(n + 1).$$

Now, dividing by 2 gives the result.   ∎

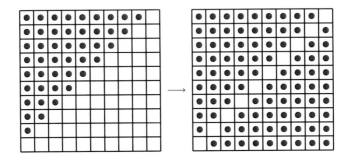

Fig. 1.12 A picture for describing the proof

Problem 1.70 Prove that

$$\sum_{k=0}^{n} 2^k \binom{n}{k} = 3^n.$$

Solution We double count strings over the alphabet $\{a, b, c\}$ of length n.

First way: There are three possibilities per character. Hence, there exist 3^n possibilities in total.

Second way: First, choose the number of times k that the letter a should be used $(0 \le k \le n)$. Then, choose the position for those characters, there are $\binom{n}{k}$ possibilities. Finally, choose for each of the remaining $n - k$ positions if they should be b or c, so there are 2^{n-k} choices. Therefore, in total, there is

$$\sum_{k=0}^{n} 2^k \binom{n}{k}$$

number of possibilities. By reversing the order of summation, we get

$$\sum_{k=0}^{n} 2^k \binom{n}{n-k} = \sum_{k=0}^{n} 2^k \binom{n}{k}.$$

This completes the proof.   ∎

Problem 1.71 If a and b are integers, the *least common multiple* of a and b, written as $[a, b]$, is defined as that positive integer d such that

(1) $a|d$ and $b|d$;
(2) If $a|c$ and $b|c$, then $d|c$.

If a and b are two positive integers, prove that

$$[a, b] = \frac{ab}{(a, b)}.$$

Solution Suppose that $d = (a, b)$. If $d = 1$, then any common multiple of a and b must be a multiple of ab too. This means that $[a, b] = ab$.

If $d > 1$, then $(a/d, b/d) = 1$. So, according to the above discussion, we have

$$\left[\frac{a}{d}, \frac{b}{d}\right] = \frac{a}{d} \cdot \frac{b}{d}.$$

Hence, we get $d^2[a/d, b/d] = ab$, and so $d[a, b] = ab$, or equivalently $(a, b)[a, b] = ab$. This completes the proof. ∎

Problem 1.72 Let a and b be two positive integers, where $b \nmid a$. Show that the Euclidean algorithm for calculating (a, b) takes at most $2([\log_2 b] + 1)$ steps (divisions).

Solution Suppose that $k = [\log_2 b] + 1$. In the notation of Euclidean algorithm, we must show that $n \leq 2k$. To do this, assume that $n \geq 2k + 1$. We claim that $r_{j+2} < r_j/2$, for each $1 \leq j \leq n - 2$. We consider the following two cases:

Case 1: If $r_{j+1} \leq r_j/2$, then $r_{j+2} < r_{j+1} \leq r_j/2$.

Case 2: If $r_{j+1} > r_j/2$, then $r_j = r_{j+1}q_{j+1} + r_{j+2}$, where $q_{j+1} = 1$. So, we have $r_{j+2} < r_j - r_{j+1} \leq r_j/2$.

Consequently, we obtain

$$1 \leq r_n < \frac{r_n - 2}{2} < \frac{r_n - 4}{4} < \ldots < \frac{r_n - 2^k}{2^k} \leq \frac{r_1}{2^k} = \frac{b}{2^k},$$

and so $k < \log_2 b$. This is a contradiction, because

$$k = [\log_2 b] + 1 > \log_2 b.$$

This completes the proof. ∎

Problem 1.73 If p is a prime, prove that

$$\binom{p}{k} = \frac{p!}{k!(p-k)!} \equiv 0 \pmod{p},$$

for each $1 \leq k \leq p - 1$.

Solution We have

$$p! = \binom{p}{k}(p-k)!k!.$$

Clearly, p divides the left side, so p must divide one of the factors on the right side. But all factors of $(p - k)!$ and $k!$ are less than p, this yields that $p \nmid (p - k)!$ and $p \nmid k!$. Therefore, we conclude that $p \mid \binom{p}{k}$. This completes the proof. ∎

Problem 1.74 (Fermat's Little Theorem) For any prime p and any integer a, prove that $a^p \equiv a \pmod{p}$. Note that this statement is equivalent to: If p is a prime and a is an integer with $(a, p) = 1$, then $a^{p-1} \equiv 1 \pmod{p}$.

Solution If $a = 0$, the result is clear. Let $a > 0$ and we apply mathematical induction on a. The result is true for $a = 1$. Suppose that the congruence is true for a, and we prove it for $a + 1$. By using Problem 1.73, we obtain

$$(a + 1)^p = \sum_{k=0}^{p} \binom{p}{k} a^k \equiv a^p + 1 \equiv a + 1 \pmod{p}.$$

This proves the congruence for positive integers. Since each integer is congruent to a positive integer modulo p, the result follows.   ■

Problem 1.75 (Wilson's Theorem) For each prime p,

$$(p - 1)! \equiv -1 \pmod{p}.$$

Solution If $p = 2$, the result is clear. So, we suppose that $p > 2$. For each $1 \le a \le p - 1$, there exists a unique $1 \le a' \le p - 1$ satisfying $aa' \equiv -1 \pmod{p}$. If $a = 1$ or $a = p - 1$, then $a = a'$. The converse statement also is true, because $a' = a$ implies that $(a - 1)(a + 1) = a^2 - 1 \equiv 0 \pmod{p}$, so that $a \equiv -1 \pmod{p}$ or $a \equiv p - 1 \pmod{p}$. The remaining numbers, i.e., $1 < a < p - 1$, can be considered in $s = (p - 3)/2$ pairs (a, a'), say $(a_1, a_1'), \ldots, (a_s, a_s')$. Therefore, we obtain $(p - 1)! \equiv 1 \times (p - 1) \times (a_1 a_1') \ldots (a_s a_s') \equiv p - 1 \equiv -1 \pmod{p}$, as desired.   ■

1.7   Supplementary Exercises

1. Let U be a set. For any two subsets of U we define $A + B = (A \setminus B) \cup (B \setminus A)$ and $A \cdot B = A \cap B$. Prove that

   (a) $A + B = B + A$;
   (b) $A + \emptyset = A$;
   (c) $A \cdot A = A$;
   (d) $A + A = \emptyset$;

   (e) $A + (B + C) = (A + B) + C$;
   (f) If $A + B = A + C$, then $B = C$;
   (g) $A \cdot (B + C) = A \cdot B = A \cdot C$.

2. Let A, B, and C be three finite sets, prove that

   $$|A \cup B \cup C| = |A| + |B| + |C| - |A \cap B| - |A \cap C| - |B \cap C| + |A \cap B \cap C|.$$

   More generally, for finite sets $A_1, A_2, \ldots, A_n$, prove that

$$\left| \bigcup_{i=1}^{n} A_i \right| = \sum_{i=1}^{n} |A_i| - \sum_{1 \le i < j \le n} |A_i \cap A_j| + \sum_{1 \le i < j < k \le n} |A_i \cap A_j \cap A_k|$$
$$+ \ldots + (-1)^{n-1} |A_1 \cap A_2 \cap \ldots \cap A_n|.$$

3. If $0 < m < n$, how many subsets are there that have exactly m elements?

4. Given sets $A = \{a, b, c, d, e, f\}$ and $B = \{a, b, c, d, e\}$. Compute $|\mathcal{P}(A) \setminus \mathcal{P}(B)|$.

5. Let X and Y be two sets having m and n elements, respectively. How many relations of X into Y exist?

6. Let $X = \mathbb{Z} \times (\mathbb{Z} \setminus \{0\})$. Define a relation R by declaring that $(a, b)R(c, d)$ if and only if $ad = bc$. Prove that the relation R is an equivalence relation.

7. Let S be the set of all integers of the form $2^m 3^n$, $m \ge 0$, $n \ge 0$. Show that there is a one to one correspondence between S and $\mathbb{N}$.

8. If X is a finite set and f is a function of X onto itself, show that f must be one to one.

9. If X is a finite set and f is a one to one function of X into itself, show that f must be onto.

10. Let $f : X \to Y$ be a function, and let $A \subset X$ and $B \subseteq Y$. Prove that

(a) If f is one to one, then $f^{-1}(f(A)) = A$;

(b) If f is onto, then $f(f^{-1}(B)) = B$.

11. Prove that the set of all rational numbers is countable.

12. Let $(X, \preceq)$ be a partially ordered set in which every totally ordered subset has a lower bound. Show that X has a minimal element.

13. A partially ordered set $(L, \preceq)$ is called a *lattice* if every subset $\{x, y\}$ with two elements has a (unique) least upper bound, denoted by $x \vee y$, and a (unique) greatest lower bound, denoted by $x \wedge y$. Prove that a lattice in which every chain has an upper bound has a unique maximal element.

14. If a set has $2n$ elements, show that it has more subsets with n elements than with any other number of elements.

15. Show that
$$\binom{n+r-1}{n} = \sum_{i=0}^{n} \binom{n-i+r-2}{n-i}.$$

Hint: Make use of Theorem 1.43.

16. Find integers n and r such that the following equation is true:
$$\binom{13}{5} + 2\binom{13}{6} + \binom{13}{7} = \binom{n}{r}.$$

17. If $a = p_1^{\alpha_1} \ldots p_k^{\alpha_k}$ and $b == p_1^{\beta_1} \ldots p_k^{\beta_k}$, where the p_i are distinct prime numbers and where each $\alpha_i \ge 0$ and $\beta_i \ge 0$, prove that

(a) $(a, b) = p_1^{\gamma_1} \ldots p_k^{\gamma_k}$, where $\gamma_i = \min\{\alpha_i, \beta_i\}$, for each i;

(b) $[a, b] = p_1^{\delta_1} \ldots p_k^{\delta_k}$, where $\delta_i = \max\{\alpha_i, \beta_i\}$, for each i.

18. Prove that $(a, b) = (a + b, [a, b])$.
19. Show that $n! + 1$ and $(n + 1)! + 1$ are relatively prime for all positive integers.
20. Find a rule for divisibility by 11.
21. Find a simple characterization of all numbers which are sums of two squares.
22. Find the least positive integer n such that $n | 2^n - 2$ but $n \nmid 3^n - 3$.
23. If $a_1 < a_2 < \ldots < a_n$ are integers, show that

$$\prod_{1 \le i \le j \le n} \frac{a_i - a_j}{i - j}$$

is an integer.

Chapter 2
Symmetries of Shapes

In this chapter, we are interested in the symmetric properties of plane figures. By a symmetry of a plane figure we mean a motion of the plane that moves the figure so that it falls back on itself.

2.1 Symmetry

One of the most important and beautiful themes unifying many areas of modern mathematics is the study of symmetry. Many of us have an intuitive idea of symmetry, and we often think about certain shapes or patterns as being more or less symmetric than others. In this chapter we sharpen the concept of "shape" into a precise definition of "symmetry".

Definition 2.1 A *transformation of the plane* is a function $f : \mathbb{R}^2 \to \mathbb{R}^2$.

Transformation involves moving an object from its original position to a new position. The object in the new position is called the image. Each point in the object is mapped to another point in the image.

A geometric shape or object is symmetric if it can be divided into two or more identical pieces that are arranged in an organized fashion. This means that an object is symmetric if there is a transformation that moves individual pieces of the object, but doesn't change the overall shape. The type of symmetry is determined by the way the pieces are organized.

Example 2.2 Symmetry occurs in nature in many ways; for example, the human form is symmetric, see Fig. 2.1.

Example 2.3 The heart carved out is an example of symmetry, see Fig. 2.2.

© The Author(s), under exclusive license to Springer Nature Singapore Pte Ltd. 2021
B. Davvaz, *A First Course in Group Theory*,
https://doi.org/10.1007/978-981-16-6365-9_2

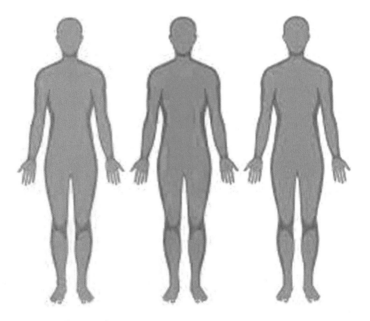

Fig. 2.1 Symmetry of human form

Fig. 2.2 The heart carved
out

Definition 2.4 The symmetry through a line $\mathcal{L}$ is a transformation of the plane which sends point P into point Q such that $\mathcal{L}$ is the midperpendicular to segment PQ. Such a transformation is also called the *axial symmetry* and $\mathcal{L}$ is called the *axis of the symmetry*. If a figure turns into itself under the symmetry through line $\mathcal{L}$, then $\mathcal{L}$ is called the axis of symmetry of this figure.

The line of symmetry can be vertical, horizontal, or diagonal. There may be one or more lines of symmetry, see Fig. 2.3.

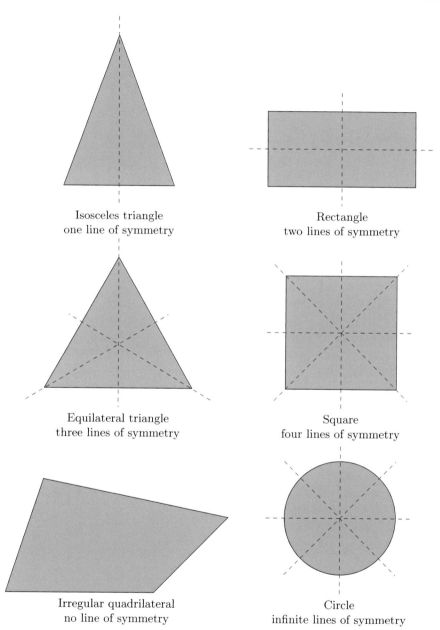

Fig. 2.3 Line of symmetry

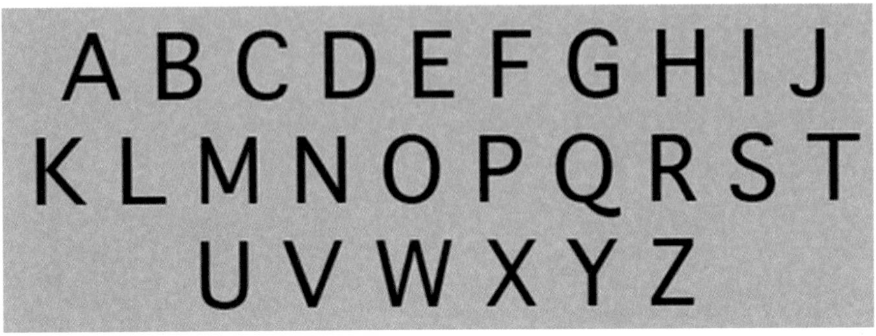

A B C D E F G H I J
K L M N O P Q R S T
U V W X Y Z

Fig. 2.4 Font Geneva

Exercises

1. Prove that the inverse of a bijective transformation is a bijective transformation.
2. The letters in Fig. 2.4 are in the font Geneva. Some of them have one line of symmetry, some have two, some have none, and some have point symmetry. The latter are invariant under a half-turn. Which ones are in the first set? The second? The third? The fourth?

2.2 Translations

Translation is a term used in geometry to describe a function that moves an object a certain distance. The object is not altered in any other way. It is not rotated, reflected, or resized. In a translation, every point of the object must be moved in the same direction and for the same distance.

Definition 2.5 A *translation* is an object from one location to another, without any change in size or orientation.

A horizontal translation refers to an object from left to right or vice versa along the x-axis (the horizontal access). A vertical translation refers to an object up or down along the y-axis (the vertical access). In many cases, a translation will be both horizontal and vertical, resulting in a diagonal object across the coordinate plane, for example, see Fig. 2.5. The trivial translation is the translation through zero distance; all other translations are non-trivial.

Definition 2.6 Let P and Q be two points in a plane. The *translation* from P to Q is transformation $f : \mathbb{R}^2 \to \mathbb{R}^2$ such that

Fig. 2.5 A translation

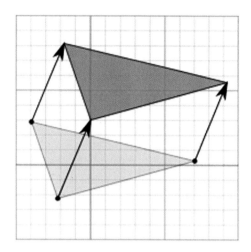

Fig. 2.6 The quadrilateral
$PQB'B$ and $AA'B'B$ are
parallelograms

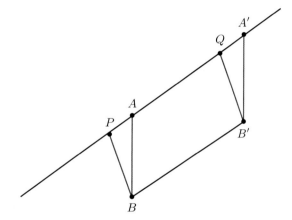

(1) $Q = f(P)$;
(2) If $P = Q$, then f is the identity;
(3) If $P \neq Q$, let A be any point on PQ and let B be any point of PQ; let $A' = f(A)$
   and $B' = f(B)$. Then quadrilaterals $PQB'B$ and $AA'B'B$ are parallelograms,
   see Fig. 2.6.

When $P \neq Q$ one can think of a translation as a slide in the direction of vector
PQ. If A is any point and $\mathcal{L}$ is the line through A parallel to PQ, then $f(A)$ is the
point on $\mathcal{L}$ whose distance from A in the direction of vector PQ is PQ.

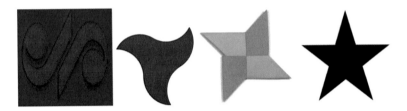

Fig. 2.7 Some shapes with rotational symmetries

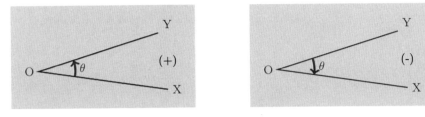

Fig. 2.8 Positively (+) and negatively (−) oriented angle XOY

Exercises

1. Prove that the composition of two translations is a translation.
2. Prove that

    (a) A composition of translations commutes;
    (b) The inverse of a translation is a translation.

2.3   Rotation Symmetries

An equilateral triangle can be rotated by 120°, 240°, or 360° angles without really changing it. If you were to close your eyes, and a friend rotated the triangle by one of those angles, then after opening your eyes you would not notice that anything had changed. In contrast, if that friend rotated the triangle by 33° or 85°, you would notice that the bottom edge of the triangle is no longer perfectly horizontal. Many other shapes that are not regular polygons also have rotational symmetries. Each shape illustrated in Fig. 2.7, for example, has rotational symmetries.

In order to define the rotation we need the notions of *"oriented angle"* and of its *"signed measure"*.

The oriented angle XOY is an angle in which we distinguish the order of its sides OX, OY (see Fig. 2.8). If the transition from OX to OY is opposite to the direction of the clock's hands, then we consider the angle as being "positively oriented", or

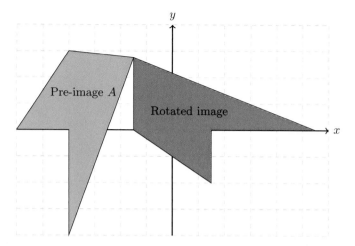

Fig. 2.9 Example of a rotation

simply a positive angle. If the transition is in the same direction as the clock's, we consider the angle as being "negatively oriented", or simply a negative angle.

Definition 2.7 A *rotation* in $\mathbb{R}^2$ is a circular movement of an object around a center of rotation.

Example 2.8 An equilateral triangle can be rotated by 120°, 240°, or 360° angles without really changing it. If you were to close your eyes, and a friend rotated the triangle by one of those angles, then after opening your eyes you would not notice that anything had changed.

Example 2.9 Figure 2.9 shows that the pre-image A is rotated 90° counterclockwise about the center point A to form the rotated image.

Exercises

1. Show that the composition of two rotations is a rotation.
2. Find the image of the ellipse $x^2/4 + y^2/9 = 1$ under the 60° rotation about $(0, 0)$.
3. A rotation of about $(-1, 0)$ is followed by a rotation of about $(1, 0)$. The first rotation is applied again after that. Analyze the composite of these three rotations.

Fig. 2.10 Beautiful reflections in nature

2.4 Mirror Reflection Symmetries

Another type of symmetry that we can find in two-dimensional geometric shapes is mirror reflection symmetry. More specifically, we can draw a line through some shapes and reflect the shape through this line without changing its appearance.

Definition 2.10 A *reflection* is defined by its axis or line of symmetry, i.e., the *mirror line*. Each point $P(x, y)$ is mapped onto the point $P'(x', y')$ which is the mirror image of (x, y) in the mirror line. This yields that PP' is perpendicular to the mirror. A reflection preserves distances.

Example 2.11 Many objects in nature appear the same on the left and right; for instance, see Figs. 2.10 and 2.11. The left half of a butterfly appears the same as the right half, and if we were to place a mirror down the center to reflect the left half, the resulting butterfly would look the same as the original, see Fig. 2.11.

Example 2.12 In Fig. 2.12, we can observe some reflections in nature.

A *glide reflection* is a composition of transformations. In a glide reflection, a translation is first performed on the figure, then it is reflected over a line, which is parallel to the direction of the previous translation. Reversing the order of combining gives the same result. Glide reflections with non-trivial translation have no fixed points. The composition of a reflection in a line and a translation in a perpendicular direction is a reflection in a parallel line. However, a glide reflection cannot be reduced like that. Thus the effect of a reflection on a line combined with a translation in one of

Fig. 2.11 Butterfly and reflection

Fig. 2.12 Some reflections in nature

Fig. 2.13 Footprints fixed
by a glide reflection

the directions of that line is a glide reflection, with a special case as just a reflection. Therefore, the only required information is the translation rule and a line to reflect over, the resulting orientation of the two figures is opposite.

Example 2.13 In your mind, picture the footprints you leave when walking in the sand. Imagine a line $\mathcal{L}$ positioned midway between your left and right footprints. In your mind, slide the entire pattern one-half step in a direction parallel to $\mathcal{L}$ then reflect in line $\mathcal{L}$. The image pattern exactly superimposes on the original pattern. This transformation is an example of a glide reflection with axis $\mathcal{L}$ (see Fig. 2.13).

Alternatively, we can think of a glide reflection with axis $\mathcal{L}$ as a reflection in line $\mathcal{L}$ followed by a translation parallel to $\mathcal{L}$.

Definition 2.14 A non-identity transformation f is an *involution* if and only if $f^2 = id$.

Note that an involution f has the property that $f = f^{-1}$.

Theorem 2.15 *A reflection is an involution.*

Proof Left as an exercise for the reader.   ■

Exercises

1. What conjectures can you make about a figure reflected in two lines?
2. Prove that a non-identity translation is not a reflection.
3. Show that the composition of translations and non-trivial rotation is a rotation.
4. Point P is reflected in two parallel lines, $\mathcal{L}$ and $\mathcal{L}'$, to form P' and P''. The distance from $\mathcal{L}$ to $\mathcal{L}'$ is 10 cm. What is the distance PP''.
5. Words such as **MOM** and **RADAR** that spell the same forward and backward, are called *palindromes*.

    (a) When reflected in their vertical midlines, **MOM** remains **MOM** but the **R**s and **D** in **RADAR** appear backward. Find at least five other words like **MOM** that are preserved under reflection in their vertical midlines.
    (b) When reflected in their horizontal midlines, **MOM** becomes **WOW**, but **BOB** remains **BOB**. Find at least five other words like **BOB** that are preserved under reflection in their horizontal midlines.

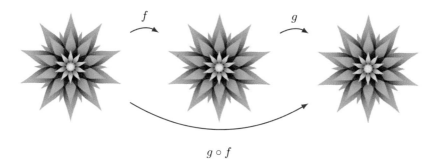

Fig. 2.14 $S(F)$ is closed under composition of functions

6. Show that any glide reflection can be written as the composition $R \circ S$ of a reflection R and a rotation S. Comment on the uniqueness of this decomposition.
7. Show that any glide reflection can be written as the product of reflections in the sides of an equilateral triangle.
8. Find all values for a and b such that $f(x, y) = (ay, x/b)$ is an involution.

2.5 Congruence Transformations

We say that two plane figures are *congruent* if they have the same shape and size. In other words, two plane figures are congruent if one figure can be moved so that it fits exactly on top of the other figure. This movement can always be affected by a sequence of translations, rotations, and reflections. Each part of one figure can be matched with a part of the other figure, and matching angles have the same size, matching intervals have the same length, and matching regions have the same area.

For instance, our reflections in a mirror have the same shape and size as we do, so we would say that we are congruent to our reflection in a mirror.

Definition 2.16 A *congruence transformation* is a transformation under which the image and pre-image are congruent.

We denote the set of all symmetries of a plane figure F by $S(F)$. The elements of $S(F)$ are distance-preserving functions $f : \mathbb{R}^2 \to \mathbb{R}^2$ such that $f(F) = F$. So we can form the composite of any two elements f and g in $S(F)$ to obtain the function $g \circ f : \mathbb{R}^2 \to \mathbb{R}^2$. Let $f, g \in S(F)$. Since f and g both map F to itself, so must $g \circ f$; and since f and g both preserve distance, so must $g \circ f$. Hence $g \circ f \in S(F)$. We describe this situation by saying that the set $S(F)$ is closed under composition of functions, see Fig. 2.14.

Example 2.17 We consider some examples of composition in $S(\square)$, the set of symmetries of the square. Any non-trivial translation alters the location of the square in

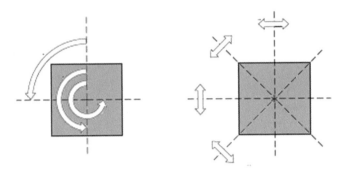

Fig. 2.15 Rotations and reflections of a square

Fig. 2.16 Initial position

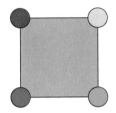

the plane and so cannot be a symmetry of the square. Therefore, we consider only rotations and reflections as potential symmetries of the square. Figure 2.15 shows our labeling for the following elements:

R_0 = Rotation of $0°$	S_0 = Reflection about a horizontal axis
R_1 = Rotation of $90°$	S_1 = Reflection about a vertical axis
R_2 = Rotation of $180°$	S_2 = Reflection about the main diagonal
R_3 = Rotation of $270°$	S_3 = Reflection about the other diagonal

We want to find $R_1 \circ S_0$ and $S_2 \circ R_1$. We consider the initial position as Fig. 2.16 to keep track of the composition of the symmetries.

Figure 2.17 shows the effect of $R_1 \circ S_0$, i.e., first S_0 and then R_1. Comparing the initial and final positions, we observe that the effect of $R_1 \circ S_0$ is to reflect the square in the diagonal from bottom left to top right. This is the symmetry that we have called S_2, and hence $R_1 \circ S_0 = S_2$.

Exercises

1. With the notation given in Example 2.17, find the following composites of symmetries of the square: $R_2 \circ S_3$, $R_2 \circ S_2$, and $R_3 \circ_3$.

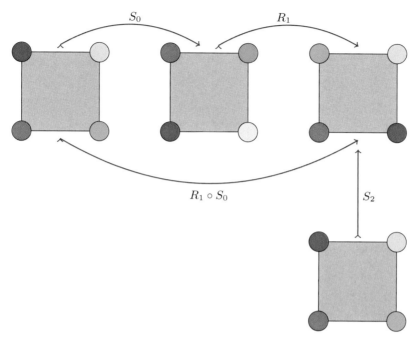

Fig. 2.17 $R_1 \circ S_0 = S_2$

2.6 Worked-Out Problems

Problem 2.18 Let ABC be a triangle with the vertices labeled clockwise such that $AC = BC$ and $\angle ACB = \pi/2$. Let S_{AB} be the reflection in the line AB, S_{AC} be the reflection in the line AC, and R be the rotation by $\pi/2$ counterclockwise around B. Identify the composition $R \circ S_{AB} \circ S_{AC}$.

Solution We can solve problems like this one using the following simple strategy. Find three points which form a triangle and see where the composition of isometries takes them. Next, it's time to guess what the isometry is. If your guess is correct for the three vertices of the triangle, then it must be correct. And this is because the theorem above guarantees that if you know what an isometry does to three corners of a triangle, then you know what the isometry does to every point in the plane. Figure 2.18 shows triangle ABC drawn on a grid of squares. Since we want to choose three points which form a triangle, we may as well choose the points A, B, and C.

- It is easy to check that $S_{AC}(A) = A$, $S_{AB}(A) = A$ and $R(A) = P$. In other words, $R \circ S_{AB} \circ S_{AC}(A) = P$;
- It is easy to check that $S_{AC}(B) = Q$, $S_{AB}(Q) = N$ and $R(N) = Q$. In other words, $R \circ S_{AB} \circ S_{AC}(B) = Q$;
- It is easy to check that $S_{AC}(C) = C$, $S_{AB}(C) = M$ and $R(M) = C$. In other words, $R \circ S_{AB} \circ S_{AC}(C) = C$.

Fig. 2.18 Triangle ABC
drawn on a grid of squares

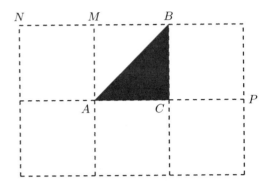

Hence, can you think of an isometry which takes A to P, B to Q, and C to C? If
you think hard enough, you should realize that it's just a rotation by π around C. So,
we have managed to deduce that the composition $R \circ S_{AB} \circ S_{AC}$ is a rotation by π
around C.   ∎

Problem 2.19 Let $ABCD$ be a rectangle with the vertices labeled counterclockwise
such that $BC = 2AB$. Suppose that

- S_{AB} is the reflection in the line AB;
- R_B is the counterclockwise rotation by $\pi/2$ about B;
- T_{DB} is the translation which takes D to B;
- G_{CD} is the glide reflection in the line CD which takes C to D.

Identify the composition $S_{AB} \circ R_B \circ T_{DB} \circ G_{CD}$.

Solution Figure 2.19 shows rectangle $ABCD$ drawn on a grid of squares. Since we
want to choose three points which form a triangle, we may as well choose the points
A, B, and C.

- It is easy to check that $G_{CD}(A) = E, T_{DB}(E) = D, R_B(D) = F$ and $S_{AB}(F) = J$.
  In other words, $S_{AB} \circ R_B \circ T_{DB} \circ G_{CD}(A) = J$;
- It is easy to check that $G_{CD}(B) = H, T_{DB}(H) = C, R_B(C) = I$, and $S_{AB}(I) = I$.
  In other words, $S_{AB} \circ R_B \circ T_{DB} \circ G_{CD}(B) = I$;
- It is easy to check that $G_{CD}(C) = D, T_{DB}(D) = B, R_B(B) = B$, and $S_{AB}(B) = B$. In other words, $S_{AB} \circ R_B \circ T_{DB} \circ G_{CD}(C) = B$.

So can you think of an isometry which takes A to J, B to I, and C to B? If you think
hard enough, you should realize that it is a rotation, although you might not be sure of
where the center lies. However, we can use the fact that if a rotation takes X to Y, then
the center of rotation must lie on the perpendicular bisector of xy. In particular, the
center of the rotation that we are interested in must lie on the perpendicular bisector
of AJ as well as the perpendicular bisector of BI. And there is only one point which
does that namely, the point O labeled in Fig. 2.19. It is now easy to deduce that the
composition must be a rotation about O by $\angle AOJ = \pi/2$ in the clockwise direction.
                                                                                         ∎

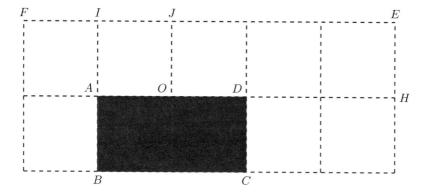

Fig. 2.19 Rectangle $ABCD$ drawn on a grid of squares

Fig. 2.20 What letter has
been folded once to make
this shape?

Fig. 2.21 A ray of light is
reflected by two
perpendicular flat mirrors

2.7 Supplementary Exercises

1. Prove that the composition of two reflections with parallel axes is a translation
   perpendicular to these axes by a distance twice that from the first axis to the
   second.
2. Prove that the composition of two reflections with axes meeting at a point is a
   rotation about that point through an angle twice that from the first axis to the
   second.

3. Prove that the composition of three reflections is a reflection if three axes are parallel or concurrent, and otherwise is a glide reflection.
4. Five reflections are composed, with axes in order the lines $x = 0$, $x + y = 6$, $y = 6$, $y = x + 2$, $y = x + 8$. Is the composition a reflection or a glide reflection? Give details.
5. What capital letters could be cut out of paper and given a single fold to produce Fig. 2.20?
6. Give an example of a bijection $f : \mathbb{R}^2 \to \mathbb{R}^2$ that preserves angles but not distances. Describe in general terms the effect of f on lines, circles, and triangles.
7. Prove that the composition of a non-trivial rotation and a reflection is glide reflection except when the axis of the reflection passes through the center of the rotation, in which case it is a reflection.
8. A ray of light is reflected by two perpendicular flat mirrors. Prove that the emerging ray is parallel to the initial incoming ray, as indicated in Fig. 2.21.

Chapter 3
Groups

The theory of groups is the oldest branch of modern algebra. The concept of a group is surely one of the central ideas of mathematics. A group is a set in which you can perform one operation with some nice properties. In this chapter we introduce the basic concepts of group theory.

3.1 A Short History of Group Theory

The concept of a group is one of the most fundamental in modern mathematics. Group theory can be considered the study of symmetry: the collection of symmetries of some object preserving some of its structure forms a group; in some sense all groups arise this way. Although permutations had been studied earlier, the theory of groups really began with Galois (1811–1832) who demonstrated that polynomials are best understood by examining certain groups of permutations of their roots. Since that time, groups have arisen in almost every branch of mathematics. There are three historical roots of group theory:

(1) The theory of algebraic equations;
(2) Number theory;
(3) Geometry.

Euler, Gauss, Lagrange, Abel, and Galois were early researchers in the field of group theory. Galois is honored as the first mathematician linking group theory and field theory, with the theory that is now called Galois theory.

Permutations were first studied by Lagrange (1770, 1771) on the theory of algebraic equations. Lagrange's main object was to find out why cubic equations could be solved algebraically. In studying the cubic, for example, Lagrange assumes the roots of a given cubic equation are x', x'', and x'''. Then, taking 1, w, and w^2 as the cube roots of unity, he examines the expression

B. Davvaz, *A First Course in Group Theory*,
https://doi.org/10.1007/978-981-16-6365-9_3

$$R = x' + wx'' + w^2 x'''$$

and notes that it takes just two different values under the six permutations of the roots x', x'', and x'''. But, he could not fully develop this insight because he viewed permutations only as rearrangements, and not as bijections that can be composed. The composition of permutations does appear in the works of Ruffini and Abbati about 1800; in 1815 Cauchy established the calculus of permutations.

Galois found that if $r_1, r_2, \ldots, r_n$ are the n roots of an equation, there is always a group of permutations of the rs such that every function of the roots invariable by the substitutions of the group is rationally known, and conversely, every rationally determinable function of the roots is invariant under the substitutions of the group. Galois also contributed to the theory of modular equations and to that of elliptic functions. His first publication on the group theory was made at the age of 18 (1829), but his contributions attracted little attention until the publication of his collected papers in 1846.

The number-theoretic strand was started by Euler and taken up by Gauss, who developed modular arithmetic and considered additive and multiplicative groups related to quadratic fields. Indeed, in 1761, Euler studied modular arithmetic. In particular, he examined the remainders of powers of a number modulo n. Although Euler's work is, of course, not stated in group-theoretic terms, he does provide an example of the decomposition of an abelian group into cosets of a subgroup. He also proves a special case of the order of a subgroup is being a divisor of the order of the group.

Gauss in 1801 was to take Euler's work much further and gives a considerable amount of work on modular arithmetic which amounts to a fair amount of theory of abelian groups. He examines orders of elements and proves (although not in this notation) that there is a subgroup for every number dividing the order of a cyclic group. Gauss also examined other abelian groups. He looked at binary quadratic forms

$$ax^2 + 2bxy + cy^2,$$

where a, b, and c are integers. Gauss examined the behavior of forms under transformations and substitutions. He partitions forms into classes and then defines a composition on the classes. Gauss proves that the order of composition of three forms is immaterial, so in modern language, the associative law holds. In fact, Gauss has a finite abelian group, and later (in 1869) Schering, who edited Gauss's works, found a basis for this abelian group.

Geometry has been studied for a very long time, so it is reasonable to ask what happened to geometry at the beginning of the nineteenth century that contributed to the rise of the group concept. Geometry had begun to lose its *metric* character with projective and non-Euclidean geometries being studied. Also the movement to study geometry in n dimensions led to an abstraction in geometry itself. The difference between metric and incidence geometry comes from the work of Monge, his student Carnot, and perhaps, most importantly, the work of Poncelet. Non-Euclidean geometry was studied by Lambert, Gauss, Lobachevsky, and, János Bolyai, among others.

Möbius in 1827, although he was completely unaware of the group concept, began to classify geometries using the fact that a particular geometry studies properties invariant under a particular group. Steiner in 1832 studied notions of synthetic geometry which were to eventually become part of the study of transformation groups.

Arthur Cayley and Augustin Louis Cauchy were among the first to appreciate the importance of the theory, and to the latter especially are due to a number of important theorems. The subject was popularized by Serret, Camille Jordan, and Eugen Netto. Other group theorists of the nineteenth century were Bertrand, Charles Hermite, Frobenius, Leopold Kronecker, and Emile Mathieu.

It was Walther von Dyck who, in 1882, gave the modern definition of a group.

The study of what are now called Lie groups, and their discrete subgroups, as transformation groups, started systematically in 1884 by Sophus Lie; followed by the works of Killing, Study, Schur, Maurer, and Cartan. The discontinuous (discrete group) theory was built up by Felix Klein, Lie, Poincaré, and Charles Emile Picard, in connection in particular with modular forms.

Other important mathematicians in this subject area include Emil Artin, Emmy Noether, Ludwig Sylow, and many others (Fig. 3.1).

3.2 Binary Operations

We are used to addition and multiplication of real numbers. These operations combine two real numbers to generate a unique single real number. So, we can look at these operations as functions on the set $\mathbb{R} \times \mathbb{R}$ defined by

$$+ : \mathbb{R} \times \mathbb{R} \to \mathbb{R}$$
$$(a, b) \to a + b$$

and

$$\cdot : \mathbb{R} \times \mathbb{R} \to \mathbb{R}$$
$$(a, b) \to a \cdot b$$

These operations are examples of binary operations. The general definition of a binary operation is as follows:

Definition 3.1 Let S be a non-empty set. A *binary operation* on S is a function $\star : S \times S \to S$ that maps each ordered pair (x, y) of S to an element $\star(x, y)$ of S. For convenience we write $a \star b$ instead of $\star(a, b)$. The set S is said to be *closed under the operation* $\star$. The pair $(S, \star)$ (or just S, if there is no fear of confusion) is called a *groupoid*.

Binary operations are usually denoted by special symbols $+, -, \cdot, \times, \star, *, \circ, \oplus, \odot, \vee, \wedge, \cup, \cap$ rather than by letters.

Example 3.2 The ordinary operations of addition "$+$", subtraction "$-$", and multiplication "$\cdot$" are binary operations on $\mathbb{Z}$, $\mathbb{Q}$, and $\mathbb{R}$. Subtraction is not a binary

(a) L. Euler (1707–1783)

(b) J. L. Lagrange (1736–1813)

(c) A. F. Möbius (1790–1868)

(d) J. Steiner (1796–1863)

(e) N. H. Abel (1802–1829)

(f) É. Galois (1811–1832)

(g) A. Cayley (1821–1895)

(h) L. Sylow (1832–1918)

(i) C. Jordan (1838–1922)

(j) F. Klein (1849–1925)

(k) E. Noether (1882–1935)

(l) E. Artin (1898–1962)

Fig. 3.1 Pictures of some famous mathematicians

operation on $\mathbb{N}$, because $3 - 7$ is not in $\mathbb{N}$. Division is not a binary operation on $\mathbb{Q}$, however, division is a binary operation on $\mathbb{Q} \setminus \{0\}$.

Example 3.3 Exponential operation $a \star b = a^b$ is a binary operation on the set $\mathbb{N}$ of natural numbers while it is not a binary operation on the set $\mathbb{Z}$ of integers.

Binary operators can be defined on arbitrary sets, not only sets of numbers.

Example 3.4 We might consider a set S of colors, and define a binary operation $\oplus$ which tells us how to combine two colors to form another color. If Red, Blue, Green, Yellow, and Purple are elements of S, then we can write Red $\oplus$ Blue $=$ Purple, since we can combine the first two colors to make the third.

Example 3.5 If X is a set, then union "$\cup$" and intersection "$\cap$" are binary operations on $\mathcal{P}(X)$.

Example 3.6 Let $\mathbb{P}$ be the set of all propositions. Then, "*and*" and "*or*" are binary operations on $\mathbb{P}$. In mathematical logic "*and*" is usually represented by $\wedge$, and "*or*" is represented by $\vee$.

Example 3.7 We noted earlier that binary operations can act not only on numbers but also on arbitrary elements, such as colors or other sets. Here we consider the set $S = \{R_0, R_1, R_2, R_3, S_0, S_1, S_2, S_3\}$ as defined in Example 2.17, a set of geometric transformations, which we can combine. We can consider the binary operation of combining these geometric transformations by doing one and then doing the other.

Definition 3.8 One way of describing a binary operation $\star$ on a set S (provided S is not too big) is to form a grid with rows and columns labeled by the elements of S, and enter the element $a \star b$ in the cell in row a and column b (for all $a, b \in S$). This is called a *multiplication table* or a *Cayley table*.

Let $\star$ be be a binary operation on a finite set $S = \{a_1, a_2, \ldots, a_n\}$ having n elements. We construct a Cayley table for $\star$ as follows:

$\star$	a_1	a_2	$\ldots$	a_n
a_1	$a_1 \star a_1$	$a_1 \star a_2$	$\ldots$	$a_1 \star a_n$
a_2	$a_2 \star a_1$	$a_2 \star a_2$	$\ldots$	$a_2 \star a_n$
$\vdots$	$\vdots$	$\vdots$	$\vdots$	$\vdots$
a_n	$a_n \star a_1$	$a_n \star a_2$	$\ldots$	$a_n \star a_n$

Let $\star$ be a binary operation on a set S. There are a number of interesting properties that a binary operation may or may not have. Specifying a list of properties that a binary operation must satisfy will allow us to define deep mathematical objects such as groups.

- $\star$ is *commutative* if $a \star b = b \star a$, for all $a, b \in S$.
- $\star$ is *associative* if $a \star (b \star c) = (a \star b) \star c$, for all $a, b, c \in S$.

Fig. 3.2 Cayley table of a
commutative binary
operation

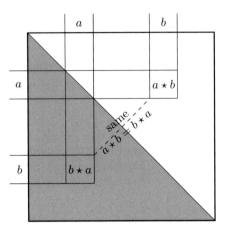

Let S be finite. The property that $a \star b = b \star a$, for all $a, b \in S$, means that the Cayley table must be symmetric across the main diagonal, see Fig. 3.2.

Example 3.9 (1) Let $S = \mathbb{R}$ and $a \star b = a$, for all $a, b \in S$. Then $3 \star 5 = 3$ and $5 \star 3 = 5$, and so $\star$ is not commutative. However, if $a, b, c \in S$, then $a \star (b \star c) = a = a \star b = (a \star b) \star c$, and hence $\star$ is associative.

(2) Let $S = \mathbb{Q} \setminus \{0\}$ and $a \star b = a/b$, for all $a, b \in S$. Then $1 \star 3 = 1/3 \neq 3 = 3/1$, and so $\star$ is not commutative. Also, we have

$$1 \star (2 \star 3) = 1 \star (2/3) = 3/2 \neq 1/6 = (1/2) \star 3 = (1 \star 2) \star 3.$$

Hence $\star$ is not associative.

Exercises

1. Is $(\mathbb{Z}, \star)$ a groupoid if $\star$ is defined, for each $x, y \in \mathbb{Z}$, by
   (a) $x \star y = \sqrt{x + y}$;
   (b) $x \star y = (x + y)^2$;
   (c) $x \star y = x - y - xy$;
   (d) $x \star y = 0$.

2. Define $\star$ on $\mathbb{Q}$ as $a \star b = ab + 1$, for all $a, b \in \mathbb{Q}$. Is $\star$ associative (prove or find a counterexample)?

3. Prove that if $\star$ is an associative and commutative binary operation on a set S, then $(a \star b) \star (c \star d) = ((d \star c) \star a) \star b$, for all $a, b, c, d \in S$.

4. Let S be a finite set containing n elements. Find the total number of binary operations on S.

5. Let S be a finite set containing n elements. Show that the total number of commutative binary operation on S is $n^{n(n-1)/2}$.

3.3 Semigroups and Monoids (Optional)

A semigroup is an algebraic structure consisting of a non-empty set together with an associative binary operation. The formal study of semigroups began in the early twentieth century. Semigroups are important in many areas of mathematics. We give here some basic definitions and very basic results concerning semigroups.

Definition 3.10 A *semigroup* is a pair $(S, \star)$ in which S is a non-empty set and $\star$ is a binary associative operation on S.

Semigroups are therefore one of the most basic types of algebraic structure.
For an element $x \in S$ we let x^n be the product of x with itself n times. So, $x^1 = x$, $x^2 = x \star x$, and $x^{n+1} = x^n \star x$ for $n \geq 1$.
Let $x_1, x_2, \ldots, x_n$ be a sequence of elements of a semigroup $(S, \star)$. We define

$$p_1 = x_1,$$
$$p_k = p_{k-1} \star x_k, \quad \text{for } k > 1.$$

Now an arbitrary product of n elements of S is determined by an expression involving n elements of S together with equal numbers of left and right parentheses that determine the order in which the product is evaluated. The general associative law ensures that the value of such a product is determined only by the order in which the elements of the semigroup occur within that product. Thus a product of n elements of S has the value $x_1 \star x_2 \star \ldots \star x_n$, where $x_1, x_2, \ldots, x_n$ are the elements to be multiplied, listed in the order in which they occur in the expression defining the product.

Example 3.11 Given four elements x_1, x_2, x_3, and x_4 of a semigroup $(S, \star)$, the products

$$((x_1 \star x_2) \star x_3) \star x_4, \quad (x_1 \star x_2) \star (x_3 \star x_4), \quad (x_1 \star (x_2 \star x_3)) \star x_4,$$
$$x_1 \star ((x_2 \star x_3) \star x_4), \quad x_1 \star (x_2 \star (x_3 \star x_4)),$$

all have the same value. Note that according to the above definition, $p_4 = ((x_1 \star x_2) \star x_3) \star x_4$.

Theorem 3.12 (General Associative Law) *Let $(S, \star)$ be a semigroup and $x_1, x_2, \ldots, x_n \in S$. Every way of inserting balanced pairs of brackets into the product $x_1 \star x_2 \star \ldots \star x_n$ give the same result.*

Proof We prove that any insertion of brackets into the product gives the same result as $x_1 \star (x_2 \star (x_3 \star \ldots \star x_n) \ldots)$. We proceed by mathematical induction on n. For

$n = 1$, the result is trivially true, for there is only one way to insert balanced pairs of brackets into the product x_1. This is the base case of the induction.

So, assume that the result holds for all $n < k$; we aim to show it is true for k. Take some bracketing of the product $x_1, x_2, \ldots, x_k$ and let y be the result. This bracketing is a product of some bracketing of $x_1, \ldots, x_j$ and some bracketing of $x_{j+1}, \ldots, x_k$ for some $1 \leq j < k$. Now, we consider the following two cases:

Case 1: Suppose that $j = 1$. By the assumption, the result of inserting brackets into $x_{j+1} \star \ldots \star x_k = x_2 \star \ldots \star x_k$ is equal to $x_2 \star (x_3 \star (\ldots x_k) \ldots)$. Thus, $y = x_1 \star (x_2 \star (x_3 \star (\ldots x_k) \ldots))$, which is the result with $n = k$.

Case 2: Suppose that $j > 1$. By the assumption, the result of the bracketing of $x_1 \star \ldots \star x_j$ is $x_1 \star (x_2 \star (\ldots x_j) \ldots)$ and the result of the bracketing of $x_{j+1} \star \ldots \star x_k$ is $x_{j+1} \star (x_{j+2} \star (\ldots x_k) \ldots)$. Consequently, we obtain

$$
\begin{aligned}
y &= \big(x_1 \star (x_2 \star (\ldots x_j) \ldots)\big)\big(x_{j+1} \star (x_{j+2} \star (\ldots x_k) \ldots)\big) \\
&= x_1 \star \big(((x_2 \star (\ldots x_j) \ldots)) \star (x_{j+1} \star (x_{j+1} \star (\ldots x_k) \ldots))\big) \text{ (by associativity)} \\
&= x_1 \star (x_2 \star (x_3 \ldots x_k) \ldots) \text{ (by assumption } n = k - 1),
\end{aligned}
$$

which is the result with $n = k$.

Therefore, by induction, the result holds for all n.   ∎

A semigroup S is *finite* if it has only a finitely many elements. A semigroup S is *commutative*, if it satisfies $x \star y = y \star x$, for all $x, y \in S$. If there exists e in S such that for all $x \in S$,

$$
e \star x = x \star e = x
$$

we say that S is a *semigroup with identity* or (more usual) a *monoid*. The element e of S is called *identity*.

Proposition 3.13 *Let e be a left identity of S and e' is a right identity of S, then $e = e'$. Consequently, a semigroup contains at most one identity.*

Proof Since e is a left identity, it follows that $e \star e' = e'$. Since e' is a right identity, it follows that $e = e \star e'$. Hence, we get $e = e \star e' = e'$, as desired.   ∎

By Proposition 3.13, the identity element is unique and we shall generally denote it by 1.

Example 3.14 Let $\mathbb{N}^0 = \mathbb{N} \cup \{0\}$ be the set of all non-negative integers. Then, $(\mathbb{N}^0, \cdot)$ is a semigroup for the usual multiplication of integers. Also, $(\mathbb{N}^0, +)$ is a semigroup, when $+$ is the ordinary addition of integers. Define $(\mathbb{N}^0, \star)$ by $n \star m = \max\{n, m\}$. Then, $(\mathbb{N}^0, \star)$ is a semigroup, since

$$
\begin{aligned}
n \star (m \star k) &= \max\{n, \max\{m, k\}\} = \max\{n, m, k\} \\
&= \max\{\max\{n, m\}, k\} = (n \star m) \star k.
\end{aligned}
$$

Example 3.15 Let S be a non-empty set. There are two simple semigroup structures on S: with the multiplication given by $x \star y = x$, for all $x, y \in S$, in this case, the

Fig. 3.3 The rectangular band

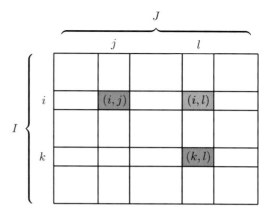

semigroup $(S, \star)$ is called the *left zero semigroup* over S. Also, fixing an element $a \in S$ and putting $x \star y = a$, for all $x, y \in S$, gives a semigroup structure on S.

Example 3.16 The *opposite semigroup* S^{op} of S is the semigroup with the same set as S but "*reversed multiplication*". That is, for $x, y \in S$, the product $x \star_{op} y$ in S^{op} is equal to the product $y \star x$ in S. It is easy to check that S^{op} is indeed a semigroup. Notice that if S is commutative, then S^{op} and S are the same semigroup.

An element x in a semigroup S is said to be *idempotent* if $x^2 = x$.

Example 3.17 Let I, J be two non-empty sets and set $T = I \times J$ with the binary operation

$$(i, j) \star (k, l) = (i, l).$$

Then, we have

$$\big((i, j) \star (k, l)\big) \star (m, n) = (i, l) \star (m, n) = (i, n),$$
$$(i, j) \star \big((k, l) \star (m, n)\big) = (i, l) \star (k, n) = (i, n),$$

for all $(i, j), (k, l)$ $(m, n) \in T$, and so the multiplication is associative. Hence, $(T, \star)$ is a semigroup called the *rectangular band* on $I \times J$. Note that $(i, j)^2 = (i, j) \star (i, j) = (i, j)$, i.e., every element is an idempotent. The name derives from the observation that if the members of $I \times J$ are pictured in a rectangular grid in the obvious fashion, then the product of two elements lies at the intersection of the row of the first member and the column of the second, see Fig. 3.3.

Example 3.18 The *direct product* $S \times T$ of two semigroups $(S, \cdot)$ and $(T, \circ)$ is defined by

$$(s_1, t_1) \star (s_2, t_2) = (s_1 \cdot s_2, t_1 \circ t_2) \ (s_1, s_2 \in S, \ t_1, t_2 \in T).$$

It is easy to show that the so defined product is associative and hence the direct product is, indeed, a semigroup. The direct product is a convenient way of combining two semigroup operations. The new semigroup $S \times T$ inherits properties of both S and T.

Example 3.19 On the Cartesian product $\mathbb{Z} \times \mathbb{Z}$ we define a binary operation as follows:

$$(a, b) \star (c, d) = \begin{cases} (a - b + c, d) & \text{if } b < c \\ (a, d) & \text{if } b = c \\ (a, d + b - c & \text{if } b > c, \end{cases}$$

for all $a, b, c, d \in \mathbb{Z}$. The set $\mathbb{Z} \times \mathbb{Z}$ with such defined operation is a semigroup.

Example 3.20 Let $(S, \star)$ be a semigroup, which is not a monoid. Find a symbol 1 such that $1 \notin S$. Now, we extend the multiplication on S to $S \cup \{1\}$ by

$$a \star b = a \star b \qquad \text{if } a, b \in S,$$
$$a \star 1 = a = 1 \star a \text{ for all } a \in S,$$
$$1 \star 1 = 1.$$

Then, $\star$ is associative. Thus, we have managed to extend multiplication in S to $S \cup \{1\}$. For an arbitrary semigroup S the monoid S^1 is defined by

$$S^1 = \begin{cases} S & \text{if } S \text{ is a monoid,} \\ S \cup \{1\} & \text{if } S \text{ is not a monoid.} \end{cases}$$

Therefore, S^1 is "S with a 1 *adjoined*" if necessary.

In a semigroup $(S, \star)$ an element z_l such that $z_l \star x = z_l$ for every $x \in S$ is called a *left zero element* of S, and an element z_r such that $x \star z_r = z_r$ for every $x \in S$ is called a *right zero element* of S. If z is both a left and a right zero element of S, then z is called a *two-sided zero element*, or simply a *zero element*, of S. A semigroup S may have any number of left (or of right) zero elements, but if it has a left zero element z_l and a right zero element z_r, then $z_l = z_l \star z_r = z_r$, whence S has a unique two-sided zero element and no other left or right zero element. Usually, we denote zero element by 0.

Example 3.21 Let $(S, \star)$ be a semigroup. Similar to Example 3.20, let 0 be an element not in S and extend the multiplication on S to $S \cup \{0\}$ by

$$0 \star x = x \star 0 = 0 \star 0 = 0,$$

for all $x \in S$. Again, this extended multiplication is associative. Thus, $S \cup \{0\}$ is a semigroup with zero element. We define

$$S^0 = \begin{cases} S & \text{if } S \text{ has a zero,} \\ S \cup \{0\} & \text{otherwise.} \end{cases}$$

Then, S^0 is called the *semigroup obtained by adjoining a zero* to S if necessary.

Theorem 3.22 *If $(S, \star)$ is a finite semigroup, then there is an idempotent element in S.*

Proof Since S is finite, it follows that $a, a^2, a^3, \ldots$ cannot be distinct elements. So, there exist integers i and j with $i < j$ such that $a^i = a^j$. This implies that $a^{i+k} = a^i$, where $k = j - i$. Now, we have

$$a^{2i+k} = a^i \star a^{i+k} = a^{2i}.$$

Next, by induction, we observe that

$$a^{ni+k} = a^{ni},$$

for all positive integer n. Also, we have

$$a^{ni+2k} = a^{ni+k} \star a^k = a^{ni} \star a^k = a^{ni+k} = a^{ni},$$
$$a^{ni+3k} = a^{ni+2k} \star a^k = a^{ni+2k} = a^{ni},$$

and so on. Therefore, we deduce that

$$a^{ni+rk} = a^{ni},$$

for all positive integer r. In particular, we have $a^{ki+ki} = a^{ki}$, or $a^{2ki} = a^{ki}$. Now, if we take $m = ki$, then a^m is an idempotent element. ∎

Let $(S, \star)$ be a semigroup. An element b of the semigroup S is called a *right divisor* of the element a of the same semigroup if there exists an element $x \in S$ such that $x \star b = a$. An element b is called a *left divisor* of a if there exists an element $y \in S$ such that $b \star y = a$. If b is a right divisor of a, we say that a is *divisible on the right* by b. If b is a left divisor of a, we say that a is *divisible on the left* by b. If the element b of S is a right divisor of the element a of the same semigroup, then the element x satisfying the equation $x \star b = a$ is called a *left inverse* of b with respect to a. The notion of *right inverse* is defined analogously. An element which is both a right inverse and a left inverse of b with respect to a is called a *two-sided inverse*, or shortly an *inverse*, of b with respect to a.

Definition 3.23 Let M be a monoid and $x \in M$. If there exists an element x' such that $x \star x' = 1$, then x' is a *right inverse* for x, and x is *right invertible*. Similarly, suppose that there exists an element x'' such that $x'' \star x = 1$, then x'' is a *left inverse* for x, and x is *left invertible*. If x is both left and right invertible, then x is *invertible*.

Proposition 3.24 *Let M be a monoid, and let $x \in M$. Suppose that x is invertible and let x' be a right inverse of x and x'' be a left inverse of x. Then, $x' = x''$.*

Proof Since x' and x'' are right and left inverses of x, respectively, it follows that $x \star x' = 1$ and $x'' \star x = 1$. Thus, we have

$$x' = 1 \star x' = (x'' \star x) \star x' = x'' \star (x \star x') = x'' \star 1 = x'',$$

as desired.   ∎

Proposition 3.24 says that right and left inverses coincide when they both exist. The existence of one does not imply the existence of the other. Thus, if x is an invertible element of a monoid M, denote the unique right and left inverse of x by x^{-1}.

Definition 3.25 A *subsemigroup* A of a semigroup S is a non-empty subset of S such that $A \star A \subseteq A$, i.e., $x, y \in A$ implies $x \star y \in A$. A *submonoid* A of a monoid M is a subsemigroup A which contains the identity 1 of M.

Theorem 3.26 *The set of invertible elements of a monoid forms a submonoid.*

Proof Let I be the set of invertible elements of a monoid M. Since $1 \in I$, it follows that I is non-empty. Suppose that x and y are two arbitrary elements of I. Since x and y are invertible, it follows that

$$(y^{-1} \star x^{-1}) \star (x \star y) = y^{-1} \star (x^{-1} \star x) \star y = y^{-1} \star y = 1,$$
$$(x \star y) \star (y^{-1} \star x^{-1}) = x \star (y \star y^{-1}) \star x^{-1} = x \star x^{-1} = 1.$$

Hence, $x \star y$ is invertible and so $x \star y \in I$. Consequently, I is a subsemigroup of M. In addition, $1 \in I$ is an identity for I. Therefore, I is a submonoid of M.   ∎

Definition 3.27 A semigroup $(S, \star)$ is *left cancellative*, if

$$c \star a = c \star b \Rightarrow a = b,$$

and $(S, \star)$ is *right cancellative*, if

$$a \star c = b \star c \Rightarrow a = b.$$

If $(S, \star)$ is both left and right cancellative, then it is *cancellative*.

Exercises

1. Let $(\mathbb{Q}, \star)$ be a groupoid, where $a \star b = a + b - ab$, for all $a, b \in \mathbb{Q}$.

   (a) Is the groupoid $(\mathbb{Q}, \star)$ a semigroup?
   (b) Is there an identity element in $(\mathbb{Q}, \star)$?
   (c) Which elements of the groupoid have inverses?

2. If $\mathbb{Q}$ is replaced by $\mathbb{Z}$ in Exercise 1, are the solutions the same?
3. A set S has n elements, what is the number of commutative binary operation of S?
4. Prove that if S is a semigroup and $e \in S$ is both a right zero and a right identity, then S is trivial.
5. In the set of all continuous functions of two variables x and y, we define in the square $0 \le x \le a, 0 \le y \le a$, the following operation, which plays an important role in the theory of integrals. The result of this operation carried out for the functions $K_1(x, y)$ and $K_2(x, y)$ is the function

$$\int_0^a K_1(x, t) K_2(t, y) dt.$$

By using the simplest properties of integrals, show that we obtain a semigroup.
6. We consider the set of functions of one variable which are absolutely integrable for $0 \le x < \infty$. In many branches of mathematics, one considers the operation in this set, the result of which for $f_1(x)$ and $f_2(x)$ is the function

$$\int_0^x f_1(t) f_2(x - t) dt.$$

Show that this operation is associative and commutative.
7. Prove that for any commutative monoid M, the set of idempotent elements of M forms a submonoid.
8. Let S be a left cancellative semigroup. Suppose that $e \in S$ is an idempotent. Prove that e is a left identity. Deduce that a cancellative semigroup can contain at most one idempotent, which must be an identity.
9. Give an example of a semigroup containing more than one idempotent but containing no identity.

3.4 Groups and Examples

In this section, we embark on the study of the algebraic object known as a group. We give a few basic definitions and then concentrate on examples of groups.

Definition 3.28 Let G be a non-empty set and $\star$ be a binary operation on G. We say that $(G, \star)$ is a *group* if it satisfies the following conditions:

(1) *Associativity Law:* $a \star (b \star c) = (a \star b) \star c$, for all $a, b, c \in G$;
(2) *Existence of Identity:* There exists an element $e \in G$ called an *identity* such that $a \star e = e \star a = a$, for all $a \in G$;
(3) *Existence of Inverse:* For each $a \in G$, there exists an element $b \in G$ such that $a \star b = b \star a = e$ (we write this element b as a^{-1} and call it the *inverse* of a in G).

Hence, a group G is a monoid G that satisfies the condition (3). In other words, a group G is a semigroup with an identity in which every element has an inverse.

Remark 3.29 In many books, there is an extra condition in the definition of group, that is, "*G is closed under* $\star$". In other words, $x \star y \in G$, for all $a, b \in G$. But according to our definition, this condition is a result of the definition of binary operation.

Corollary 3.30 *If G is a group, then*

(1) The identity element of G is unique;
(2) Every $a \in G$ has a unique inverse in G.

Proof Since any group is a semigroup and a monoid, the result follows from Propositions 3.13 and 3.24. ∎

Definition 3.31 The number of elements in a group G is called the *order* of G and is denoted by $|G|$. If $|G|$ is finite, then G is said to be a *finite group*; otherwise G is an *infinite group*.

Before starting to look into the nature of groups, we look at some examples.

Example 3.32 (*Additive Groups*) The set of integers $\mathbb{Z}$, the set of rational numbers $\mathbb{Q}$, the set of real numbers $\mathbb{R}$, and the set of complex numbers $\mathbb{C}$ are all groups under ordinary addition.

Example 3.33 If E is the set of even integers, then E is a group under addition of integers.

Example 3.34 The sets $\mathbb{Q}^* = \mathbb{Q} \setminus \{0\}$ and $\mathbb{R}^* = \mathbb{R} \setminus \{0\}$ are groups under ordinary multiplication.

Example 3.35 Let G be the set of real numbers of the form $a + b\sqrt{2}$, where $a, b \in \mathbb{Q}$ and are not simultaneously zero. Then, G is a group under the usual multiplication of real numbers.

Example 3.36 Let $\mathbb{C}^*$ be the set of all non-zero complex numbers, i.e.,

$$\mathbb{C}^* = \{a + bi \mid a, b \in \mathbb{R} \text{ and } (a \neq 0 \text{ or } b \neq 0)\}.$$

Recall that $i^2 = -1$. We define multiplication of complex numbers as follows:

$$(a + bi) \cdot (c + di) = (ac - bd) + (ad + bc)i.$$

This is a binary operation on $\mathbb{C}^*$. Note that both $ac - bd$ and $ad + bc$ cannot be zero. Suppose that $a + bi, c + di$ and $e + fi$ are arbitrary elements of $\mathbb{C}^*$. Then, we have

$$\big((a + bi) \cdot (c + di)\big) \cdot (e + fi)$$
$$= \big((ac - bd) + (ad + bc)i\big) \cdot (e + fi)$$
$$= \big((ac - bd)e - (bc + ad)f\big) + \big((bc + ad)e + (ac - bd)f\big)i$$
$$= \big(a(ce - df) - b(de + cf)\big) + \big(b(ce - df) + a(de + cf)\big)i$$
$$= (a + bi) \cdot \big((ce - df) + (de + cf)\big)$$
$$= (a + bi) \cdot \big((c + di) \cdot (e + fi)\big).$$

Hence, the multiplication · is associative on $\mathbb{C}^*$. Clearly, $1 = 1 + 0i \in \mathbb{C}^*$ is the identity element in $\mathbb{C}^*$. Now, we check the existence of inverse. Assume that $a + bi \in \mathbb{C}^*$. Then not both a and b are zero, and so $a^2 + b^2 \neq 0$. Consequently, we have

$$\frac{a}{a^2 + b^2} - \frac{b}{a^2 + b^2}i \in \mathbb{C}^*.$$

Moreover, we see that

$$\Big(\frac{a}{a^2 + b^2} - \frac{b}{a^2 + b^2}i\Big)(a + bi) = (a + bi)\Big(\frac{a}{a^2 + b^2} - \frac{b}{a^2 + b^2}i\Big) = 1.$$

Therefore, we proved that $(\mathbb{C}^*, \cdot)$ is a group and we call this group the *multiplicative group of non-zero complex numbers*.

Example 3.37 The set $K_4 = \{e, a, b, c\}$ under the binary operation defined by the following Cayley table is a group.

·	e	a	b	c
e	e	a	b	c
a	a	e	c	b
b	b	c	e	a
c	c	b	a	e

This group is known as *Klein's four group*.

Definition 3.38 A group G is said to be *abelian* (or *commutative*) if $a \star b = b \star a$, for all $a, b \in G$.

All the previous examples of groups are abelian. Now, we give some examples of non-abelian groups.

Example 3.39 The group D_3 is the *symmetry group of an equilateral triangle* with vertices on the unit circle, at angles 0, $2\pi/3$, and $4\pi/3$, that is, it is the set of all reflection, rotation, and combinations of these, that leave the shape and position of this triangle fixed. Figure 3.4 shows the effect of the sixth element of D_3. Note that r, s and t can be equally well considered as a rotation of $\theta = \pi$ in $\mathbb{R}^3$ about the axes r, s and t. Then, these six operations form a group. The Cayley table for this group is:

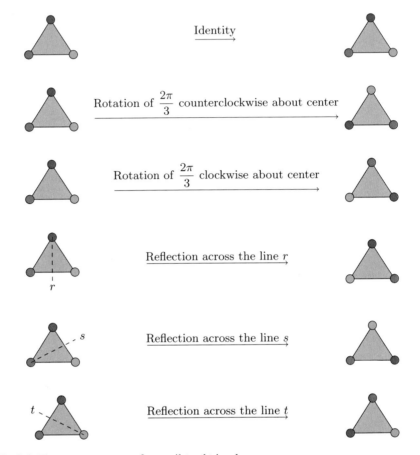

Fig. 3.4 The symmetry group of an equilateral triangle

·	R_0	R_1	R_2	S_0	S_1	S_2
R_0	R_0	R_1	R_2	S_0	S_1	S_2
R_1	R_1	R_2	R_0	S_1	S_2	S_0
R_2	R_2	R_0	R_1	S_2	S_0	S_1
S_0	S_0	S_2	S_1	R_0	R_2	R_1
S_1	S_1	S_0	S_2	R_1	R_0	R_2
S_2	S_2	S_1	S_0	R_2	R_1	R_0

Example 3.40 The group D_4 is the *symmetry group of a square* with vertices on the unit circle, at angles 0, $\pi/2$, π, and $3\pi/2$. Figure 3.5 shows the eight symmetries of square. Let

$$D_4 = \{R_0,\ R_1,\ R_2,\ R_3,\ S_0,\ S_1,\ S_2,\ S_3\}$$

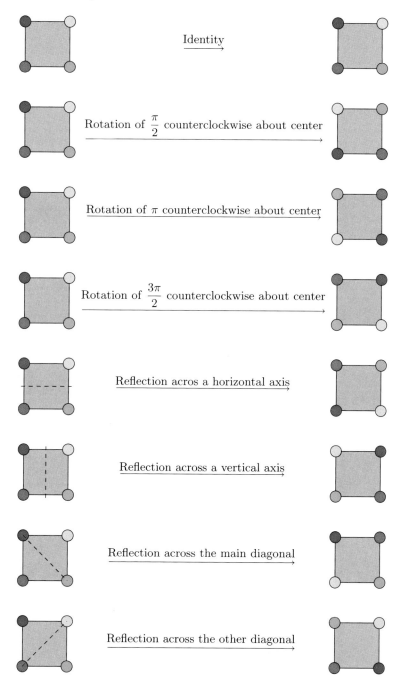

Fig. 3.5 The symmetry group of a square

be the symmetry group of a square. The Cayley table for this group is:

·	R_0	R_1	R_2	R_3	S_0	S_1	S_2	S_3
R_0	R_0	R_1	R_2	R_3	S_0	S_1	S_2	S_3
R_1	R_1	R_2	R_3	R_0	S_1	S_2	S_3	S_0
R_2	R_2	R_3	R_0	R_1	S_2	S_3	S_0	S_1
R_3	R_3	R_0	R_1	R_2	S_3	S_0	S_1	S_2
S_0	S_0	S_3	S_2	S_1	R_0	R_3	R_2	R_1
S_1	S_1	S_0	S_3	S_2	R_1	R_0	R_3	R_2
S_2	S_2	S_1	S_0	S_3	R_2	R_1	R_0	R_3
S_3	S_3	S_2	S_1	S_0	R_3	R_2	R_1	R_0

Example 3.41 Let G be the set of all 2×2 matrices $\begin{bmatrix} a & b \\ c & d \end{bmatrix}$ where a, b, c, and d are real numbers such that $ad - bc \neq 0$. For the binary operation in G we use the following multiplication:

$$\begin{bmatrix} a & b \\ c & d \end{bmatrix} \cdot \begin{bmatrix} a' & b' \\ c' & d' \end{bmatrix} = \begin{bmatrix} aa' + bc' & ab' + bd' \\ ca' + dc' & cb' + dd' \end{bmatrix}.$$

Clearly, the entries of this 2×2 matrix are real. In order to see that this matrix belongs to G we must show that

$$(aa' + bc')(cb' + dd') - (ab' + bd')(ca' + dc') \neq 0.$$

A short computation gives that

$$(aa' + bc')(cb' + dd') - (ab' + bd')(ca' + dc') = (ad - bc)(a'd' - b'c') \neq 0,$$

because both

$$\begin{bmatrix} a & b \\ c & d \end{bmatrix} \text{ and } \begin{bmatrix} a' & b' \\ c' & d' \end{bmatrix}$$

are in G. It is not difficult to check that the associative law holds in G. The element $I_2 = \begin{bmatrix} 1 & 0 \\ 0 & 1 \end{bmatrix}$ belongs to G, and it acts as an identity element relative to the binary operation of G.

Now, suppose that $\begin{bmatrix} a & b \\ c & d \end{bmatrix}$ be an arbitrary element of G. Then, we have $ad - bc \neq 0$, and so the matrix

$$\begin{bmatrix} \dfrac{d}{ad - bc} & \dfrac{-b}{ad - bc} \\ \dfrac{-c}{ad - bc} & \dfrac{a}{ad - bc} \end{bmatrix}$$

makes sense. Moreover, we obtain

$$\left(\frac{d}{ad-bc}\right)\left(\frac{a}{ad-bc}\right) - \left(\frac{-b}{ad-bc}\right)\left(\frac{-c}{ad-bc}\right) = \frac{ad-bc}{(ad-bc)^2} = \frac{1}{ad-bc} \neq 0.$$

Thus, the matrix

$$\begin{bmatrix} \dfrac{d}{ad-bc} & \dfrac{-b}{ad-bc} \\ \dfrac{-c}{ad-bc} & \dfrac{a}{ad-bc} \end{bmatrix}$$

belongs to G. An easy computation shows that

$$\begin{bmatrix} a & b \\ c & d \end{bmatrix} \cdot \begin{bmatrix} \dfrac{d}{ad-bc} & \dfrac{-b}{ad-bc} \\ \dfrac{-c}{ad-bc} & \dfrac{a}{ad-bc} \end{bmatrix} = \begin{bmatrix} 1 & 0 \\ 0 & 1 \end{bmatrix} = \begin{bmatrix} \dfrac{d}{ad-bc} & \dfrac{-b}{ad-bc} \\ \dfrac{-c}{ad-bc} & \dfrac{a}{ad-bc} \end{bmatrix} \cdot \begin{bmatrix} a & b \\ c & d \end{bmatrix}.$$

This means that this element of G is the inverse of $\begin{bmatrix} a & b \\ c & d \end{bmatrix}$. Therefore, we conclude that G is a group.

Example 3.42 Let $G = \{(a, b) \mid a, b \in \mathbb{R} \text{ and } a > 0\}$. We define the following operation on G as follows:

$$(a, b) \star (c, d) = (ac, bc + d),$$

for all (a, b), $(c, d) \in G$. Obviously, if $b \neq 0$ and $c \neq 0$, then $bc + d \neq 0$. So, $\star$ is a binary operation on G. Now, we verify associativity. Suppose that (a, b), (c, d), and (e, f) are arbitrary elements of G. Then, we have

$$\begin{aligned} \big((a, b) \star (c, d)\big) \star (e, f) &= (ac, bc + d) \star (e, f) \\ &= (ace, bce + de + f) \\ &= (a, b) \star (ce, de + f) \\ &= (a, b) \star \big((c, d) \star (e, f)\big). \end{aligned}$$

Moreover, since $(a, b) \star (1, 0) = (1, 0) \star (a, b) = (a, b)$, it follows that $(1, 0)$ is the identity element. Finally, since

$$(a, b) \star \left(\frac{1}{a}, -\frac{b}{a}\right) = \left(\frac{1}{a}, -\frac{b}{a}\right) \star (a, b) = (1, 0),$$

it follows that $(1/a, -b/a)$ is the inverse of (a, b). Therefore, $(G, \star)$ is a group. But

$$(2, 3) \star (1, 4) = (2, 7) \text{ and } (1, 4) \star (2, 3) = (2, 11).$$

This shows that G is not abelian.

Example 3.43 Let G be the set of all functions $T_{a,b} : \mathbb{R} \to \mathbb{R}$ defined by $T_{a,b}(x) = ax + b$, for each $x \in \mathbb{R}$, where $a, b \in \mathbb{R}$ and $a \neq 0$, i.e.,

$$G = \{T_{a,b} : \mathbb{R} \to \mathbb{R} \mid a, b \in \mathbb{R} \text{ and } a \neq 0\}.$$

Consider the product of elements of G as the compositions of functions. Hence, if $T_{a,b}$ and $T_{c,d}$ are elements in G, then

$$
\begin{aligned}
\left(T_{a,b} \circ T_{c,d}\right)(x) = T_{a,b}\left(T_{c,d}(x)\right) &= aT_{c,d}(x)b \\
&= a(cx + d) + b = (ac)x + (ad + b) \\
&= T_{ac,ad+b}(x),
\end{aligned}
$$

for all $x \in \mathbb{R}$. Therefore, we conclude that

$$T_{a,b} \circ T_{c,d} = T_{ac,ad+b}.$$

This yields that $T_{ac,ad+b} \in G$, i.e., G is closed under the composition of functions. For all elements $T_{a,b}$, $T_{c,d}$ and $T_{e,f}$ in G, we see that

$$
\begin{aligned}
\left(T_{a,b} \circ T_{c,d}\right) \circ T_{e,f} &= T_{ac,ad+b} \circ T_{e,f} \\
&= T_{ace,acf+ad+b} \\
&= T_{a,b} \circ T_{ce,cf+d} \\
&= T_{a,b} \circ \left(T_{c,d} \circ T_{e,f}\right).
\end{aligned}
$$

So, the associativity law holds. The element $T_{1,0}$ is the identity function and $T_{1,0} \circ T_{a,b} = T_{a,b} \circ T_{1,0} = T_{1,0}$. Finally, what is $T_{a,b}^{-1}$? We must find real numbers $x \neq 0$ and y such that $T_{a,b} \circ T_{x,y} = T_{x,y} \circ T_{a,b} = T_{1,0}$, or equivalently

$$T_{ax,ay+b} = T_{ax,bx+y} = T_{1,0}.$$

This implies that $ax = 1$ and $ay + b = bx + y = 0$. Remember $a \neq 0$, so $a = 1/a$ and $y = -b/a$. Thus, $T_{a^{-1},-a^{-1}b}$ is the inverse of $T_{a,b}$. Consequently, $(G, \circ)$ is a group. Since $T_{1,2} \circ T_{3,4} \neq T_{3,4} \circ T_{1,2}$, it follows that G is not abelian.

It is a little clumsy to keep writing the $\star$ for the product in G, and from now on we shall write the product $a \star b$ simply as ab, for all $a, b \in G$.

Theorem 3.44 *If G is a group, then*

(1) For every $a \in G$, $(a^{-1})^{-1} = a$;
(2) For all $a, b \in G$, $(ab)^{-1} = b^{-1}a^{-1}$.

Proof (1) For any $a \in G$, we have

$$aa^{-1} = e = a^{-1}a \quad \text{and} \quad (a^{-1})^{-1}a^{-1} = e = a^{-1}(a^{-1})^{-1}.$$

Since the inverse of a^{-1} is unique, it follows that $(a^{-1})^{-1} = a$.

(2) This item is proved by the equalities

$$(ab)(b^{-1}a^{-1}) = a(bb^{-1})a^{-1} = aea^{-1} = aa^{-1} = e,$$
$$(b^{-1}a^{-1})(ab) = b^{-1}(a^{-1}a)b = b^{-1}eb = b^{-1}b = e,$$

and the uniqueness of the inverse. ∎

In part (2), notice the change in the order of the factors from ab to $b^{-1}a^{-1}$. If in a group the binary operation is written additively, then $a + b$ (for $a, b \in G$) is called the *sum* of a and b, and the identity element is denoted by 0. Also, the inverse of $a \in G$ is denoted by $-a$. We write $a - b$ for $a + (-b)$. Abelian groups are frequently written additively.

Note that $a^n = \underbrace{aa \dots a}_{n \text{ times}}$. If $n = -m$ is a negative integer, then we define $a^n = (a^{-1})^m$; also, we define $a^0 = e$. The formulas $a^m a^n = a^{m+n}$ and $(a^m)^n = a^{mn}$ hold for any element a of G and any pair of integers m and n. One must be careful with this notation when dealing with a specific group whose binary operation is addition. In this case, the definitions and group properties expressed in multiplicative notation must be translated to additive notation. Table 3.1 shows the common notation and corresponding terminology for groups under multiplication and groups under addition.

Corollary 3.45 *In any group G both the cancellation laws hold, i.e.,*

(1) $ab = ac$ implies $b = c$;
(2) $ba = ca$ implies $b = c$.

Proof It is enough to multiply a^{-1} on the left or right side. ∎

Theorem 3.46 *Any finite semigroup in which both cancellation laws hold is a group.*

Proof Suppose that G is a finite semigroup with n elements in which both cancellation laws hold and let

$$G = \{x_1, \ x_2, \ \dots, x_n\}.$$

Suppose that $a \in G$ is arbitrary and fixed. Then, the elements $ax_1, ax_2, \ \dots, ax_n$ are distinct, because $ax_i = ax_j$ (for some i and j) implies that $x_i = x_j$ (the left cancellation law). Consequently, we have

Table 3.1 Notation and corresponding terminology for groups under multiplication and addition

Multiplicative group		Additive group	
$a \cdot b$ or ab	Multiplicative	$a + b$	Addition
e or 1	Identity	0	Zero
a^{-1}	Multiplicative inverse of a	na	Multiple of a
a^n	Power of a	$a - b$	Difference

$$G = \{ax_1, \ ax_2, \ \ldots, ax_n\}.$$

Hence, for each $x_i \in G$, there exists $x_j \in G$ such that

$$x_i = ax_j. \tag{3.1}$$

In particular, there exists $x_k \in G$ such that $a = ax_k$. Then, $ax_i = (ax_k)x_i = a(x_k x_i)$. Now, by the left cancellation law, we conclude that $x_i = x_k x_i$, for each $1 \le i \le n$.

Similarly, by considering the elements $x_1 a, \ x_2 a, \ \ldots, \ x_n a$ and using the right cancellation law, we can find an element $x_l \in G$ such that $x_i = x_i x_l$, for each $1 \le i \le n$. So, we get $x_l = x_k x_l = x_k$. Therefore, x_k is the identity element in G. As usual, we set $x_k = e$.

Taking $x_i = e$ in (3.1), then we can say that there exists $x_m \in G$ such that

$$e = ax_m.$$

In a similar way, by considering the elements $x_1 a, \ x_2 a, \ \ldots, \ x_n a$, we can find an element $a_r \in G$ such that

$$e = x_r a.$$

Thus, we can write

$$x_r = x_r e = x_r(ax_m) = (x_r a)x_m = ex_m = x_m.$$

Hence, $x_m a = ax_m = e$, which implies that $x^{-1} = x_m$. This shows that every element in G has its inverse in G. Therefore, G is a group. ∎

Corollary 3.47 *A finite semigroup G is a group if and only if G satisfies both cancellation laws.*

Proof The proof follows immediately from Corollary 3.45 and Theorem 3.46. ∎

Theorem 3.48 *Let G be a non-empty set together with a binary operation. Then, G is a group if and only if*

(1) $a(bc) = (ab)c$, for all $a, b \in G$;
(2) Given any $a, b \in G$, the equations $ax = b$ and $ya = b$ have unique solutions in G.

Proof Let G be a group. Since $a(a^{-1}b) = (aa^{-1})b = eb = b$, it follows that $x = a^{-1}b$ is a solution of the equation $ax = b$ in G. Similarly, $y = ba^{-1}$ is a solution of the equation $ya = b$ in G. The uniqueness of the solution follows from the cancellation law.

Suppose that G is a non-empty set together with a binary operation satisfying (1) and (2). In order to show that G is a group we need to show that G contains an identity and each element of G has an inverse.

Let $a \in G$ be an arbitrary element. By (2), since $ax = a$ has solution in G, it follows that there is an element $e \in G$ such that $ae = a$.

Now, given $b \in G$, then by (2) there exists $y \in G$ such that $b = ya$. Then, $be = (ya)e = y(ae) = ya = b$, and so $be = b$, for each $b \in G$. Similarly, since the equation $ya = a$ has solution in G, it follows that there exists $e' \in G$ such that $e'a = a$. Again, given $b \in G$, by (2) there exists $z \in G$ such that $b = az$. Then, $e'b = e'(az) = (e'a)z = az = b$, and so $e'b = b$, for each $b \in G$. In particular, since $be = b$, for each $b \in G$, it follows that $e'e = e'$. Also, since $e'b = b$, for each $b \in G$, it follows that $e'e = e$. Consequently, $e = e'$ and we get $be = b = eb$, for each $b \in G$. This shows that e is the identity element of G.

Next, we prove the existence of inverse. Let $a \in G$. Then, by (2), there exist $a', a'' \in G$ such that $aa' = e$ and $a''a = e$. Hence, we obtain

$$a' = ea' = (a''a)a' = a''(aa') = a''e = a'',$$

and we conclude that $a'a = e = aa'$. This yields that $a^{-1} = a$. ∎

Theorem 3.49 *Let G be a non-empty set together with a binary operation. Then, G is a group if and only if*

(1) $a(bc) = (ab)c$, for all $a, b \in G$;
(2) There exists an element $e \in G$ such that $ae = a$, for all $a \in G$;
(3) For every $a \in G$, there exists $a' \in G$ such that $aa' = e$.

Proof If G is a group, then by Definition 3.28, the statements (1), (2), and (3) hold.

Now, assume that G is a non-empty set together with a binary operation satisfying (1), (2), and (3). Let $a \in G$. Then, by (3), there is $a' \in G$ such that $aa' = e$. Again, by (3), there exists $a'' \in G$ such that $a'a'' = e$. Hence, we get

$$\begin{aligned} a'a &= (a'a)e = (a'a)(a'a'') = a'(a(a'a'')) = a'((aa')a'') \\ &= a'(ea'') = (a'e)a'' = a'a'' = e. \end{aligned}$$

Consequently, we have $aa' = e = a'a$. In addition, we find that

$$ea = (aa')a = a(a'a) = a(a'a) = ae = a,$$

and so $ea = ae = a$. Therefore, we deduce that G is a group. ∎

Theorem 3.50 *Let G be a non-empty set together with a binary operation. Then, G is a group if and only if*

(1) $a(bc) = (ab)c$, for all $a, b \in G$;
(2) There exists an element $e \in G$ such that $ea = a$, for all $a \in G$;
(3) For every $a \in G$, there exists $a' \in G$ such that $a'a = e$.

Proof The proof is similar to the proof of Theorem 3.49. ∎

The following example shows that the conclusion of Theorems 3.49 and 3.50 fails, if a semigroup S contains a right identity e and each element $a \in G$ may possess a left inverse.

Example 3.51 Let S be a set having at least two elements. For any $x, y \in G$, we define $xy = x$. It is easy to check that S together with this binary operation is a semigroup. Let b and c be two distinct elements of S. Then, for each $a \in S$, we have $ab = ac$, while $b \neq c$. Hence, S does not satisfy cancellation law, and so S is not a group. Now, let e be any fixed element of S. We have

(1) For each $a \in S$, $ae = a$. This means that e is a right identity element;
(2) For each $a \in G$, since $ea = e$, it follows that e is a left inverse of a with respect to e.

Consequently, S has the right identity e and each element of S has a left inverse in S but we know that G is not a group.

We leave it to readers to construct a similar example of a semigroup S which has a left identity and right inverse for each of its elements but it fails to be a group.

Definition 3.52 Let a and b be elements of a group G. We say that a and b are *conjugate* in G (and call b a conjugate of a) if there exists $x \in G$ such that $b = x^{-1}ax$.

We write, for this, $a \sim_{Conj} b$ and refer to this relation as conjugacy.

Theorem 3.53 *Conjugacy is an equivalence relation on G.*

Proof Since $a = e^{-1}ae$, for each $a \in G$, it follows that $a \sim_{Conj} a$, i.e., $\sim_{Conj}$ is reflexive.

If $a \sim_{Conj} b$, then there exists $x \in G$ such that $b = x^{-1}ax$. Hence, $a = (x^{-1})^{-1} bx^{-1}$. Since $x^{-1} \in G$, it follows that $b \sim_{Conj} a$, i.e., $\sim_{Conj}$ is symmetric.

Suppose that $a \sim_{Conj} b$ and $b \sim_{Conj} c$, where $a, b, c \in G$. Then, $b = x^{-1}ax$ and $c = y^{-1}by$, for some $x, y \in G$. Substituting for b in the expression for c we obtain $c = y^{-1}x^{-1}axy$. This shows that $c = (xy)^{-1}a(xy)$. Since $xy \in G$, it follows that $a \sim_{Conj} c$, i.e., $\sim_{Conj}$ is transitive. ∎

Theorem 3.53 shows that we can divide a group G into equivalence classes under the relation of conjugacy. Each such equivalence class is called a *conjugate class*.

Exercises

1. Prove that the set of all rational numbers of the form $3^m 6^n$, where m and n are integers, is a group under multiplication.
2. Let $G = \{a \in \mathbb{R} \mid -1 < a < 1\}$. Define a binary operation $\star$ on G by

$$a \star b = \frac{a + b}{1 + ab},$$

for all $a, b \in G$. Show that $(G, \star)$ is a group.

3. Let G be the set of all rational numbers except -1. Show that $(G, \star)$ is a group where
$$a \star b = a + b + ab,$$
for all $a, b \in G$.

4. If a and b are elements of a group G, prove that $ab^n a^{-1} = (aba^{-1})^n$, for each integer n.

5. Show that if every element of the group G is its own inverse, then G is abelian.

6. If G is a group of even order, prove it has an element $a \neq e$ satisfying $a^2 = e$.

7. Prove that a finite semigroup G with identity is a group if and only if G contains only one idempotent.

8. Let G be a group.

   (a) If G has three elements, then show that it must be abelian;
   (b) Do part (a) if G has four elements;
   (c) Do part (a) if G has five elements.

9. Consider the set $\{0, 1, 2, 3, 4, 5, 6, 7\}$. Suppose that there is a binary operation $\star$ on G that satisfies the following two conditions:

   (a) $a \star b \leq a + b$, for all $a, b \in G$;
   (b) $a \star a = 0$, for all $a \in G$.

   Construct Cayley table for G. (This group sometimes called *Nim group*).

10. Find an example which shows it is impossible to have $(ab)^{-2} \neq b^{-2}a^{-2}$.

11. In a finite group, show that the number of non-identity elements that satisfy the equation $x^5 = e$ is a multiple of 5. If the stipulation that the group be finite is omitted, what can you say about the number of non-identity elements that satisfy the equation $x^5 = e$?

12. For each positive integer n, prove that the number of groups of order n is finite.

13. Let G be a group in which $(ab)^n = a^n b^n$ for some fixed integer $n > 1$ and for all $a, b \in G$. For all $a, b \in G$, prove that

   (a) $(ab)^{n-1} = b^{n-1}a^{n-1}$;
   (b) $a^n b^{n-1} = b^{n-1}a^n$;
   (c) $(aba^{-1}b^{-1})^{n(n-1)} = e$.

14. In Theorem 3.46 show by an example that if one just assumed one of the cancellation laws, then the conclusion need not follow.

15. Show that in Theorem 3.46 infinite examples exist, satisfying the conditions which are not groups.

16. Show that the equation $x^2ax = a^{-1}$ has a solution for x in a group G if and only if a is the cube of some element in G.

3.5   Turning Groups into Latin Squares (Optional)

The name "Latin square" was inspired by mathematical papers by Leonhard Euler (1707–1783), who used Latin characters as symbols, but any set of symbols can be used.

Definition 3.54 A *Latin square* is an $n \times n$ table filled with n different symbols in such a way that each symbol occurs exactly once in each row and exactly once in each column.

Example 3.55 The following are two examples of Latin squares.

1	2	3
2	3	1
3	1	2

a	b	c	d	e
b	a	e	c	d
c	d	b	e	a
d	e	a	b	c
e	c	d	a	b

Theorem 3.56 *Cayley table of any finite group is a Latin square.*

Proof Take a row indexed by the group element a. Suppose that two elements in this row are equal. In other words, there are two columns c and d such that $ac = ad$. If we multiply by a^{-1} on the left, this gives us $c = d$, i.e., these were in fact the same columns, and therefore, there are no repetitions in this row. The same logic tells us that there are also no repetitions in any column; therefore, this is a Latin square. ∎

A natural question is: which Latin squares are Cayley tables of groups?

Many of the group axioms for the operation · of a finite group can be checked by referring to the Cayley table for the group. From the table, one can quite easily check the existence of an identity element and observe that each group element occurs exactly once in each row and each column. The only group property that is difficult to check directly from the properties of the Cayley table is associativity. It must be shown that $(ab)c = a(bc)$ for all elements a, b, and c of the group, and to check this property case by case is quite tedious. If the group contains n elements, there are n^3 ordered triple (a, b, c) that must be considered with $(ab)c$ and $a(bc)$ checked for equality in each case.

The following theorem may simplify the verification of the associative law for multiplication.

Theorem 3.57 *Let G be a finite group. Let r and s be in the array representing the Cayley table for G such that the column containing r and the row containing s intersect at the identity element e. Then, rs is the element at the fourth corner of the rectangle formed by r, s, and e.*

Proof The identity element e occurs exactly once in each row and in each column, so suppose that e of the theorem is in the row headed by q and in the column headed by b. Then, r occurs exactly once in the column headed by b, say in the row headed by p; so s occurs exactly once in the row headed by q, say in the column headed by a.

$$
\begin{array}{c|cc}
\cdot & a & b \\
\hline
p & & r \\
q & s & e
\end{array}
$$

Thus, $pb = r$, $qa = s$, and $qb = e$. We wish to show that the fourth corner pa of the rectangle formed by r, s, and e is rs. We have

$$
rs = (pb)(qa) = p(bq)a.
$$

The last equality is a result of the associativity law present in the group G. However, qb is the identity; thus bq is the identity element.

It follows that $rs = pea = pa$. ∎

The rule stated in Theorem 3.57 is called the *rectangle rule*.

Since we wish to emphasize that the rectangle rule follows from the associative law, Theorem 3.57 could be stated as follows:

Suppose that G is a system with identity element e such that each group element occurs exactly once in each row and in each column of the Cayley table; then the rectangle rule is a consequence of Gs having the associative property.

Suppose that we number the rows and columns from 1 to n and let a_{ij} be the element in the ith row and jth column. Suppose that $r = a_{ik}$ and $s = a_{jt}$. Then, $pa = a_{it}$ since pa is the same row as r in the same column as s. Also, $e = a_{jk}$ since e is in the same column as r and in the same row as s. The rectangle rule $pa = rs$ now becomes the following:

If $e = a_{jk}$, then $a_{it} = a_{ik}a_{jt}$, for all i and t. The table

$$
\begin{array}{c|cc}
\cdot & a & b \\
\hline
p & rs & r \\
q & s & e
\end{array}
$$

now takes the form

$$
\begin{array}{c|cc}
\cdot & a & b \\
\hline
p & a_{it} & a_{ik} \\
q & a_{jt} & a_{jk}
\end{array}
$$

Theorem 3.58 *Suppose that G is a system with identity element e and with the following properties:*

(1) The Cayley table is a Latin square. In other words, each element of G occurs exactly once in each row and in each column;
(2) The rectangle rule holds.

Then, the operation obeys the associative law.

Fig. 3.6 Rectangle *A*

Fig. 3.7 Rectangle *B*

Fig. 3.8 Rectangles *A* and *B* together

Fig. 3.9 Rectangles *A B* and *C* together

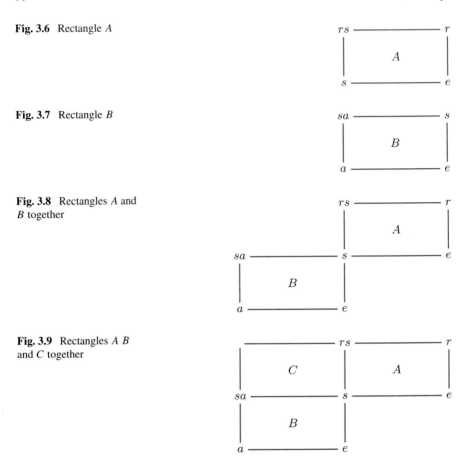

Proof Let *r*, *s*, and *a* be elements of *G*. By (1), *e* occurs in each column, so pick an *e* in the table. By (1), we can locate the *r* in the column containing *e* and the *s* in the row containing *e*. By (2), we know that the element in the fourth corner of the rectangle *A* is *rs*, see Fig. 3.6.

By (1), we can find the *e* in the column containing the above mentioned *s*. By (1), *a* can be located in the row containing this *e*. By (2), the element in the fourth corner of rectangle *B* is *sa*, see Fig. 3.7. Rectangles *A* and *B* fit together to form Fig. 3.8. We see that a new rectangle *C* is formed with three of the corners labeled *sa*, *s*, and *rs*, see Fig. 3.9. By (2), the element in the fourth corner of rectangle *C* union *A* is *r*(*sa*). By (2), the element in the fourth corner of rectangle *C* union *B* is (*rs*)*a*. But these corners are the same. Consequently, we obtain *r*(*sa*) = (*rs*)*a*. ∎

Example 3.59 Consider the following table:

Fig. 3.10 Rectangle related
to Example 3.59

·	e	a	b	c	d
e	e	a	b	c	d
a	a	e	d	b	c
b	b	c	e	d	a
c	c	d	a	e	b
d	d	b	c	a	e

This table is a Latin square, but rectangle rule does not hold as $bb \neq d$ in the rectangle in Fig. 3.10.

The operation does not have the associative property, because $(ab)c = a$ and $a(bc) = c$.

Another method of verifying the associative law is to use permutations. We shall study this method in the future.

Exercises

1. Show that the number of $n \times n$ Latin squares is 1, 2, 12, 576 for $n = 1, 2, 3, 4$, respectively.
2. Which of the following Latin squares are Cayley tables of groups?

1	2	3	4	5
2	5	4	1	3
3	1	2	5	4
4	3	5	2	1
5	4	1	3	2

1	2	3	4	5
2	4	1	5	3
3	1	5	2	4
4	5	2	3	1
5	3	4	1	2

3. Suppose the following is the Cayley table of a group G. Fill in the blank entries.

·	e	a	b	c	d
e	e	–	–	–	–
a	–	b	–	–	e
b	–	c	d	e	–
c	–	d	–	a	b
d	–	–	–	–	–

3.6 Subgroups

In general, we shall not be interested in arbitrary subsets of a group G because they do not reflect that G has an algebraic structure imposed on it. So, we are interested in special subsets of a group, called "subgroups". The concept of subgroup is one of the most basic ideas in group theory.

Definition 3.60 A non-empty subset H of a group G is called a *subgroup* of G if relative to the operation in G, H itself forms a group.

Let e be the identity element of a group G. Then, trivially G and $\{e\}$ are subgroups of G. These subgroups are called *trivial subgroups* of G. A subgroup H of G is called a *proper subgroup* if it is different from G as well as from $\{e\}$. We denote the relation "H is a subgroup of G" by $H \leq G$.

It would be useful to have some criterion for deciding whether a given subset of a group is a subgroup.

Theorem 3.61 (Two-Step Subgroup Test) *A non-empty subset H of the group G is a subgroup if and only if*

(a) $a, b \in H$ implies $ab \in H$;
(b) $a \in H$ implies $a^{-1} \in H$.

Proof If H is a subgroup of G, then it is obvious that (1) and (2) must hold.

Conversely, suppose that H is a subset of G for which (1) and (2) hold. In order to establish that H is a subgroup, all that is needed is to verify that $e \in H$ and that the associative law holds for elements of H. Since the associative law does hold for G, it holds all the more so for H, which is a subset of G. Now, if $a \in H$, then by (2), $a^{-1} \in H$. Next, since $a \in H$ and $a^{-1} \in H$, by (1), it follows that $e = aa^{-1} \in H$. This yields that H is a subgroup of G.. ∎

Theorem 3.62 (One-Step Subgroup Test) *A non-empty subset H of the group G is a subgroup if and only if*

$$a, b \in H \text{ implies } ab^{-1} \in H.$$

Proof Let H be a subgroup of G and let $a, b \in H$. By Theorem 3.61 (2), we conclude that $b^{-1} \in H$. Now, since $a, b^{-1} \in H$, by Theorem 3.61 (1), it follows that $ab^{-1} \in H$.

Conversely, let H be a non-empty subset of G such that $ab^{-1} \in H$, for all $a, b \in H$. Since H is non-empty, it follows that there exists $x \in H$. So, $e = xx^{-1} \in H$. Now, suppose that a and b are two arbitrary elements of H. As $a, e \in H$, $ea^{-1} \in H$, i.e., $a^{-1} \in H$. Next, since $a, b \in H$, it follows that $b^{-1} \in H$ and $a(b^{-1})^{-1} \in H$. This implies that $ab \in H$. Therefore, by Theorem 3.61, we conclude that H is a subgroup of G. ∎

Theorem 3.63 (Finite Subgroup Test) *If H is a non-empty finite subset of a group G, then H is a subgroup of G if*

$$a, b \in H \text{ implies } ab \in H.$$

Proof In light of Theorem 3.61, we need to prove that $a^{-1} \in H$ whenever $a \in H$. Assume that a is an arbitrary element of H. If $a = e$, then $a^{-1} = a$ and we are done. If $a \neq e$, then the set $\{a, a^2, a^3, \ldots\}$ is a subset of H. Since H is finite, there must be repetition in this set of elements, i.e., for some positive integers i and j, $a^i = a^j$ with $i > j > 0$. Then, by the cancellation law in G, $a^{i-j} = e$, and since $a \neq e, i - j > 1$. Hence, we can write $a^{i-j} = aa^{i-j-1} = e$, and consequently $a^{i-j-1} = a^{-1}$. But $i - j - 1 \geq 1$ implies $a^{i-j-1} \in H$ and we are done. ∎

Example 3.64 Let $\mathbb{R}$ be the group of all real numbers under addition, and let $\mathbb{Z}$ be the set of all integers. Then, $\mathbb{Z}$ is a subgroup of $\mathbb{R}$.

Example 3.65 Let $\mathbb{R}^*$ be the group of all non-zero real numbers under ordinary multiplication, and let $\mathbb{Q}^+$ be the set of positive rational numbers. Then, $\mathbb{Q}^+$ is a subgroup of $\mathbb{R}^*$.

Example 3.66 Let $\mathbb{C}^*$ be the group of all non-zero complex numbers under multiplication, and let $H = \{a + bi \mid a^2 + b^2 = 1\}$. Then, H is a subgroup of $\mathbb{C}^*$.

Example 3.67 Let G be an abelian group. Then, $H = \{x \in G \mid x^2 = e\}$ is a subgroup of G.

The following remark is clear.

Remark 3.68 If H is a subgroup of G and K is a subgroup of H, then K is a subgroup of G.

Example 3.69 Suppose that $H = \{1, -1, i, , -i\}$ and $K = \{1, -1\}$, where $i^2 = -1$. Then, H is a subgroup of $\mathbb{C}^*$, the group of non-zero complex numbers under multiplication, and K is a subgroup of H. So, we can say that K is a subgroup of $\mathbb{C}^*$.

Example 3.70 Consider the group of integers $\mathbb{Z}$ with ordinary addition. Then, the set

$$n\mathbb{Z} = \{\ldots, -3n, -2n, -n, 0, n, 2n, 3n, \ldots\}$$

for each positive integer n is a subgroup of $\mathbb{Z}$.

Example 3.71 Let G be the group defined in Example 3.41. Let H be the subset of G consisting of all 2×2 matrices $\begin{bmatrix} a & b \\ c & d \end{bmatrix}$ such that $ad - bc = 1$. Then, as easily verified, H is a subgroup of G.

Example 3.72 Let $C_{12} = \{e = a^0, a^1, a^2, \ldots, a^{11}\}$. We define a binary operation on C_{12} as follows:

$$a^i a^j = \begin{cases} a^{i+j} & \text{if } i + j < 12 \\ a^{i+j-12} & i + j \geq 12. \end{cases}$$

Fig. 3.11 Hasse diagram for
the family of subgroups of
C_{12}

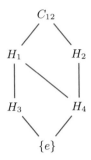

It is easy to see that C_{12} is a group under the above multiplication. Moreover, all of
the subgroups of C_{12} are as follows:

$$\{e\},\ G,$$
$$H_1 = \{e,\ a^2,\ a^4,\ a^6,\ a^8,\ a^{10}\},$$
$$H_2 = \{e,\ a^3,\ a^6,\ a^9\},$$
$$H_3 = \{e,\ a^4,\ a^8\}$$
$$H_4 = \{e,\ a^6\}.$$

In Fig. 3.11, we illustrate Hasse diagram for subgroups of C_{12}.

Example 3.73 Let G be any group and H be a subgroup of G. For $a \in G$, let

$$a^{-1}Ha = \{a^{-1}ha \mid h \in H\}.$$

We assert that $a^{-1}Ha$ is a subgroup of G. Let x and y be two arbitrary elements of
$a^{-1}Ha$. Then, $x = a^{-1}ha$ and $y = a^{-1}h'a$, for some $h, h' \in H$. Thus, we obtain

$$xy^{-1} = (a^{-1}ha)(a^{-1}h'a)^{-1} = (a^{-1}ha)(a^{-1}h'^{-1}a) = a^{-1}hh'^{-1}a.$$

Since $h, h' \in H$ and H is a subgroup of G, it follows that $hh'^{-1} \in H$. This shows
that $xy^{-1} \in a^{-1}Ha$, and so $a^{-1}Ha$ is a subgroup of G.

Theorem 3.74 *If G is a group and $\{H_i \mid i \in I\}$ is a non-empty family of subgroups
of G, then*

$$\bigcap_{i \in I} H_i$$

is a subgroup of G.

Proof Suppose that a and b are two elements of $\bigcap_{i \in I} H_i$. Then, $a, b \in H_i$, for
all $i \in I$. Since for each $i \in I$, H_i is a subgroup of G, it follows that $ab^{-1} \in H_i$,
for all $i \in I$. Hence, we conclude that $ab^{-1} \in \bigcap_{i \in I} H_i$. This shows that $\bigcap_{i \in I} H_i$ is a
subgroup of G.   ∎

In particular, the intersection of two subgroups of a group is a subgroup. The following example shows that the union of two subgroups of a group is not a subgroup, in general.

Example 3.75 Suppose that $G = \mathbb{Z}$. We know that $H_1 = 2\mathbb{Z}$ and $H_2 = 3\mathbb{Z}$ are subgroups of G. We have

$$2 \in H_1 \subseteq H_1 \cup H_2 \text{ and } 3 \in H_2 \subseteq H_1 \cup H_2,$$

while $2 + 3 = 5 \notin H_1 \cup H_2$. So, $H_1 \cup H_2$ cannot be a subgroup of G.

Definition 3.76 (*Center of a Group*) The *center* $Z(G)$ of a group G is the subset of elements in G that commute with every element of G, i.e.,

$$Z(G) = \{a \in G \mid ax = xa, \text{ for all } x \in G\}.$$

Theorem 3.77 *The center of a group G is a subgroup of G.*

Proof We use Theorem 3.61 to prove this result. Since $e \in Z(G)$, it follows that $Z(G)$ is non-empty. Now, assume that a and b be two arbitrary elements of $Z(G)$. Then, we have

$$(ab)x = a(bx) = a(xb) = (ax)b = (xa)b = x(ab),$$

for all $x \in G$. Thus, $ab \in Z(G)$. Also, we have

$$xa^{-1} = exa^{-1} = (a^{-1}a)xa^{-1} = a^{-1}(ax)a^{-1} = a^{-1}(xa)a^{-1}$$
$$= a^{-1}x(aa^{-1}) = a^{-1}xe = a^{-1}x,$$

for all $x \in G$. This shows that $a^{-1} \in Z(G)$ whenever $a \in Z(G)$. ∎

Corollary 3.78 *If G is an abelian group, then $Z(G) = G$.*

Definition 3.79 Let X be a fixed non-empty subset of a group G. The *centralizer of X in G, $C_G(X)$*, is the set of all elements in G that commute with elements of X, i.e.,

$$C_G(X) = \{a \in G \mid ax = xa, \text{ for all } x \in X\}.$$

Theorem 3.80 *The centralizer of a non-empty subset of a group G is a subgroup of G.*

Proof The proof is similar to the proof of Theorem 3.77 and it is left to the reader. ∎

If $X = \{x\}$, then we write $C_G(x)$ instead of $C_G(\{x\})$. In this case, $C_G(x)$ is the set of all elements in G that commute with x.

Corollary 3.81 *We have*
$$Z(G) = \bigcap_{x \in G} C_G(x).$$

Proof It is straightforward.   ∎

Theorem 3.82 *Let H_1 and H_2 be two subgroups of G. Then, $H_1 \cup H_2$ is a subgroup of G if and only if $H_1 \subseteq H_2$ or $H_2 \subseteq H_1$.*

Proof If $H_1 \subseteq H_2$ or $H_2 \subseteq H_1$, then it is clear that $H_1 \cup H_2$ is a subgroup of G.

Conversely, let $H_1 \cup H_2$ be a subgroup of G. We prove by contradiction method. Suppose that $H_1 \nsubseteq H_2$ and $H_2 \nsubseteq H_1$. Then, there exist $x \in H_1 - H_2$ and $y \in H_2 - H_1$. So, we conclude that $x^{-1} \in H_1$, $y^{-1} \in H_2$ and $x, y \in H_1 \cup H_2$. Since $H_1 \cup H_2$ is a subgroup, it follows that $xy \in H_1 \cup H_2$. This implies that $xy \in H_1$ or $xy \in H_2$. If $xy \in H_1$, then $y = x^{-1}(xy) \in H_1$, and this is a contradiction. Similarly, if $xy \in H_2$, then $x = (xy)y^{-1} \in H_2$, and this is again a contradiction. Therefore, we deduce that $H_1 \subseteq H_2$ or $H_2 \subseteq H_1$.   ∎

Corollary 3.83 *If $\{H_i \mid i \in I\}$ is a chain of subgroups of a group G, then $\bigcup\limits_{i \in I} H_i$ is a subgroup of G.*

Exercises

1. If G is an abelian group and if $H = \{a \in G \mid a^2 = e\}$, show that H is a subgroup of G.
2. Give an example of a non-abelian group for which the H in Exercise 1 is not a subgroup.
3. Let G be an abelian group, fix a positive integer n. Let $H = \{a^n \mid a \in G\}$. Show that H is a subgroup of G.
4. In Example 3.43, let $H = \{T_{a,b} \in G \mid a \text{ is rational}\}$. Show that H is a subgroup of G.
5. If H is a subgroup of G, let
$$N = \bigcap_{x \in G} x^{-1} H x.$$

   Prove that N is a subgroup of G such that $a^{-1} N a = N$, for all $a \in G$.
6. If H is a subgroup of G such that $a^{-1} H a \subseteq H$, for all $a \in G$, prove that actually $a^{-1} H a = H$.
7. Let H and K be two subgroups of G such that $a^{-1} H a = H$ and $a^{-1} k a = K$, for all $a \in G$. If $H \cap K = \{e\}$, prove that $hk = kh$, for any $h \in H$ and $k \in K$.
8. For a non-empty subset X of a group G, define
$$X^k = \Big\{ \prod_{i=1}^{k} x_i \mid x_i \in X \Big\},$$

for any positive integer k. Prove that if G has n element, then X^n is a subgroup of G.

9. Let G be a group, and let $a \in Z(G)$. In a Cayley table for G, how does the row headed by a compare with the column headed by a? Is the converse of your answer true?

10. Show that $x \in Z(G)$ if and only if $C_G(x) = G$.

11. Let $x, y \in G$ and let $xy = z$ if $z \in Z(G)$. Show that x and y commute.

12. If G is a group and $a, c \in G$, prove that

$$C_G(a^{-1}xa) = a^{-1}C_G(x)a.$$

13. Let G be a group and X be a non-empty subset of G. The set

$$N_G(X) = \{a \in G \mid a^{-1}Xa = X\}$$

is called the *normalizer of X in G*. If H is a subgroup of G, prove that

(a) $N_G(H)$ is a subgroup of G;
(b) $H \leq N_G(H)$.

14. Give an example of a group G and a subgroup H such that $N_G(H) \neq C_G(H)$. Is there any relation between $N_G(H)$ and $C_G(H)$?

3.7 Worked-Out Problems

Problem 3.84 Let G be the group defined in Example 3.41. Find the center of G.

Solution Suppose that $\begin{bmatrix} a & b \\ c & d \end{bmatrix}$ is an arbitrary element in the center of G. So, this matrix commutes with all elements of G. Since $\begin{bmatrix} 0 & 1 \\ 1 & 0 \end{bmatrix} \in G$, it follows that

$$\begin{bmatrix} a & b \\ c & d \end{bmatrix} \cdot \begin{bmatrix} 0 & 1 \\ 1 & 0 \end{bmatrix} = \begin{bmatrix} 0 & 1 \\ 1 & 0 \end{bmatrix} \cdot \begin{bmatrix} a & b \\ c & d \end{bmatrix},$$

and so $a = d$ and $b = c$. Similarly, since $\begin{bmatrix} 1 & 1 \\ 0 & 1 \end{bmatrix} \in G$, it follows that

$$\begin{bmatrix} a & b \\ c & d \end{bmatrix} \cdot \begin{bmatrix} 1 & 1 \\ 0 & 1 \end{bmatrix} = \begin{bmatrix} 1 & 1 \\ 0 & 1 \end{bmatrix} \cdot \begin{bmatrix} a & b \\ c & d \end{bmatrix}.$$

Hence, we get

$$\begin{bmatrix} a & a+b \\ b & b+a \end{bmatrix} = \begin{bmatrix} a+b & b+a \\ b & a \end{bmatrix}.$$

This implies that $b = 0$. Therefore, each element of the center of G is in the form

$$\begin{bmatrix} a & 0 \\ 0 & a \end{bmatrix},$$

where $a \neq 0$. On the other hand, it is easy to check that for each non-zero real number a, the matrix $\begin{bmatrix} a & 0 \\ 0 & a \end{bmatrix}$ belongs to the center of G. Consequently, we have

$$Z(G) = \left\{ \begin{bmatrix} a & 0 \\ 0 & a \end{bmatrix} \mid a \in \mathbb{R},\ a \neq 0 \right\},$$

and we are done.   ∎

Problem 3.85 Let $S = \mathbb{N}^0 \times \mathbb{N}^0$. On S we define a binary operation

$$(a, b) \star (c, d) = \big(a - b + \max\{b, c\}, d - c + \max\{b, c\}\big).$$

Prove that $(S, \star)$ is a monoid with identity $(0, 0)$. This monoid is called *Bicycle semigroup/monoid*.

Solution Suppose that $(a, b), (c, d) \in S$. Then, we have $\max\{b, c\} - b \geq 0$ and $\max\{b, c\} - c \geq 0$. So, we get $a - b + \max\{b, c\} \geq a$ and $d - c + \max\{b, c\} \geq d$. Consequently, $(a - b + \max\{b, c\}, d - c + \max\{b, c\}) \in S$, and hence the multiplication is closed. Clearly, $(0, 0) \in S$ and for any $(a, b) \in S$ we have

$$(0, 0) \star (a, b) = \big(0 - 0 + \max\{0, a\}, b - a + \max\{0, a\}\big)$$
$$= (a, b) = (a, b) \star (0, 0).$$

Hence, $(0, 0)$ is the identity element of S. Now, we verify the associativity. Assume that $(a, b), (c, d)$ and (e, f) are arbitrary elements of S. Then, we have

$$\big((a, b) \star (c, d)\big) \star (e, f)$$
$$= \big(a - b + \max\{b, c\}, d - c + \max\{b, c\}\big) \star (e, f)$$
$$= \big(a - b - d + c + \max\{d - c + \max\{b, c\}, e\},$$
$$\quad f - e + \max\{d - c + \max\{b, c\}, e\}\big)$$

and

$$(a, b) \star \big((c, d) \star (e, f)\big)$$
$$= (a, b) \star \big(c - d + \max\{d, e\}, f - e + \max\{d, e\}\big)$$
$$= \big(a - b + \max\{b, c - d + \max\{d, e\}\},$$
$$\quad f - e - c + d + \max\{b, c - d + \max\{d, e\}\}\big).$$

Now, we must show that

$$a - b - d + c + \max\{d - c + \max\{b, c\}, e\} = a - b + \max\{b, c - d + \max\{d, e\}\},$$
(3.2)

and

$$f - e + \max\{d - c + \max\{b, c\}, e\} = f - e - c + d + \max\{b, c - d + \max\{d, e\}\}.$$
(3.3)

Equations (3.2) and (3.3) are the same, and so we only need to show that

$$-d + c + \max\{d - c + \max\{b, c\}, e\} = \max\{b, c - d + \max\{d, e\}\}.$$

But this equality is equivalent to

$$\max\{\max\{b, c\}, c - d + e\} = \max\{b, c - d + \max\{d, e\}\},$$

and this equality holds, because

$$\begin{aligned}
\max\{b, c - d + \max\{d, e\}\} &= \max\{b, \max\{c - d + d, c - d + e\}\} \\
&= \max\{b, \max\{c, c - d + e\}\} \\
&= \max\{b, c, c - d + e\} \\
&= \max\{\max\{b, c\}, c - d + e\}.
\end{aligned}$$

Therefore, the multiplication is associative and hence S is a monoid.   ■

Problem 3.86 Let G be a group and $a_0 \in G$. Define a new binary operation $*$ on G such that $(G, *)$ is a group with a_0 as its identity.

Solution We define $*$ on G as follows:

$$x * y = xa_0^{-1}y,$$

for all $x, y \in G$. Let x, y, and z be arbitrary elements of G. Then, we can write

$$\begin{aligned}
(x * y) * z &= (xa_0^{-1}y) * z = xa_0^{-1}ya_0^{-1}z = xa_0^{-1}(ya_0^{-1}z) \\
&= x * (ya_0^{-1}z) = x * (y * z).
\end{aligned}$$

Hence, the associative law holds.

For any $x \in G$, $x * a_0 = xa_0^{-1}a_0 = xe = x$, where e is the identity of G. Similarly, $a_0 * x = a_0a_0^{-1}x = ex = x$. This shows that a_0 is the identity element of G with respect to $*$.

Next, let $x \in G$. Then, we have $x * (a_0x^{-1}a_0) = xa_0^{-1}a_0x^{-1}a_0 = a_0$. Also, we have $(a_0x^{-1}a_0) * x = a_0x^{-1}a_0a_0^{-1}x = a_0$. Hence, $a_0x^{-1}a_0$ is the inverse of x with respect to $*$.

Therefore, $(G, *)$ is a group with a_0 as its identity.   ■

Problem 3.87 Let G be a group and m, n be relatively prime positive integers such that

$$a^m b^m = b^m a^m \quad \text{and} \quad a^n b^n = b^n a^n.$$

for all $a, b \in G$. Prove that G is abelian.

Solution Since $(m, n) = 1$, it follows that there exist integers x and y such that $mx + ny = 1$. Then, we have

$$
\begin{aligned}
(a^m b^n)^{mx} &= (a^m b^n)(a^m b^n) \dots (a^m b^n) \\
&= a^m \big((b^n a^m)(b^n a^m) \dots (b^n a^m)\big) b^n \\
&= a^m (b^n a^m)^{mx} (b^n a^m)^{-1} b^n \\
&= a^m (b^n a^m)^{mx} a^{-m} b^{-n} b^n \\
&= a^m (b^n a^m)^{mx} a^{-m} \\
&= a^m \big((b^n a^m)^x\big)^m a^{-m} \\
&= \big((b^n a^m)^x\big)^m a^m a^{-m} \\
&= (b^n a^m)^{mx}.
\end{aligned}
$$

This shows that

$$(a^m b^n)^{mx} = (b^n a^m)^{mx}. \tag{3.4}$$

In a similar way, we see that

$$(a^m b^n)^{ny} = (b^n a^m)^{ny}. \tag{3.5}$$

From (3.4) and (3.5), we obtain

$$a^m b^n = (a^m b^n)^{mx+ny} = (b^n a^m)^{mx+ny} = b^n a^m.$$

Finally, we can write

$$
\begin{aligned}
ab &= a^{mx+ny} b^{mx+ny} = a^{mx}(a^{ny} b^{mx}) b^{ny} \\
&= a^{mx} b^{mx} a^{ny} b^{ny} = b^{mx} a^{mx} b^{ny} a^{ny} \\
&= b^{mx} b^{ny} a^{mx} a^{ny} = b^{mx+ny} a^{mx+ny} = ba.
\end{aligned}
$$

This completes the proof.   ∎

Problem 3.88 If G is a group in which

$$(ab)^k = a^k b^k \tag{3.6}$$

for three constructive integers k and for all $a, b \in G$, show that G is abelian.

Solution Suppose that (3.6) is true for n, $n + 1$, and $n + 2$, where n is an integer, i.e.,

$$
\begin{aligned}
(ab)^n &= a^n b^n, \\
(ab)^{n+1} &= a^{n+1} b^{n+1}, \\
(ab)^{n+2} &= a^{n+2} b^{n+2}.
\end{aligned}
$$

Then, we obtain

$$a^{n+1}b^{n+1} = (ab)^{n+1} = (ab)^n ab = a^n b^n ab.$$

By cancellation law, we get

$$ab^n = b^n a. \tag{3.7}$$

On the other hand, we have

$$a^{n+2}b^{n+2} = (ab)^{n+2} = (ab)^n (ab)^2 = a^n b^n abab.$$

Again, by cancellation law, we obtain

$$a^2 b^{n+1} = b^n aba. \tag{3.8}$$

From (3.7) and (3.8), we have

$$a^2 b^{n+1} = b^n aba = ab^n ba = ab^{n+1} a.$$

This yields that

$$ab^{n+1} = b^{n+1} a. \tag{3.9}$$

Now, from (3.7) and (3.9), we conclude that $ab^{n+1} = b^{n+1}a = bb^n a = bab^n$. This shows that $ab = ba$. ∎

3.8 Supplementary Exercises

1. Let G be the set of all real 2×2 matrices $\begin{bmatrix} a & b \\ c & d \end{bmatrix}$, where $ad - bc \neq 0$ is a rational number. Prove that G forms a group under matrix multiplication.

2. Let G be the set of all real 2×2 matrices $\begin{bmatrix} a & b \\ 0 & d \end{bmatrix}$, where $ad \neq 0$. Prove that G forms a group under matrix multiplication. Is G abelian?

3. Let G be the set of all real 2×2 matrices $\begin{bmatrix} a & 0 \\ 0 & a^{-1} \end{bmatrix}$, where $a \neq 0$. Prove that G is an abelian group under matrix multiplication.

4. A semigroup S is called *regular* if for each $y \in S$ there exists $a \in S$ such that $yay = y$. Let S be a semigroup with at least three elements and $x \in S$ is such that $S \setminus \{x\}$ is a group. Prove that S is regular if and only if $x^2 = x$.

5. Let $x_1, x_2, \ldots, x_n$ be elements of a group G. Prove that there are

$$\frac{(2n-2)!}{n!(n-1)!}$$

ways of evaluating the product of $x_1, x_2, \ldots, x_n$ in that order. All these ways yield the same group element which can thus be denoted unambiguously by $x_1 x_2 \ldots x_n$.

6. Let G be a finite group. Show that there are an odd number of elements x of G such that $x^3 = e$. Show that there are an even number of elements y of G such that $y^2 \neq e$.

7. Let G be a group with an odd number of elements. Prove that for each $a \in G$, the equation $x^2 = a$ has a unique solution.

8. Show that the conclusion of Problem 3.88 does not follow if we assume the relation $(ab)^k = a^k b^k$ for just two constructive integers.

9. Let G be a group, $a, b \in G$ and m, n be two integers.

   (a) If $b^{-1}ab = a^m$, show that $b^{-1}a^n b = a^{mn}$ and $b^{-n}ab^n = a^{m^n}$;
   (b) If $a^{-1}b^2 a = b^3$ and $b^{-1}a^2 b = a^3$, prove that $a = b = e$.

10. Let G be the set of all positive rational numbers of the form

$$\frac{a^2 + b^2}{c^2 + d^2},$$

where $a, b, c,$ and d are integers. Is G a group, being ordinary multiplication?

11. A group G satisfies the *ascending chain condition (ACC) on subgroups* if every ascending sequence
$$H_1 \leq H_2 \leq \ldots$$

of subgroups must eventually be constant, that is, if there is an $n > 0$ such that $H_{n+k} = H_n$ for all $k > 0$. A group G satisfies the *maximal condition on subgroups* if every non-empty family of subgroups has a maximal member. Prove that a group G satisfies the maximal condition on subgroups if and only if it satisfies the ascending chain condition on subgroups.

12. Let G be a group of order n and let m be an integer relatively prime to n. Show that if $x^m = y^m$, then $x = y$. Hence show that for each $z \in G$ there is a unique $x \in G$ such that $x^m = z$.

Chapter 4
Cyclic Groups

The simplest type of groups are group of integers modulo n and cyclic groups. Cyclic groups are groups in which every element is a power of some fixed element. In this chapter, we examine cyclic groups in detail and determine their important characteristics. We observe that a cyclic subgroup with generator a is the smallest subgroup containing the set $A = \{a\}$. Can we extend subgroups generated by sets with more than one element? We answer this question in this chapter too.

4.1 Group of Integers Modulo n

Let $a, b \in \mathbb{Z}$ and n be a positive integer. Already, we defined

$$a \equiv b(\mathrm{mod}\ n) \ \Leftrightarrow \ n|a - b.$$

This relation is named as congruence modulo n. In Lemma 1.62, we proved that this relation is an equivalence relation. The equivalence class of $a \in \mathbb{Z}$ is the set

$$\bar{a} = \{b \in \mathbb{Z} \mid a \equiv b(\mathrm{mod}\ n)\}.$$

The integers modulo n partition $\mathbb{Z}$ into n different equivalence classes. We denote the set of these equivalence classes by $\mathbb{Z}_n$. Thus,

$$\mathbb{Z}_n = \{\bar{0}, \bar{1}, \ldots, \overline{n-1}\}.$$

For instance, if $n = 5$, then

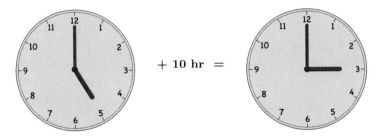

Fig. 4.1 5+10=3

$$\overline{0} = \{\ldots, -15,\ -10,\ -5,\ 0,\ 5,\ 10,\ 15, \ldots\},$$
$$\overline{1} = \{\ldots, -14,\ -9,\ -4,\ 1,\ 6,\ 11,\ 16, \ldots\},$$
$$\overline{2} = \{\ldots, -13,\ -8,\ -3,\ 2,\ 7,\ 12,\ 17, \ldots\},$$
$$\overline{3} = \{\ldots, -12,\ -7,\ -2,\ 3,\ 8,\ 13,\ 18, \ldots\},$$
$$\overline{4} = \{\ldots, -11,\ -6,\ -1,\ 4,\ 9,\ 14,\ 19, \ldots\},$$

and

$$\mathbb{Z}_5 = \{\overline{0},\ \overline{1},\ \overline{2},\ \overline{3},\ \overline{4}\}.$$

Theorem 4.1 $(\mathbb{Z}_n, +)$ *is an abelian group, where $+$ is the addition modulo n, i.e.,*
$\overline{a} + \overline{b} = \overline{a + b}$, *for all $\overline{a}, \overline{b} \in \mathbb{Z}_n$.*

Proof First we show that $+$ is well defined on $\mathbb{Z}_n$. If $\overline{a_1} = \overline{a_2}$ and $\overline{b_1} = \overline{b_2}$, then there
exist integers q_1 and q_2 such that $a_1 - a_2 = q_1 n$ and $b_1 - b_2 = q_2 n$. So, we have

$$(a_1 + b_1) - (a_2 + b_2) = a_1 - a_2 + b_1 - b_2 = q_1 n + q_2 n = (q_1 + q_2) n.$$

This implies that $a_1 + b_1 \equiv a_2 + b_2 (\text{mod } n)$, or equivalently $\overline{a_1 + b_1} = \overline{a_2 + b_2}$.
Clearly, addition on $\mathbb{Z}_n$ is associative and commutative. $\overline{0}$ is the additive identity
for $\mathbb{Z}_n$ and each $\overline{a} \in \mathbb{Z}_n$ has an additive inverse $\overline{-a}$ in $\mathbb{Z}_n$. Note that $\overline{-a} = \overline{n - a}$. ∎

Sometimes, when no confusion can arise, we use $0, 1, \ldots, n - 1$ to indicate the
equivalence classes $\overline{0}, \overline{1}, \ldots, \overline{n - 1}$.

Example 4.2 For $n = 12$, we have $3 + 6 = 9, 8 + 9 = 5, 6 + 7 = 1$ and $5 + 10 = 3$. A familiar use of modular arithmetic is in the 12-hour clock (see Fig. 4.1, in which
the day is divided into two 12-hour periods).

Example 4.3 The Cayley table for $(\mathbb{Z}_5, +)$ is as follows:

+	$\bar{0}$	$\bar{1}$	$\bar{2}$	$\bar{3}$	$\bar{4}$
$\bar{0}$	$\bar{0}$	$\bar{1}$	$\bar{2}$	$\bar{3}$	$\bar{4}$
$\bar{1}$	$\bar{1}$	$\bar{2}$	$\bar{3}$	$\bar{4}$	$\bar{0}$
$\bar{2}$	$\bar{2}$	$\bar{3}$	$\bar{4}$	$\bar{0}$	$\bar{1}$
$\bar{3}$	$\bar{3}$	$\bar{4}$	$\bar{0}$	$\bar{1}$	$\bar{2}$
$\bar{4}$	$\bar{4}$	$\bar{0}$	$\bar{1}$	$\bar{2}$	$\bar{3}$

We can also multiply elements of $\mathbb{Z}_n$. Given two elements $\bar{a}$ and $\bar{b}$ in $\mathbb{Z}_n$, we define

$$\bar{a} \cdot \bar{b} = \overline{ab}.$$

This is well defined and with respect to it, $\mathbb{Z}_n$ becomes a commutative monoid.

A multiplicative inverse for $\bar{a} \in \mathbb{Z}_n$ is an element $\bar{b} \in \mathbb{Z}_n$ such that $\bar{a} \cdot \bar{b} = \bar{1}$. An element $\bar{a} \in \mathbb{Z}_n$ is a *unit* if it has a multiplicative inverse in $\mathbb{Z}_n$. In other words, the integer a is a unit modulo n, meaning that $ab \equiv 1 \pmod{n}$ for some integer b.

By the above operation we do not obtain a group. For instance, $\bar{0}$ does not have a multiplicative inverse.

Lemma 4.4 $\bar{a}$ *is a unit in* $\mathbb{Z}_n$ *if and only if* $(a, n) = 1$.

Proof If $\bar{a}$ is a unit, then $ab - 1 = qn$ for some integers b and q, and so $ab + qn = 1$. Now, by Theorem (1.51) we conclude that $(a, n) = 1$.

Conversely, if $(a, n) = 1$, then by Theorem (1.51) there exist $x, y \in \mathbb{Z}$ such that $1 = ax + by$. Thus, $\bar{x}$ is a multiplicative inverse of $\bar{a}$. ∎

Let U_n denote the set of units of $\mathbb{Z}_n$.

Theorem 4.5 *For each integer* $n \geq 1$*, the set* U_n *forms an abelian group under multiplication modulo* n*, with identity element* $\bar{1}$*.*

Proof We first show that the product of two units $\bar{a}$ and $\bar{b}$ is also a unit. If $\bar{a}$ and $\bar{b}$ are units, then they have inverses $\bar{c}$ and $\bar{d}$, respectively, such that

$$\bar{a} \cdot \bar{c} = \overline{ac} = \bar{1} \text{ and } \bar{b} \cdot \bar{d} = \overline{bd} = \bar{1}.$$

Hence, we obtain that

$$\overline{ab} \cdot \overline{cd} = \overline{abcd} = \overline{acbd} = \overline{ac} \cdot \overline{bd} = \bar{1} \cdot \bar{1} = \bar{1}.$$

Consequently, $\overline{ab}$ has the inverse $\overline{cd}$, and is a unit. Moreover, for each $\bar{a}, \bar{b}, \bar{c} \in U_n$, we have

$$\bar{a} \cdot \bar{b} = \overline{ab} = \overline{ba} = \bar{b} \cdot \bar{a},$$
$$\bar{a} \cdot (\bar{b} \cdot \bar{c}) = \bar{a} \cdot \overline{bc} = \overline{a(bc)} = \overline{(ab)c} = \overline{ab} \cdot \bar{c} = (\bar{a} \cdot \bar{b}) \cdot \bar{c}.$$

Table 4.1 A short table of values $\varphi(n)$

n	1	2	3	4	5	6	7	8	9	10	11	12	13	14	15
$\varphi(n)$	1	1	2	2	4	2	6	4	6	4	10	4	12	6	8

The identity element of U_n is $\bar{1}$, since $\bar{a} \cdot \bar{1} = \bar{a} = \bar{1} \cdot \bar{a}$, for all $\bar{a} \in U_n$. Finally, if $\bar{a} \in U_n$, then by the definition of U_n, there exists $\bar{c} \in \mathbb{Z}_n$ such that $\bar{a} \cdot \bar{c} = \bar{1} = \bar{c} \cdot \bar{a}$. This yields that $\bar{c} \in U_n$ and $\bar{c}$ is the inverse of $\bar{a}$.   ∎

Definition 4.6 Let n be a positive integer. The *Euler function* φ is defined to be the number of positive integers not exceeding n which are relatively prime to n.

Table 4.1 is a short table of values $\varphi(n)$:

Corollary 4.7 *For each integer $n \geq 1$, $|U_n| = \varphi(n)$.*

Proof It follows from Lemma 4.4 and Theorem 4.5.   ∎

Example 4.8 In $\mathbb{Z}_{14}$, the group of units is $U_{14} = \{1, 3, 5, 9, 11, 13\}$. The operation in U_{14} is multiplication modulo 14. The following is Cayley table for U_{14}:

$\cdot$	$\bar{1}$	$\bar{3}$	$\bar{5}$	$\bar{9}$	$\bar{11}$	$\bar{13}$
$\bar{1}$	$\bar{1}$	$\bar{3}$	$\bar{5}$	$\bar{9}$	$\bar{11}$	$\bar{13}$
$\bar{3}$	$\bar{3}$	$\bar{9}$	$\bar{1}$	$\bar{13}$	$\bar{5}$	$\bar{11}$
$\bar{5}$	$\bar{5}$	$\bar{1}$	$\bar{11}$	$\bar{3}$	$\bar{13}$	$\bar{9}$
$\bar{9}$	$\bar{9}$	$\bar{13}$	$\bar{3}$	$\bar{11}$	$\bar{1}$	$\bar{5}$
$\bar{11}$	$\bar{11}$	$\bar{5}$	$\bar{13}$	$\bar{1}$	$\bar{9}$	$\bar{3}$
$\bar{13}$	$\bar{13}$	$\bar{11}$	$\bar{9}$	$\bar{5}$	$\bar{3}$	$\bar{1}$

Example 4.9 If p is prime, then all positive integers smaller than p are relatively prime to p. Hence, $U_p = \{\bar{1}, \bar{2}, \ldots, \overline{p - 1}\}$.

Theorem 4.10 *Let $p > 1$ be an integer. Then the following statements are equivalent.*

(1) p is prime;
(2) If $\bar{a} \cdot \bar{b} = \bar{0}$ in $\mathbb{Z}_p$, then $\bar{a} = \bar{0}$ or $\bar{b} = \bar{0}$.

Proof $(1 \Rightarrow 2)$ Since p is prime, it follows that $U_p = \mathbb{Z}_p - \{0\}$ is a group. Assume that $\bar{a} \cdot \bar{b} = \bar{0}$ in $\mathbb{Z}_p$. If $\bar{a} = \bar{0}$, then there is nothing to prove. If $\bar{a} \neq \bar{0}$, then $\bar{a} \in U_p$. So, there exists $\bar{x} \in U_p$ such that $\bar{x} \cdot \bar{a} = \bar{1}$. Therefore, we get

$$\bar{0} = \bar{x} \cdot \bar{0} = \bar{x} \cdot (\bar{a} \cdot \bar{b}) = (\bar{x} \cdot \bar{a}) \cdot \bar{b} = \bar{1} \cdot \bar{b} = \bar{b}.$$

$(2 \Rightarrow 1)$ Suppose that $p = ab$ for some integers a and b. We show $a = \pm 1$ or $\pm p$. Since $p = ab$, it follows that $\overline{ab} = \overline{p} = \overline{0}$. This implies that $\overline{a} \cdot \overline{b} = \overline{0}$ in $\mathbb{Z}_p$. Now, by (2) we conclude that $\overline{a} = \overline{0}$ or $\overline{b} = \overline{0}$. If $\overline{a} = \overline{0}$, then $p|a$, or equivalently $a = kp$ for some integer k. Hence, we have $p = ab = kpb$. This implies that $1 = kb$. Since k and b are integers, it follows that the only possibilities are $b = \pm 1$, and so $a = \pm p$. If $\overline{b} = \overline{0}$, then a similar argument shows that $a = \pm 1$. Therefore, p is prime. ∎

Exercises

1. Solve equation $x^2 + \overline{4}x + \overline{8} = \overline{0}$ in $\mathbb{Z}_{11}$.
2. Find the number of distinct solutions to the equation $x^2 + \overline{(-1)} = \overline{0}$ in $\mathbb{Z}_5$, $\mathbb{Z}_{13}$, and in $\mathbb{Z}_9$.
3. Find, if possible, a multiplicative inverse for $\overline{8}$ in each of $\mathbb{Z}_5$, $\mathbb{Z}_{21}$, and $\mathbb{Z}_{264}$.
4. Find a positive integer n and three elements $\overline{a}$, $\overline{b}$, and $\overline{b}$ in $\mathbb{Z}_n$ such that none of $\overline{a} \cdot \overline{b}, \overline{b} \cdot \overline{c}$, and $\overline{c} \cdot \overline{a}$ is equal to $\overline{0}$, yet $\overline{a} \cdot \overline{b} \cdot \overline{c} = \overline{0}$.
5. Given $\overline{a}$, $\overline{b} \in \mathbb{Z}_n$, we say that $\overline{b}$ is a *square root* of $\overline{a}$ if $\overline{b} \cdot \overline{b} = \overline{a}$.

   (a) Find all square roots of elements in $\mathbb{Z}_{17}$, if exist;
   (b) If p is any prime, show that a has at most two square roots modulo n;
   (c) Give an example that shows that it is possible for a number to have more than two square roots.

4.2 Cyclic Groups

This section contains a few observations about cyclic groups. The structure of cyclic groups is relatively simple. We examine cyclic groups in detail and determine many of their important characteristics.

If G is a group, $a \in G$ and $k \in \mathbb{Z}$, then we define

$$a^n = \begin{cases} \underbrace{aa \ldots a}_{k \text{ times}} & \text{if } k > 0 \\ e & \text{if } k = 0 \\ \underbrace{a^{-1}a^{-1} \ldots a^{-1}}_{-k \text{ times}} & \text{if } k < 0. \end{cases}$$

Lemma 4.11 *If G is a group and $a \in G$, then $H = \{a^n \mid n \in \mathbb{Z}\}$ is a subgroup of G.*

Proof It is straightforward. ∎

Fig. 4.2 An examples of
string art to show that 5 is a
generator of $\mathbb{Z}_{12}$

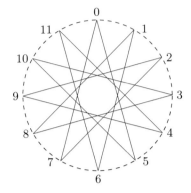

Definition 4.12 Let G be a group. A subgroup H of G is *cyclic* if $H = \{a^n \mid n \in \mathbb{Z}\}$, for some $a \in H$. In this case we say that H is the *cyclic subgroup generated by a*.

When this happens, we write $H = \langle a \rangle$.

Definition 4.13 A group G is called *cyclic* if there is an element $a \in G$ such that $G = \langle a \rangle$. Such an element a is called a *generator* of G.

Example 4.14 The set of integers under addition is an example of an infinite group which is cyclic and is generated by both 1 and -1.

Example 4.15 If $K_4 = \{e, a, b, c\}$ is the Klein's 4-group, then $H = \{e, a\}$ is a cyclic subgroup of K_4; however, K_4 is not a cyclic group.

Example 4.16 The set $\mathbb{Z}_n = \{0, 1, \ldots, n - 1\}$ for $n \geq 1$ is a cyclic group under addition modulo n. Both 1 and -1 are generators.

Depending on which n we are given, $\mathbb{Z}_n$ may have many generators.

Example 4.17 $\mathbb{Z}_{12} = \langle 1 \rangle = \langle 5 \rangle = \langle 7 \rangle = \langle 11 \rangle$. To verify for instance that $\mathbb{Z}_{12} = \langle 5 \rangle$ we can use a *string art*. Figure 4.2 is an example of string art that illustrates how 5 generates $\mathbb{Z}_{12}$. Twelve tacks are placed along a circle and numbered. A string is tied to tack 0 and then looped around every fifth tack. As a result, the numbers of the tacks that are reached are exactly the ordered numbers of 5 modulo 12. Note that if every seventh tack were used, the same artwork would be obtained. If every third tack were connected, as in Fig. 4.3, the resulting loop would only use four tacks, and so 3 does not generate $\mathbb{Z}_{12}$.

Example 4.18 An *nth root of unity* is a complex number z which satisfies the equation $z^n = 1$ for some positive integer n. Let $\omega_n = e^{2\pi i/n}$ be an nth root of unity. All the nth roots of unity form a group under multiplication. It is a cyclic group, generated by ω_n, which is called a *primitive root of unity*. The term "primitive" exactly refers to being a generator of the cyclic group, namely an nth root of unity is primitive when there is no positive integer k smaller than n such that $\omega_n^k = 1$. See Fig. 4.4 for $n = 5$.

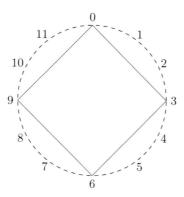

Fig. 4.3 An example of string art to show that 3 is not a generator of $\mathbb{Z}_{12}$

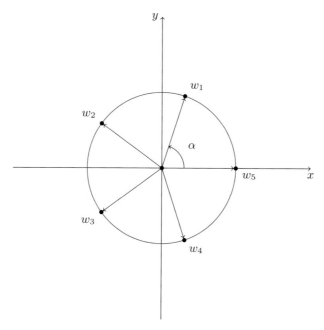

Fig. 4.4 5th roots of unity, where $\alpha = \dfrac{2\pi}{5}$

Definition 4.19 The *order of an element* a in a group G is the least positive integer n such that $a^n = e$. In additive notation this would be $na = 0$. If no such integer exist, then we say a has *infinite order*. The order of an element a is denoted by $o(a)$.

Note that the critical part of Definition 4.19 is that the order is the *least* positive integer with the given property. The terminology *order* is used both for groups and group elements, but it is usually clear from the context which one is considered.

So, to find the order of an element a we need only compute the series of products a, a^2, a^3, ... until we first reach the identity. If the identity never appears in the series, then a has infinite order.

Definition 4.20 A *torsion group* is a group all of whose elements have finite order. On the other hand, a group is said to be *torsion-free* if all its non-identity elements have infinite order.

Lemma 4.21 *Let $a \neq e$ be an element of a group G. If n is the order of a and $a^k = e$ for some integer k, then n divides k.*

Proof By the Division algorithm, there exist integers q and r such that $k = qn + r$ and r is strictly smaller than n. So, we have

$$e = a^k = a^{nq+r} = a^{nq} a^r.$$

Since a has order n, it follows that $a^{nq} = (a^n)^q = e$. This implies that $a^r = e$, which contradicts our choice of n as the smallest positive integer for which a power of a is the identity, unless $r = 0$. In this case $k = nq$ and we conclude that n must divide k.   ∎

Theorem 4.22 *If $a \in G$ has order n, then*

$$o(a^k) = \frac{n}{(n, k)}.$$

Proof Suppose that $(n, k) = d$. Then, there exist integers u and v such that $k = ud$, $n = vd$ and $(u, v) = 1$. Now, we have

$$(a^k)^v = (a^{ud})^v = a^{uvd} = a^{un} = (a^n)^u = e.$$

Moreover, if $(a^k)^m = e$, then $n|km$. So, $vd|udm$, or equivalently $v|um$. Since $(u, v) = 1$, by Euclid's lemma we conclude that $v|m$. Therefore, we obtain $o(a^k) = n/d$. ∎

We now turn to some results about cyclic groups.

Theorem 4.23 *Every cyclic group is abelian.*

Proof Suppose that G is a cyclic group. By definition, there exists $a \in G$ such that $G = \langle a \rangle$. We need to show that $xy = yx$, for all $x, y \in G$. Now, x and y are of the form x^m and y^n, for some integers m and n, respectively. Moreover, we have

$$xy = a^m a^n = a^{m+n} = a^{n+m} = a^n a^m = yx,$$

as desired.   ∎

Theorem 4.24 *Every subgroup of a cyclic group is also cyclic.*

Proof Let $G = \langle a \rangle$ be our cyclic group and suppose that H is a subgroup of G. We must show that H is cyclic. If it consists of the identity alone, then it is trivial subgroup and there is nothing to prove. So, we may assume that $H \neq \{e\}$. We consider the set

$$S = \{s \in \mathbb{N} \mid a^s \in H\}.$$

Since H is non-trivial, it follows that S is non-empty. But being a non-empty subset of positive integers, S must have the least positive integer, call it k. We next claim that $H = \langle a^k \rangle$. In order to prove this claim, it suffices to let h be an arbitrary element of H and show that $h \in \langle a^k \rangle$. Since $h \in G = \langle a \rangle$, it follows that $h = a^m$, for some integer m. We now invoke the Division algorithm to m and k to obtain integers q and r such that $m = kq + r$, where $0 \leq r < k$. Therefore, we have

$$h = a^m = a^{kq+r} = a^{kq} a^r.$$

This implies that

$$a^r = a^{-kq} a^m.$$

As a^k belongs to H, so does a^{-kq}. By closure property of a subgroup, we know that $a^{-kq} a^m = a^r$ belongs to H. As r is strictly smaller than k, it contradicts the choice of k as the smallest element of S, unless $r = 0$. This proves that $h = a^{qk} = (a^k)^q \in \langle a^k \rangle = H$. ∎

Remark 4.25 Theorem 4.24 tells us that all subgroups of additive group $\mathbb{Z}$ are of the form $n\mathbb{Z} = \{nk \mid k \in \mathbb{Z}\}$.

Theorem 4.26 *Let G be a finite cyclic group generated by a.*

(1) *The generator a has finite order;*
(2) *If $o(a) = n$, then the elements a^k with $k = 0, 1, \ldots, n - 1$ are all distinct and*

$$G = \{e, a, a^2, \ldots, a^{n-1}\}.$$

Proof Since G is cyclic, it follows that

$$G = \langle a \rangle = \{\ldots, a^{-2}, a^{-1}, e, a, a^2, \ldots\}.$$

Since G is finite, it follows that for some integers k and m we have $a^m = a^k$. This implies that $a^{m-k} = e$ and so a has finite order.

If $o(a) = n$, then we claim that $G = \{e, a, a^2, \ldots, a^{n-1}\}$. We see that for $0 \leq m < k \leq n - 1$, the elements a^m and a^k are distinct as otherwise $a^{m-k} = e$ with $|m - k| < n$. Consequently, $\{e, a, a^2, \ldots, a^{n-1}\}$ is a subgroup of G containing n elements. Now, if x is an arbitrary element of G, then $x = a^j$, for some integer j. If $j > 0$, then applying the Division algorithm, we obtain $a^j = a^r$, where $0 \leq r < n - 1$. If $j < 0$ and $s = -j$, then we can make $a^s = a^r$, for some $r < n$ and $a^j = a^{-s} = (a^s)^{-1} = (a^r)^{-1} = a^{n-r}$. This yields that $G = \{e, a, a^2, \ldots, a^{n-1}\}$.

Therefore, we conclude that the cyclic group generated by an element of order n has precisely n elements. ∎

Corollary 4.27 *Let G be a group and a be an element of G. Then,*

$$o(a) = |\langle a \rangle|.$$

Corollary 4.28 *Let G be a finite group of order n. Then, G is cyclic if and only if there exists an element of G of order n.*

Corollary 4.29 *If G is an infinite cyclic group, then G has an infinite number of subgroups.*

Proof If $G = \langle a \rangle$, then $\langle a^n \rangle \neq \langle a^m \rangle$, for each integer $n \neq m$. ∎

Theorem 4.30 *Let G be a cyclic group of order n generated by a and let k be a positive integer. If $d = (n, k)$, then x^k and x^d generate the same cyclic subgroup of G.*

Proof If $d = (n, k)$, we show that a^k and a^d are both powers of each other. Since $d \,|\, k$, it follows that $a^k = (x^d)^{k/d}$. On the other hand, since $d = (k, n)$, it follows that $d = kr + ns$, for some $r, s \in \mathbb{Z}$. This implies that $a^d = a^{kr+ns} = a^{kr}a^{ns} = a^{kr} = (a^k)^r$. This yields that a^k and a^d generate the same subgroup of G. ∎

Theorem 4.31 *Let G be a cyclic group of order n. Then, for each m dividing n, G has a unique subgroup of order m, namely $\langle a^{n/m} \rangle$.*

Proof If $k = n/m$, then $(a^k)^m = (a^{n/m})^m = a^n = e$ and no smaller positive integer power of a^k could be e. So, $\langle a^k \rangle$ is a subgroup of order m. Next, we show that $\langle a^k \rangle$ is the only subgroup of order m. In order to do this, let H be an arbitrary subgroup of G of order m. By the proof of Theorem 4.24, we have $H = \langle a^d \rangle$, where d is the smallest positive integer such that a^d is in H. We apply the Division algorithm to obtain integers q and r such that $n = dq + r$ and $0 \leq r < d$. Then, we have $e = a^n = a^{dq+r} = (a^d)^q a^r$. This implies that $a^r = (a^d)^{-q} \in H$. Consequently, $r = 0$ and so $n = dq$. Moreover, we have

$$m = |H| = |\langle a^d \rangle| = \frac{n}{d}.$$

Therefore, we obtain $d = n/m = k$, i.e., $H = \langle a^d \rangle = \langle a^k \rangle$. This completes the proof. ∎

Theorem 4.32 (Generators of a Finite Cyclic Group) *Let G be a cyclic group of order n generated by a. Then, $G = \langle a^k \rangle$ if and only if k and n are relatively prime.*

Proof Suppose that k and n are relatively prime. Then, we may write $1 = kx + ny$, for some integers x and y. Then, we have

$$a = a^{kx+ny} = a^{kx}a^{ny} = a^{kx} = (a^k)^x.$$

This implies that a belongs to $\langle a^k \rangle$, and so all powers of a belong to $\langle a^k \rangle$. Thus, we conclude that $G = \langle a^k \rangle$ and a^k is a generator of G.

Conversely, suppose that k and n are not relatively prime. Then, there exists an integer $d > 1$ such that $d|n$ and $d|k$. Hence, there exist integers m and r such that $n = md$ and $k = dr$. So, we get

$$(a^k)^m = a^{drm} = a^{rn} = (a^n)^r = e.$$

Since $d > 1$, it follows that $m < n$. This shows that a^k is not a generator of G. ∎

Corollary 4.33 *Let G be a cyclic group of order n generated by a. Then, G has $\phi(n)$ generators.*

Corollary 4.34 *Let k be a positive integer. Then, k is a generator of $\mathbb{Z}_n$ if and only if k and n are relatively prime.*

Theorem 4.35 *Let G be a cyclic group of order n generated by a. If d is a positive divisor of n, then the number of elements of order d is $\phi(d)$.*

Proof By Theorem 4.31, there exists exactly one subgroup of order d, say $\langle b \rangle$. Then, every element of order d generates $\langle b \rangle$. On the other hand, by Theorem 4.32, an element b^k generates $\langle b \rangle$ if and only if $(k, d) = 1$. The number of such elements is $\varphi(d)$. ∎

Theorem 4.36 *Every infinite cyclic group has exactly two generators.*

Proof Suppose that a is a generator of the infinite cyclic group G. Then, we have $G = \{\ldots, a^{-2}, a^{-1}, e, a, a^2, \ldots\}$. Now, if $a^k \in G$ is another generator of G, then $G = \{\ldots, a^{-2k}, a^{-k}, e, a^k, a^{2k}, \ldots\}$. Since $a^{k+1} \in G$, it follows that there exists an integer m such that $a^{k+1} = a^{km}$. This implies that $a^{k(1-m)+1} = e$. Since G is infinite, it follows that $k(m-1) + 1 = 0$, or equivalently $k(m-1) = 1$. This yields that $k = \pm 1$. Consequently, if a is a generator of G, then another generator of G is only a^{-1}. ∎

For any finite group G, let $\theta_G(d)$ be the number of elements of G of order d. Then, it is clear that

$$|G| = \sum_{d=1}^{|G|} \theta_G(d).$$

Theorem 4.37 *In a finite group (not necessary cyclic), $\theta_G(d)$ is divisible by $\varphi(d)$.*

Proof Let G be a finite group. If G has no element of order d, then $\varphi(d)|0 = \theta_G(d)$ and we are done. So, suppose that there is $a \in G$ such that $o(a) = d$. By Theorem 4.35, $\langle a \rangle$ has $\varphi(d)$ elements of order d. Now, if all elements of order d belong to $\langle a \rangle$, we are done. Hence, suppose that $x \in G$ such that $o(x) = d$ and $x \in \langle a \rangle$.

Then, $\langle x \rangle$ has $\varphi(d)$ elements of order d too. Next, if $\langle a \rangle$ and $\langle x \rangle$ have no elements of order d in common, we have found $2\varphi(d)$ elements of order d. Note that if there is $y \in \langle a \rangle \cap \langle x \rangle$ with $o(y) = d$, then we obtain $\langle a \rangle = \langle x \rangle = \langle y \rangle$, and this is a contradiction. Continuing, we conclude that the number of elements of order d is a multiple of $\varphi(d)$.   ∎

Exercises

1. Draw the Hasse diagram for a cyclic group of order 30.
2. Show that the elements of finite order in an abelian group G form a subgroup of G.
3. Let U_{24} be the group of invertible elements in $\mathbb{Z}_{24}$. Find all cyclic subgroups of U_{24}.
4. Find an example of a non-cyclic group, all of whose proper subgroups are cyclic.
5. Find a collection of distinct subgroups $\langle a_1 \rangle, \langle a_2 \rangle, \ldots, \langle a_n \rangle$ with the property that $\langle a_1 \rangle \leq \langle a_2 \rangle \leq \ldots \leq \langle a_n \rangle$ with n as large as possible.
6. Let G be a group and $a, b \in G$. Prove that

   (a) $o(a) = o(a^{-1})$;
   (b) $o(ab) = o(ba)$;
   (c) $o(a) = o(g^{-1}ag)$, for all $g \in G$;
   (d) If $ab = ba$, and $o(a)$ and $o(b)$ are relatively prime, then $o(ab) = o(a)o(b)$.

7. If a cyclic group has an element of infinite order, how many elements of finite order does it have?
8. Give an example of a group that has exactly 6 subgroups (including the trivial subgroup and the group itself). Generalize to exactly n subgroups for any positive integer n.
9. If G is an abelian group and contains cyclic subgroups of orders 4 and 5, what other sizes of cyclic subgroups must G contain? Generalize.
10. If G is an abelian group and contains a pair of cyclic subgroups of order 2, show that G must contain a subgroup of order 4. Must this subgroup be cyclic?
11. Let G be an abelian group. Show that the elements of finite order in G form a subgroup. This subgroup is called the *torsion subgroup* of G.
12. Find the torsion subgroup of the multiplicative group $\mathbb{R}^*$ of non-zero real numbers.
13. Find the torsion subgroup T of the multiplicative group $\mathbb{C}^*$ of non-zero complex numbers.
14. Let G be an abelian group of order mn such that m and n are relatively prime. If there exists $a, b \in G$ such that $o(a) = m$ and $o(b) = n$, prove that G is cyclic.
15. Show that both U_{25} and U_{27} are cyclic groups.
16. Prove that U_{2^n} ($n \geq 3$) is not cyclic.
17. Let a and b be elements of a group. If $o(a) = m$, $o(b) = n$, and m and n are relatively prime, show that $\langle a \rangle \cap \langle b \rangle = \{e\}$.

18. Suppose that $o(x) = n$. Find a necessary and sufficient condition on r and s such that $\langle x^r \rangle \le \langle x^s \rangle$.
19. Suppose that $o(x) = n$. Show that $\langle x^r \rangle = \langle x^s \rangle$ if and only if $(n, r) = (n, s)$.
20. Suppose that a is a group element such that $o(a^{28}) = 10$ and $o(a^{22}) = 20$. Determine $o(a)$.
21. Determine the Hasse diagram for subgroups of $\mathbb{Z}_{p^2 q}$, where p and q are distinct.
22. Give an example of a group and elements $a, b \in G$ such that $o(a)$ and $o(b)$ are finite but $o(ab)$ is not of finite order.

4.3 Generating Sets

We are interested in answering the following questions: What is the smallest subgroup of a group G containing elements $x_1, \ldots, x_n \in G$? How can we describe an arbitrary element in this subgroup? Or, more generally, what is the smallest subgroup of a group G containing a subset A of G and how can we describe an arbitrary element in this subgroup?

Definition 4.38 Let G be a group and A be a subset of G. Let $\{H_i \mid i \in I\}$ be the family of all subgroups of G which contain A. Then

$$\bigcap_{i \in I} H_i$$

is called the *subgroup of G generated by the set A* and denoted by $\langle A \rangle$.

Indeed, $\langle A \rangle$ is the smallest subgroup of G that contains A. The elements of A are the generators of the subgroup $\langle A \rangle$, which may also be generated by other subsets. If $A = \{x_1, \ldots, x_n\}$, then we write $\langle x_1, \ldots, x_n \rangle$ instead of $\langle A \rangle$. If $G = \langle x_1, \ldots, x_n \rangle$, then G is said to be *finitely generated*.

Remark 4.39 If $A = \{a\}$, then $G = \langle a \rangle$ is a cyclic group.

Theorem 4.40 *If G is a group and A is a non-empty subset of G, then*

$$\langle A \rangle = \{x_1^{\alpha_1} \ldots x_n^{\alpha_n} \mid x_i \in A, \ \alpha_i \in \mathbb{Z}, \ 1 \le i \le n, \ n \in \mathbb{N}\}.$$

Proof Let

$$H = \{x_1^{\alpha_1} \ldots x_n^{\alpha_n} \mid x_i \in A, \ \alpha_i \in \mathbb{Z}, \ 1 \le i \le n, \ n \in \mathbb{N}\}.$$

Clearly, H is a non-empty subset of G. Suppose that x and y are two arbitrary elements of H. Then, there exist $x_1, \ldots, x_n, y_1, \ldots, y_m \in A$ and $\alpha_1, \ldots, \alpha_n, \beta_1, \ldots, \beta_m \in \mathbb{Z}$ such that

$$x = x_1^{\alpha_1} \ldots x_n^{\alpha_n} \quad \text{and} \quad y = y_1^{\beta_1} \ldots y_m^{\beta_m}.$$

Hence, we obtain

$$xy^{-1} = x_1^{\alpha_1} \ldots x_n^{\alpha_n} y_m^{-\beta_m} \ldots y_1^{-\beta_1} \in H.$$

Thus, H is a subgroup of G. Moreover, $A \subseteq H$. Consequently, by the definition of $\langle A \rangle$, we conclude that $\langle A \rangle \subseteq H$.

On the other hand, since $\langle A \rangle$ is a subgroup of G, it follows that all elements of the form $x_1^{\alpha_1} \ldots x_n^{\alpha_n}$ (for $x_i \in A$ and $\alpha_i \in \mathbb{Z}$) belong to $\langle A \rangle$. This yields that $H \subseteq \langle A \rangle$. Therefore, we deduce that $\langle A \rangle = H$. ∎

Remark 4.41 If G is an additive group and A is a non-empty subset of G, then

$$\langle A \rangle = \{\alpha_1 x_1 + \ldots \alpha_n x_n \mid x_i \in A, \ \alpha_i \in \mathbb{Z}, \ 1 \le i \le n, \ n \in \mathbb{N}\}.$$

Example 4.42 Every finite group is finitely generated since $G = \langle G \rangle$.

Example 4.43 The additive group $\mathbb{Q}$ is not finitely generated: a finite set of rational numbers has a common denominator, say m, and the subgroup of $\mathbb{Q}$ generated by these rational numbers (their integral multiples and sums thereof) will only give rise to rational numbers with denominators dividing m. Not all rationales have such denominators (try $\frac{1}{m+1}$), so $\mathbb{Q}$ does not have a finite set of generators as an additive group.

Example 4.44 A finitely generated group is at most countable, so an uncountable group is not finitely generated. For example, the group of real numbers under addition is not finitely generated.

If G is a group and $\{H_i \mid i \in I\}$ is a family of subgroups of G, then $\bigcup_{i \in I} H_i$ is not a subgroup of G in general. The subgroup $\langle \bigcup_{i \in I} H_i \rangle$ generated by the set $\bigcup_{i \in I} H_i$ is called the *subgroup generated by groups* $\{H_i \mid i \in I\}$. If H and K are subgroups of G, then the subgroup $\langle H \cup K \rangle$ generated by H and K is called the *join* of H and K and is denote by $H \vee K$.

Theorem 4.45 *If G is a group, A is a subgroup of $Z(G)$, the center of G, and $z \in G$, then $\langle A \cup \{z\} \rangle$, the subgroup generated by A and z, is abelian.*

Proof Let $H = \langle A \cup \{z\} \rangle$ and y be an arbitrary element of H. First, we show that $y = az^k$, for some $a \in A$ and integer k. By Theorem 4.40, we can write $y = a_1 a_2 \ldots a_n$, where for each $1 \le i \le n$, $a_i \in A$, $a_i = z$ or $a_i = z^{-1}$. Since A is a subgroup of $Z(G)$, it follows that $a_1, a_2, \ldots, a_n$ commute. We set all a_is that belong to A on the left side and hence other elements will be z or z^{-1}. If a is the product of a_is that are in A, then $y = az^k$ for some integer k.

Now, suppose that x and y are two arbitrary elements of H. Then, we have

$$x = a_1 z^{k_1} \quad \text{and} \quad y = a_2 z^{k_2},$$

for some $a_1, a_2 \in A$ and integers k_1, k_2. Thus, we obtain

$$xy = (a_1 z^{k_1}) a_2 z^{k_2}$$
$$= a_2 (a_1 z^{k_1}) z^{k_2}$$
$$= a_2 a_1 z^{k_2} z^{k_1}$$
$$= a_2 z^{k_2} a_1 z^{k_1}$$
$$= yx.$$

This yields that H is an abelian subgroup of H. ∎

Exercises

1. In $(\mathbb{Z}, +)$, determine the subgroup generated by

   (a) $\{4, 6\}$;
   (b) $\{4, 5\}$.

2. Let G be a group and X be a non-empty subset of G. For every $a \in G$, prove that $\langle a^{-1} X a \rangle = a^{-1} \langle X \rangle a$.
3. Prove that group generated by $G \setminus H$ equals G, where H is a proper subgroup of G.
4. Let G be an abelian group and $A = \{a_1, \ldots, a_n\}$ be a finite subset of G. If the order of each element of A is finite, prove that $\langle A \rangle$ is a finite subgroup of G.
5. Suppose that G is a group containing subgroups H and K. Show that $H \cup K = H \vee K$ if and only if $H \cap K \in \{H, K\}$.
6. Show that there is a group G containing subgroups H, K, L such that $H \cup K \cup L = H \vee K \vee L$ but $H \cap K \cap L \notin \{H, K, L\}$.

4.4 Worked-Out Problems

Problem 4.46 Let G be an abelian group and $x, y \in G$ be of finite orders.

(1) Show that $o(xy) | [o(x), o(y)]$, where $[o(x), o(y)]$ stands for least common multiple;
(2) Give an example to illustrate that $o(xy) \neq [o(x), o(y)]$, in general.

Solution (1) Assume that x and y are two elements of orders m and n, respectively. Let $[m, n] = r$. Then, we have $r = mm' = nn'$, and hence

$$x^r y^r = (x^m)^{m'} (y^n)^{n'} = e.$$

Now, suppose that $o(xy) = k$, then k is the smallest positive integer k such that $(xy)^k = e$. Since G is abelian, it follows that

$$(xy)^r = x^r y^r = e \text{ and } (xy)^k = x^k y^k = e.$$

This shows that $(xy)^k = (xy)^r = e$. Since $o(xy) = k$, we conclude that $k|r$. This completes the proof.

(2) For this part, let x be a non-identity element of G and let $y = x^{-1}$. Then, it is clear that $1 = o(e) = o(xx^{-1}) = o(xy) \neq [o(x), o(y)]$. ∎

Problem 4.47 Prove that a group is finite if and only if it has only finitely many subgroups.

Solution Suppose that G is a finite group of order n. Since the number of subsets of G is 2^n, it follows that the number of subgroups of G is less than 2^n.

Conversely, suppose that the number of subgroups of G is finite. Let $a \in G$ be an arbitrary element. We claim that $\langle a \rangle$ is finite. If $\langle a \rangle$ is an infinite group, then the order of a is infinite. Now, for each positive integer k, let $H_k = \langle a^k \rangle$. We prove H_ks are distinct. Indeed, if $H_i = H_j$, for some positive integers m and n, then $a^i \in \langle a^j \rangle$ and $a^j \in \langle a^i \rangle$. So, there exist integers r and s such that $a^i = a^{jr}$ and $a^j = a^{is}$. This yields that $a^{i-jr} = e$ and $a^{j-is} = e$. Since the order of a is infinite, it follows that $i - jr = 0$ and $j - is = 0$, or $i|j$ and $j|i$. Hence, $i = j$. Consequently, H_ks are distinct. This means that G has many infinitely subgroups, and it is a contradiction. Therefore, $\langle a \rangle$ is finite, for every $a \in G$. Since

$$G = \bigcup_{a \in G} \langle a \rangle$$

and the number of subgroups of G is finite, we conclude that

$$G = \langle a_1 \rangle \cup \langle a_2 \rangle \cup \ldots \cup \langle a_n \rangle.$$

Since the finite union of finite sets is finite, it follows that G is finite. ∎

4.5 Supplementary Exercises

1. Let $G = \{a_1, a_2, \ldots, a_n\}$ be a finite abelian group of order n and $x = a_1 a_2 \ldots a_n$ be the product of all the elements in G. Show that $x^2 = e$.
2. Suppose that the elements x, y in the group G satisfy the relation $xyx^{-1} = y^2$, where y is a non-identity element.

   (a) Show that $x^5 y x^{-5} = y^{32}$;
   (b) If $o(x) = 5$, compute $o(y)$.

3. Let G be the collection of all rational numbers x which satisfy $0 \le x < 1$. Show that the binary operation

$$x \oplus y = \begin{cases} x + y & \text{if } 0 \le x + y < 1 \\ x + y - 1 & \text{if } x + y \ge 1. \end{cases}$$

makes G into an infinite abelian group all of whose elements have finite order.

4. Let G be a group and $x \in G$ such that $o(x) = mn$, where m and n are positive integers and relatively prime. Show that one can write $x = ab$, where $o(a) = m$, $o(b) = n$ and $ab = ba$. Moreover, prove the uniqueness of such a representation.

5. Give an example of an infinite group that has exactly two elements of order 4.

6. For every integer n greater than 2, prove that the group U_{n^2-1} is not cyclic.

7. Let G be an abelian and suppose that G has elements of orders m and n, respectively. Prove that G has an element whose order is the least common multiple of m and n.

8. Let G be a finite group. Prove that the following conditions are equivalent:

    (a) G is cyclic and $|G| = p^n$, where p is prime and n is a non-negative integer;
    (b) If A and B are subgroups of G, then $A \leq B$ or $B \leq A$.

9. Prove that the additive group of all rational numbers is a torsion-free group and, further, that it can be represented as a union of an ascending chain of cyclic subgroups.

10. Suppose that G is a finitely generated group and that $G = \langle Y \rangle$, where Y is not necessarily a finite set. Prove that there is a finite subset X of Y such that $G = \langle X \rangle$.

11. Prove that a group G is finitely generated if and only if any increasing sequence $H_1 \leq H_2 \leq \dots$ of subgroups of G stabilizes, i.e., $H_i = H_j$ for $i, j \geq k$ starting from some k.

Chapter 5
Permutation Groups

In this chapter, we construct some groups whose elements are called permutations. Often, an action produced by a group element can be regarded as a function, and the binary operation of the group can be regarded as function composition. The symmetric group on a set is the group consisting of all bijections from the set to itself with function composition as the group operation. These groups will provide us with examples of finite non-abelian groups.

5.1 Inverse Functions and Permutations

In this section, we study certain groups of functions called permutation groups.

Theorem 5.1 *If* $f : X \rightarrow Y$, $g : Y \rightarrow Z$ *and* $h : Z \rightarrow W$ *are functions, then their compositions are associative, i.e.,* $(h \circ g) \circ f = h \circ (g \circ f)$.

Proof It is straightforward. ∎

Definition 5.2 For any non-empty set X, the *identity function* is the function $id_X : X \rightarrow X$ defined by $id_X(x) = x$, for all $x \in X$.

Clearly, if $f : X \rightarrow Y$ is any function, then $f \circ id_X = f$ and $id_X \circ f = f$.

Definition 5.3 Let $f : X \rightarrow Y$ be a function. We say that f has an *inverse function* if there exists a function $g : Y \rightarrow X$ such that $f \circ g = id_Y$ and $g \circ f = id_X$.

Theorem 5.4 *If a function* $f : X \rightarrow Y$ *has an inverse, then this inverse is unique.*

Proof Suppose that g and h are both inverses for f. Then, we have $g \circ f = h \circ f = id_X$ and $f \circ g = f \circ h = id_Y$. Thus, we obtain

© The Author(s), under exclusive license to Springer Nature Singapore Pte Ltd. 2021
B. Davvaz, *A First Course in Group Theory*,
https://doi.org/10.1007/978-981-16-6365-9_5

$$h = h \circ id_Y = h \circ (f \circ g) = (h \circ f) \circ g = id_X \circ g = g.$$

This yields that the inverse of f is unique.   ■

The inverse of f is denoted by f^{-1}.

Theorem 5.5 *A function* $f : X \to Y$ *has an inverse if and only if* f *is a bijection.*

Proof Assume that f is a bijection. We define a function $g : Y \to X$ as follows:

$$g(y) = x \iff f(x) = y.$$

Since f is one to one, it follows that g is a function. Now, by the definition, $f \circ g = id_Y$ and $g \circ f = id_X$.

Conversely, suppose that f has an inverse f^{-1}. First, we show that f is one to one. If $f(x_1) = f(x_2)$, then

$$x_1 = id_X(x_1) = f^{-1}(f(x_1)) = f^{-1}(f(x_2)) = id_X(x_2) = x_2,$$

hence f is one to one. In order to show that f is onto, take any $y \in Y$. Then,

$$y = id_Y(y) = f \circ f^{-1}(y) = f(f^{-1}(y)) = f(x),$$

where $x = f^{-1}(y)$. So, f is onto.   ■

Theorem 5.6 *If* $f : X \to Y$ *and* $g : Y \to Z$ *are bijections, then so is the function* $g \circ f$.

Proof Suppose that f and g have inverse functions f^{-1} and g^{-1}, respectively. Then, we obtain $(g \circ f)(f^{-1} \circ g^{-1}) = id_Z$ and $(f^{-1} \circ g^{-1})(g \circ f) = id_X$. Hence, the inverse of $g \circ f$ is $f^{-1} \circ g^{-1}$. Consequently, by Theorem 5.5, $g \circ f$ is a bijection.   ■

Definition 5.7 Let X be a non-empty set. A bijective function from X to itself is called a *permutation* of X.

For an arbitrary non-empty set X we define S_X to be the set of all permutations.

Theorem 5.8 *The set* S_X *of all permutations of* X *is a group under composition of functions.*

Proof We check the group axioms for S_X. By Theorem 5.6, if $f, g \in S_X$, then $f \circ g \in S_X$. The associativity axioms holds by Theorem 5.1. The identity element is id_X. Finally, the definition of an inverse function shows that if f^{-1} is the inverse of f, then f is the inverse of f^{-1}. Consequently, f^{-1} is a bijection.   ■

Since the composition of functions is not commutative, it follows that S_X is not abelian, for $|X| \geq 3$.

Let us make a small example to understand better the connection between the intuition and the formal definition.

Example 5.9 Let $X = \{\bullet, \blacksquare, \triangle\}$. Then, the permutations that belong to S_X are:

$$id_X : \begin{cases} \bullet \to \bullet \\ \blacksquare \to \blacksquare \\ \triangle \to \triangle \end{cases} \qquad\qquad f_1 : \begin{cases} \bullet \to \blacksquare \\ \blacksquare \to \bullet \\ \triangle \to \triangle \end{cases}$$

$$f_2 : \begin{cases} \bullet \to \blacksquare \\ \blacksquare \to \triangle \\ \triangle \to \bullet \end{cases} \qquad\qquad f_3 : \begin{cases} \bullet \to \triangle \\ \blacksquare \to \blacksquare \\ \triangle \to \bullet \end{cases}$$

$$f_4 : \begin{cases} \bullet \to \bullet \\ \blacksquare \to \triangle \\ \triangle \to \blacksquare \end{cases} \qquad\qquad f_5 : \begin{cases} \bullet \to \triangle \\ \blacksquare \to \bullet \\ \triangle \to \blacksquare \end{cases}$$

Exercises

1. Define $f : \mathbb{N} \to \mathbb{N}$ by $f(1) = 2$, $f(2) = 3$, $f(3) = 4$, $f(4) = 7$, $f(7) = 1$, and $f(n) = n$ for any other $n \in \mathbb{N}$. Show that $f \circ f \circ f \circ f \circ f = id_{\mathbb{N}}$. What is f^{-1} in this case?

2. Let $f : \mathbb{Z} \to \mathbb{Z}$ be a function. For each of the following cases, find a left and a right inverses if exist.

   (a) $f(x) = \begin{cases} x & \text{if } x \text{ is even} \\ 2x + 1 & \text{if } x \text{ is odd} \end{cases}$

   (b) $f(x) = \begin{cases} x/3 & \text{if } x \equiv 0 \pmod{3} \\ x + 1 & \text{otherwise.} \end{cases}$

3. Let $f : \mathbb{Z} \to \mathbb{Z}$ be defined by $f(x) = ax + b$, where a and b are integers. Find the necessary and sufficient conditions on a and b such that $f \circ f = id_{\mathbb{Z}}$.

4. Let $f : X \to X$ be a function such that $f(f(x)) = x$, for all $x \in X$. Prove that f is a symmetric relation on X.

5. A function $f : X \to Y$ is said to be *left cancellable* if for any set Z and for any mappings g and h from Z to X such that $f \circ g = f \circ h$, then $g = h$. Prove that a function $f : X \to Y$ is left cancellable if and only if f is one to one.

6. A function $f : X \to Y$ is said to be *right cancellable* if for any set Z and for any mappings g and h from Y to Z such that $g \circ f = h \circ f$, then $g = h$. Prove that a function $f : X \to Y$ is right cancellable if and only if f is onto.

7. Given two sets X and Y we declare $X \prec Y$ (X is smaller than Y) if there is a mapping of Y onto X but no mapping of X onto Y. Prove that if $X \prec Y$ and $Y \prec Z$, then $X \prec Z$.

8. If X is a finite set and f is a one to one function of X, show that for some positive integer n,

$$\underbrace{f \circ f \circ \ldots \circ f}_{n \text{ times}} = id_X.$$

9. If X has m elements in Exercise 8, find a positive integer n (in terms of m) that works simultaneously for all one to one mappings of X into itself.
10. If $a \in X$ and $H = \{f \in S_X \mid f(a) = a\}$, show that H is a subgroup of S_X.
11. Let X be an infinite set and let H be the set of all permutations $f \in S_X$ such that $f(a) \neq a$ for at most a finite number of $a \in X$.

    (a) Prove that H is a subgroup of S_X;
    (b) Show that if $f \in S_X$, then $f^{-1}Hf = H$.

12. If X has three or more elements, show that we can find $f, g \in S_X$ such that $f \circ g \neq g \circ f$.
13. Observe that for any positive integer x, we have $x = 2^m(2n + 1)$, for some non-negative integers m and n. This means that we can define $f : \mathbb{N} \to (\mathbb{N} \cup \{0\}) \times (\mathbb{N} \cup \{0\})$ such that $f(x) = (m, n)$, as indicated above. Prove that f is one to one and onto.
14. **(Schröder–Bernstein Theorem).** Let X and Y be two sets such that

    (a) For a subset A of X, there is a one to one correspondence between A and Y;
    (b) For a subset B of Y, there is a one to one correspondence between B and X.

    Prove that there exists a one to one correspondence between A and Y.
15. Let G be a group and let a be a fixed element of G. Show that the map $f_a : G \to G$, given by $f_a(x) = ax$, for $x \in G$, is a permutation of the set G.

5.2 Symmetric Groups

In this section we briefly introduce some basic concepts and constructions that we will need later. Permutations are usually studied as combinatorial objects, we will observe that they have a natural group structure.

Definition 5.10 The group S_X is called the *symmetric group* or *permutation group* on the set X.

The group of permutations of the set $X = \{1, \ldots, n\}$ is denoted by S_n.

Theorem 5.11 *The order of S_n is equal to $n!$.*

Proof We count how many permutations of $\{1, 2, \ldots, n\}$ exist. We have to fill the boxes

with numbers $1, 2, \ldots, n$ with no repetitions. For box 1, we have n possible choices. When one number has been chosen, for box 2, we have $n - 1$ choices, and so on.

Consequently, we have

$$n(n-1)(n-2)\ldots 2\cdot 1 = n!$$

permutations and so the order of S_n is $|S_n| = n!$.   ∎

We can describe a permutation $\sigma \in S_n$ in several ways. A convenient notation for specifying a given permutation $\sigma \in S_n$ is

$$\begin{pmatrix} 1 & 2 & 3 & \ldots & n \\ a_1 & a_2 & a_3 & \ldots & a_n \end{pmatrix},$$

where a_k is the image of k under σ, for each $0 \le k \le n$. In this case, we write $k\sigma = a_k$. Accordingly, regarding this notation one must be absolutely sure as to what convention is being followed in writing the product of two permutations. If $\tau, \sigma \in S_n$, then we reiterate that $\sigma\tau$ will always mean: *first apply σ and then τ*.

Example 5.12 Let

$$\sigma = \begin{pmatrix} 1 & 2 & 3 & 4 & 5 \\ 2 & 3 & 1 & 5 & 4 \end{pmatrix} \text{ and } \tau = \begin{pmatrix} 1 & 2 & 3 & 4 & 5 \\ 1 & 3 & 5 & 4 & 2 \end{pmatrix}$$

be two permutations in S_5. Then, we have

$$\sigma\tau = \begin{pmatrix} 1 & 2 & 3 & 4 & 5 \\ 3 & 5 & 1 & 2 & 4 \end{pmatrix}, \quad \tau\sigma = \begin{pmatrix} 1 & 2 & 3 & 4 & 5 \\ 2 & 1 & 4 & 5 & 3 \end{pmatrix},$$

$$\sigma^{-1} = \begin{pmatrix} 1 & 2 & 3 & 4 & 5 \\ 3 & 1 & 2 & 5 & 4 \end{pmatrix}, \quad \tau^2 = \begin{pmatrix} 1 & 2 & 3 & 4 & 5 \\ 1 & 5 & 2 & 4 & 3 \end{pmatrix}.$$

There is another notation commonly used to specify permutations. It is called cycle notation.

Definition 5.13 Let $1 \le k \le n$ and let $a_1, a_2, \ldots, a_k$ be k disjoint integers between 1 and n. The *cycle* $(a_1\ a_2\ \ldots\ a_k)$ denotes the permutation of S_n that sends

$$a_1 \to a_2,$$
$$a_2 \to a_3,$$
$$\vdots$$
$$a_{k-1} \to a_k,$$
$$a_k \to a_1,$$

and leaves the remaining $n - k$ numbers fixed. We say that the *length of the cycle* $(a_1\ a_2\ \ldots\ a_k)$ is k.

It is clear that our choice of starting point for the cycle is not important. Thus, $(a_1\ a_2\ \ldots\ a_k) = (a_2\ \ldots\ a_k\ a_1)$. The inverse of a cycle is a cycle. More precisely,

Fig. 5.1 An illustration of cycle notation

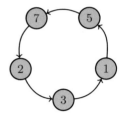

$$(a_1 \ a_2 \ \ldots \ a_k)^{-1} = (a_k \ a_{k-1} \ \ldots \ a_1).$$

Example 5.14 As an illustration of cycle notation, let us consider the permutation

$$\begin{pmatrix} 1 & 2 & 3 & 4 & 5 & 6 & 7 \\ 5 & 3 & 1 & 4 & 7 & 6 & 2 \end{pmatrix}.$$

This assignment of values could be presented schematically as in Fig. 5.1.

Example 5.15 The cycle (3 4 1 6) means the permutation where $3 \to 4$, $4 \to 1$, $1 \to 6$, $6 \to 3$, and all the other elements are fixed. So, $(3 \ 4 \ 1 \ 6) \in S_7$ corresponds to

$$(3 \ 4 \ 1 \ 6) = \begin{pmatrix} 1 & 2 & 3 & 4 & 5 & 6 & 7 \\ 6 & 2 & 4 & 1 & 5 & 3 & 7 \end{pmatrix}.$$

Example 5.16 Suppose that (1 3 4 2) and (2 5 3) are two cycles in S_5. Then

$$\begin{aligned}
(1 \ 3 \ 4 \ 2)(2 \ 5 \ 3) &= \begin{pmatrix} 1 & 2 & 3 & 4 & 5 \\ 3 & 1 & 4 & 2 & 5 \end{pmatrix} \begin{pmatrix} 1 & 2 & 3 & 4 & 5 \\ 1 & 5 & 2 & 4 & 3 \end{pmatrix} \\
&= \begin{pmatrix} 1 & 2 & 3 & 4 & 5 \\ 2 & 1 & 4 & 5 & 3 \end{pmatrix} \\
&= (1 \ 2)(3 \ 4 \ 5).
\end{aligned}$$

Example 5.17 We may write S_3, in Example 5.9, as

$$S_3 = \{id, \ (1 \ 2), \ (1 \ 3), \ (2 \ 3), \ (1 \ 2 \ 3), \ (1 \ 3 \ 2)\}.$$

The following is Cayley table for S_3.

·	id	$(1\,2)$	$(1\,3)$	$(2\,3)$	$(1\,2\,3)$	$(1\,3\,2)$
id	id	$(1\,2)$	$(1\,3)$	$(2\,3)$	$(1\,2\,3)$	$(1\,3\,2)$
$(1\,2)$	$(1\,2)$	id	$(1\,2\,3)$	$(1\,3\,2)$	$(1\,3)$	$(2\,3)$
$(1\,3)$	$(1\,3)$	$(1\,3\,2)$	id	$(1\,2\,3)$	$(2\,3)$	$(1\,2)$
$(2\,3)$	$(2\,3)$	$(1\,2\,3)$	$(1\,3\,2)$	id	$(1\,2)$	$(1\,3)$
$(1\,2\,3)$	$(1\,2\,3)$	$(2\,3)$	$(1\,2)$	$(1\,3)$	$(1\,3\,2)$	id
$(1\,3\,2)$	$(1\,3\,2)$	$(1\,3)$	$(2\,3)$	$(1\,2)$	id	$(1\,2\,3)$

Definition 5.18 Two cycles $(a_1\ a_2\ \ldots\ a_k)$ and $(b_1\ b_2\ \ldots\ b_l)$ are *distinct* if $\{a_1, a_2, \ldots, a_k\} \cap \{b_1, b_2, \ldots, b_l\} = \emptyset$.

Lemma 5.19 *If $\sigma = (a_1\ a_2\ \ldots\ a_k)$ and $\tau = (b_1\ b_2\ \ldots\ b_l)$ are distinct, then $\sigma\tau = \tau\sigma$.*

Proof Let $1 \le i \le k - 1$. Since $a_i \notin \{b_1, \ldots, b_l\}$, it follows that $a_i\tau = a_i$. Hence, we get

$$a_i(\tau\sigma) = a_i\sigma = a_{i+1}.$$

Also, since $a_{i+1} \notin \{b_1, \ldots, b_l\}$, it follows that

$$a_i(\sigma\tau) = a_{i+1}\tau = a_{i+1}.$$

Similar arguments for each $1 \le j \le l$ show that

$$b_j(\tau\sigma) = b_{j+1}\sigma = b_{j+1} = b_j\tau = b_j(\sigma\tau)$$

and

$$a_k(\sigma\tau) = a_1\tau = a_1 = a_k\sigma = a_k(\tau\sigma),$$
$$b_l(\sigma\tau) = b_l\tau = b_1 = b_1\sigma = b_l(\tau\sigma).$$

Finally, if $j \in \{a_1, \ldots, a_l, b_1, \ldots, b_l\}$, then

$$j(\tau\sigma) = (j\tau)\sigma = j\sigma = j = j\tau = (j\sigma)\tau = j(\sigma\tau).$$

Therefore, for each $j \in \{1, \ldots, n\}$ we have $j(\sigma\tau) = j(\tau\sigma)$. This yields that $\sigma\tau = \tau\sigma$. ∎

Let $\sigma \in S_n$. For each $x, y \in \{1, 2, \ldots, n\}$, we define the relation

$$x \equiv_\sigma y \;\Leftrightarrow\; x = x\sigma^k \text{ for some integer } k.$$

Lemma 5.20 *The relation $\equiv_\sigma$ is an equivalence relation.*

Proof Indeed, we have

(1) $x \equiv_\sigma x$ since $x = x\sigma^0$.

(2) If $x \equiv_\sigma y$, then $y = x\sigma^k$. Hence, $x = y\sigma^{-k}$. This implies that $y \equiv_\sigma x$.

(3) If $x \equiv_\sigma y$ and $y \equiv_\sigma z$, then $y = x\sigma^j$ and $z = y\sigma^k$ for some integers j, k. Hence, we obtain

$$z = y\sigma^k = x\sigma^j\sigma^k = x\sigma^{j+k}.$$

This implies that $x \equiv_\sigma z$.   ∎

This equivalence relation induces a decomposition of $\{1, 2, \ldots, n\}$ into disjoint subsets, namely the equivalence classes. Suppose that m_x is the smallest positive integer such that $x\sigma^{m_x} = x$. Then, the equivalence class of x under σ consists of the numbers $x, x\sigma, x\sigma^2, \ldots, x\sigma^{m_x-1}$.

Theorem 5.21 *Every permutation in S_n can be written as a cycle or as a product of disjoint cycles. Up to reordering the factors, this is unique.*

Proof Let σ be any permutation in S_n. Then, its cycles are of the form $(x \; x\sigma \; x\sigma^2 \; \ldots \; x\sigma^{m_x-1})$. Since the cycles of σ are disjoint, it follows that the image of $a \in \{1, 2, \ldots, n\}$ under σ is the same as the image of a under the product, δ, of all distinct cycles of σ. Consequently, σ and δ have the same effect on each element of $\{1, 2, \ldots, n\}$. Therefore, $\sigma = \delta$. In this way, by Lemma 5.19, we observe that every permutation can be uniquely expressed as a product of disjoint cycles.   ∎

This factorization is called the *cycle decomposition* of σ. The *cycle structure* of σ is the number of cycles of each length in the cycle decomposition of σ. For each $k = 1, \ldots, n$ assume that m_k denote the number of cycles of length k. Then, we say that σ has cycle structure

$$\underbrace{1, \ldots, 1}_{m_1}, \underbrace{2, \ldots, 2}_{m_2}, \ldots, \underbrace{n, \ldots, n}_{m_n}.$$

As notation for cycle type, we abbreviate this to $1^{m_1}, 2^{m_2}, \ldots, n^{m_n}$.

Example 5.22 The permutation

$$\sigma = (1 \; 2)(3 \; 5 \; 6)(4 \; 8)$$

in S_8 has cycle structure consisting of one cycle of length 1, two cycles of length 2, and one cycle of length 3.

Theorem 5.23 *The number of permutations in S_n of cycle structure of the form $1^{m_1}, 2^{m_2}, \ldots, n^{m_n}$ is equal to*

$$\frac{n!}{m_1! \ldots m_n! 1^{m_1} 2^{m_2} \ldots n^{m_n}}.$$

Proof A permutation of the given cycle structure is produced by filling the integers $1, 2, \ldots, n$ into the following boxes:

There exist $n!$ ways of doing this. But some of these ways give the same permutation of S_n. We try to count them.

(1) There exist $m_1!$ permutations of cycles of length 1, $m_2!$ permutations of cycles of length 2, $m_3!$ permutations of cycles of length 3, and so on. So, we must divide by $m_1! \ldots m_n!$.
(2) Each cycle of length 2 can be written in two ways, i.e., $(a\ b) = (b\ a)$. Similarly, each cycle of length 3 can be written in three ways, i.e., $(a\ b\ c) = (b\ c\ a) = (c\ a\ b)$, and so on. So we must divide by $1^{m_1} 2^{m_2} \ldots n^{m_n}$.

This completes the proof. ∎

Definition 5.24 A cycle of length 2 is called a *transposition*.

Corollary 5.25 *Any cycle in S_n is a product of transpositions.*

Proof If $(a_1\ a_2\ \ldots\ a_k)$ is an arbitrary cycle in S_n, then

$$(a_1\ a_2\ \ldots\ a_k) = (a_1\ a_2)(a_1\ a_3) \ldots (a_1\ a_n),$$

as desired. ∎

Theorem 5.26 *Every permutation in S_n ($n \geq 2$) is either a transposition or a product of transpositions. In other words, S_n is generated by transpositions.*

Proof First, note that the identity permutation can be expressed as $(1\ 2)(1\ 2)$, so it is product of transpositions. By Theorem 5.21, we know that every permutation is a cycle or a product of cycles. By Corollary 5.25, since each cycle is a product of transpositions, it follows that S_n is generated by transpositions. ∎

Theorem 5.27 *The symmetric group S_n is generated by $n - 1$ transpositions $(1\ 2)$, $(1\ 3)$, $\ldots$, $(1\ n)$.*

Proof By Theorem 5.26, we know that S_n is generated by transpositions. Now, if $(a\ b)$ be an arbitrary transposition, then

$$(a\ b) = (1\ a)(1\ b)(1\ a).$$

This yields the desired result. ∎

Theorem 5.28 *The symmetric group S_n is generated by $n - 1$ transpositions $(1\ 2)$, $(2\ 3)$, $\ldots$, $(n - 1\ n)$.*

Proof By Theorem 5.26, it suffices to show that each transposition $(a\ b)$ in S_n is a product of transpositions of the form $(i\ i+1)$, where $i < n$. Suppose that $a < b$. For the proof we use mathematical induction on $b - a$ that $(a\ b)$ is a product of transpositions $(i\ i+1)$. This is obvious when $b - a = 1$, because $(a\ b) = (a\ a+1)$ is one of the transpositions we want in the desired generating set. Now, suppose that $b - a = k > 1$ and the theorem is true for all transpositions moving a pair of integers whose difference is less than k. We have

$$(a\ b) = (a\ a+1)(a+1\ b)(a\ a+1).$$

The transpositions $(a\ a+1)$ and $(a\ a+1)$ lie in our desired generating set. For the transposition $(a+1\ b)$ we have $b - (a+1) = k-1 < k$. So, by assumption, $(a+1\ b)$ is a product of transpositions of the form $(i\ i+1)$, so $(a\ b)$ is as well. ∎

Lemma 5.29 *A cycle of length m has order m.*

Proof It is straightforward. ∎

Theorem 5.30 *The order of a permutation written in disjoint cycle form is the least common multiple of the lengths of the cycles.*

Proof Suppose that $\sigma \in S_n$ and $\sigma = \sigma_1 \sigma_2 \ldots \sigma_k$, where the σ_i $(i = 1, \ldots, k)$ are disjoint cycles of length m_i. Let m be the least common multiple of $m_1, m_2, \ldots, m_k$. Since $m_i | m$ for each $1 \leq i \leq k$, it follows that

$$\sigma^m = (\sigma_1 \sigma_2 \ldots \sigma_k)^m = \sigma_1^m \sigma_2^m \ldots \sigma_k^m = id,$$

where id is the identity permutation in S_n. Consequently, the order of σ is at most m.

Now, suppose that $\sigma^r = id$. This implies that $\sigma_1^r \sigma_2^r \ldots \sigma_k^r = id$. Since σ_i $(i = 1, \ldots, k)$ are disjoint, it follows that $\sigma_i^r = id$. Since σ_i is of order m_i, it follows that $m_i | r$. This yields that $m | r$. Therefore, we conclude that σ is of order m. ∎

Example 5.31 We want to determine the number of permutations in S_7 of order 3. By Theorem 5.30, it is enough to count the number of permutations of the form

(1) $(a\ b\ c)$,
(2) $(a\ b\ c)(x\ y\ z)$.

For the first case, there exist $7 \cdot 6 \cdot 5$ such triples. But this product counts the permutation $(a\ b\ c)$ three times. Thus, the number of permutations of the form (1) is equal to 70.

For the second case, there exist 70 ways to create the first cycle and $\frac{4 \cdot 3 \cdot 2}{3} = 8$ to create the second cycle. So, we have $70 \times 8 = 560$ ways. But this product counts $(a\ b\ c)(x\ y\ z)$ and $(x\ y\ z)(a\ b\ c)$ as distinct while they are equal permutations. Consequently, the number of permutations in S_7 of the form (2) is 280.

Therefore, we have $70 + 280 = 350$ permutations of order 3 in S_7.

Lemma 5.32 *Let σ be any permutation in S_n and let*

$$\sigma = (a_1 \ \ldots \ a_i)(b_1 \ \ldots \ b_j) \ldots (c_1 \ \ldots \ c_k)$$

be the cycle decomposition of σ. Then, for each $\tau \in S_n$, we have

$$\tau^{-1}\sigma\tau = (a_1\tau \ \ldots \ a_i\tau)(b_1\tau \ \ldots \ b_j\tau) \ldots (c_1\tau \ \ldots \ c_k\tau) \qquad (5.1)$$

which is a product of disjoint cycles.

Proof We have

$$\begin{cases} (a_1\tau)\tau^{-1}\sigma\tau = a_1\sigma\tau = a_2\tau, \\ \quad\vdots \\ (a_{i-1}\tau)\tau^{-1}\sigma\tau = a_{i-1}\sigma\tau = a_i\tau, \\ (a_i\tau)\tau^{-1}\sigma\tau = a_i\sigma\tau = a_1\tau, \\ (b_1\tau)\tau^{-1}\sigma\tau = b_1\sigma\tau = b_2\tau, \\ \quad\vdots \\ (b_{j-1}\tau)\tau^{-1}\sigma\tau = b_{j-1}\sigma\tau = b_j\tau, \\ (b_j\tau)\tau^{-1}\sigma\tau = b_j\sigma\tau = b_1\tau, \\ \quad\vdots \\ (c_1\tau)\tau^{-1}\sigma\tau = c_1\sigma\tau = c_2\tau, \\ \quad\vdots \\ (c_{k-1}\tau)\tau^{-1}\sigma\tau = c_{k-1}\sigma\tau = c_k\tau, \\ (c_k\tau)\tau^{-1}\sigma\tau = c_k\sigma\tau = c_1\tau. \end{cases}$$

Moreover, if σ do not move integer d, then $\tau^{-1}\sigma\tau$ do not move $d\tau$. So, we see that the right side of (5.1) acts on every integer of $\{1, \ldots, n\}$ in the same way as $\tau^{-1}\sigma\tau$. This completes the proof. ∎

Now we are ready to cut down the size of a generating set for S_n to two.

Theorem 5.33 *The symmetric group S_n is generated by the transposition $(1\ 2)$ and the cycle $(1\ 2 \ \ldots \ n)$.*

Proof By Theorem 5.28, it is enough to show that the products of $(1\ 2)$ and $(1\ 2 \ \ldots \ n)$ give all transpositions of the form $(i\ i+1)$. We may take $n \geq 3$. Suppose that $\tau = (1\ 2 \ \ldots \ n)$. Then, by Lemma 5.32, we have

$$\tau^{-1}(1\ 2)\tau = (1\tau\ 2\tau) = (2\ 3),$$

and more generally for $i = 2, \ldots, n-1$, we obtain

$$\tau^{-(i-1)}(1\ 2)\tau^{i-1} = (1\tau^{i-1}\ 2\tau^{i-1}) = (i\ i+1).$$

This completes the proof. ∎

Theorem 5.34 *Two permutations in S_n are conjugate if and only if they have the same cycle structure up to ordering.*

Proof Suppose that σ and δ are conjugate. Then, there exists $\tau \in S_n$ such that $\tau^{-1}\sigma\tau = \delta$. Now, by Lemma 5.32, we conclude that σ and δ have the same cycle structure.

Conversely, suppose that

$$\sigma = (a_1 \ \ldots \ a_i)(b_1 \ \ldots \ b_j) \ldots (c_1 \ \ldots \ c_k),$$
$$\delta = (a'_1 \ \ldots \ a'_i)(b'_1 \ \ldots \ b'_j) \ldots (c'_1 \ \ldots \ c'_k),$$

be two permutations of S_n with the same cycle structure. Now, we define τ to be the permutation of S_n which sends

$$a_1 \to a'_1, \ \ldots \ a_i \to a'_i,$$
$$b_1 \to b'_1, \ \ldots \ b_j \to b'_j,$$
$$\vdots$$
$$c_1 \to c'_1, \ \ldots \ c_k \to c'_k.$$

By Lemma 5.32, $\tau^{-1}\sigma\tau$ and δ are the same permutation. ∎

Definition 5.35 Two permutations σ and τ in S_n are said to be *similar* if there exists a one to one correspondence between the cycles of σ and τ such that the corresponding cycles have same length.

Corollary 5.36 *Two permutations in S_n are similar if and only if they are conjugate.*

Definition 5.37 A group G is *centerless* if $Z(G) = \{e\}$.

Theorem 5.38 S_n *is centerless if* $n \geq 3$.

$$Z(S_n) = \{id\}.$$

This means that the center of symmetric group is the subgroup comprising only the identity permutation.

Proof Suppose that σ is a non-identity permutation in $Z(S_n)$ and let $\sigma = \sigma_1\sigma_2\ldots\sigma_k$, where σ_is are distinct cycles with lengths l_is such that $l_k \leq \cdots \leq l_2 \leq l_1$. We consider the following two cases:

Case 1: Let $\sigma_1 = (a_1 \ a_2 \ \ldots \ a_m)$ with $m \geq 3$. Since $\sigma \in Z(S_n)$, it follows that $\sigma(a_1 \ a_2) = (a_1 \ a_2)\sigma$. Since σ_is are distinct, it follows that $\sigma_1(a_1 \ a_2) = (a_1 \ a_2)\sigma_1$, or equivalently

$$(a_1 \ a_2 \ \ldots \ a_m)(a_1 \ a_2) = (a_1 \ a_2 \ \ldots \ a_m)(a_1 \ a_2).$$

This is a contradiction, because in the left side of the equality we have $a_m \to a_1$ while in the right side we have $a_m \to a_2$.

Case 2: Let $\sigma_1 = (a_1\ a_2)$. Since $n \geq 3$, it follows that there exists a_3 such that $a_3 \neq a_1$ and $a_3 \neq a_2$. Since $\sigma \in S_n$, it follows that $\sigma(a_1\ a_2\ a_3) = (a_1\ a_2\ a_3)\sigma$, or equivalently

$$(a_1\ a_2)\sigma_2 \ldots \sigma_k(a_1\ a_2\ a_3) = (a_1\ a_2\ a_3)(a_1\ a_2)\sigma_2 \ldots \sigma_k.$$

Since a_1 and a_2 do not appear in cycles $\sigma_2, \ldots, \sigma_k$, it follows that in the left side of the last equality $a_1 \to a_3$ while in the right side $a_1 \to a_1$. This is again a contradiction. Therefore, we conclude that $Z(S_n) = \{id\}$.   ∎

Exercises

1. Let

$$\alpha = \begin{pmatrix} 1 & 2 & 3 & 4 & 5 & 6 & 7 & 8 \\ 8 & 6 & 4 & 1 & 5 & 7 & 2 & 3 \end{pmatrix} \text{ and } \beta = \begin{pmatrix} 1 & 2 & 3 & 4 & 5 & 6 & 7 & 8 \\ 2 & 1 & 5 & 8 & 6 & 3 & 7 & 4 \end{pmatrix}.$$

   Compute each of the following:

   (a) α^{-1};
   (b) $\alpha\beta\alpha^{-1}$;
   (c) $\alpha^3\beta$;
   (d) $\alpha\beta^{-2}$.

2. Write each of the following permutations as a product of distinct cycles:

   (a) $(3\ 4\ 5\ 6)(4\ 3)(1\ 2\ 3)$;
   (b) $(1\ 2)(2\ 3)(2\ 4)(1\ 3\ 5)$.

3. Give the Cayley table for the cyclic subgroup of S_5 generated by

$$\sigma = \begin{pmatrix} 1 & 2 & 3 & 4 & 5 \\ 2 & 4 & 5 & 1 & 3 \end{pmatrix}.$$

4. Determine eight elements in S_6 that commute with $(1\ 2)(5\ 6)(3\ 4)$. Do they form a subgroup of S_6?
5. Find five subgroups of S_5 of order 24.
6. Find the number of permutations in the set $\{\sigma \in S_5 \mid 2\sigma = 5\}$.
7. How many elements of order 6 are there in the symmetric group S_{11}?
8. Find all powers of the cycle $\sigma = (x_1\ x_2\ \ldots\ x_n)$.
9. Find all permutations in the symmetric group S_n which commute with the cycle $(x_1\ x_2\ \ldots\ x_n)$, where $x_1\ x_2\ \ldots\ x_n$ is a permutation of the numbers $1, 2, \ldots, n$.
10. Count the number of elements of S_n having at least one fixed point.

11. Given the permutations $\alpha = (1\ 2)(3\ 4)$ and $\beta = (1\ 3)(5\ 6)$. Find a permutation γ such that $\gamma^{-1}\alpha\gamma = \beta$. Is γ unique?

12. Prove that the permutations

$$\alpha = \begin{pmatrix} 1 & 2 & 3 & 4 & 5 & 6 \\ 2 & 5 & 3 & 6 & 1 & 4 \end{pmatrix} \quad \text{and} \quad \beta = \begin{pmatrix} 1 & 2 & 3 & 4 & 5 & 6 \\ 5 & 3 & 4 & 2 & 1 & 6 \end{pmatrix}$$

are conjugate in the symmetric group S_6, and find the number $\gamma \in S_6$ such that $\gamma^{-1}\alpha\gamma = \beta$.

13. Show that the permutations in S_9 which send the numbers 2, 5, 7 among themselves form a subgroup of S_9. What is the order of this subgroup?

14. If n is at least 3, show that for some $f \in S_n$, f cannot be expressed in the form $f = g^3$, for any $g \in S_n$.

15. What is the smallest positive integer n such that S_n has an element of order greater than $2n$?

16. Show that in S_7, the equation $x^2 = (1\ 2\ 3\ 4)$ has no solutions but the equation $x^3 = (1\ 2\ 3\ 4)$ has at least two.

17. Let X be the set $\mathbb{Z}_{31}$, and let $f : X \to X$ be the permutation $f(x) = 2x$. Decompose this permutation into disjoint cycles.

18. Let X be the set $\mathbb{Z}_{29}$, and let $f : X \to X$ be the permutation $f(x) = x^3$. Decompose this permutation into disjoint cycles.

19. Find the cycle decomposition of the permutation induced by the action of complex conjugation on the set of roots of $x^5 - x + 1$.

20. In S_4, find the subgroup generated by $(1\ 2\ 3)$ and $(1\ 2)$. Also, for this subgroup, find the corresponding subgroup $\sigma^{-1}H\sigma$, for $\sigma = (1\ 4)$.

21. Find necessary and sufficient conditions on the pair i and j in order that $\langle (1\ 2\ \ldots\ n), (i\ j) \rangle = S_n$.

22. Show that for all $1 < i \leq n$, we have $\langle (2\ 3\ \ldots\ n), (1\ i) \rangle = S_n$.

23. Determine a permutation $\sigma \in S_n$ such that for every $1 \leq i, j \leq n$,

$$i \leq j \Rightarrow i\sigma \leq j\sigma.$$

24. Find the maximum possible order for a permutation in S_n for $n = 5$, $n = 6$, $n = 7$, $n = 10$, and $n = 15$.

25. A permutation is called *regular* if it can be decomposed into disjoint cycles of the same length. Prove that every power of a cycle of length n in S_n is a regular permutation. Prove that the length of each of the disjoint cycles in this decomposition divides n.

26. Prove that every regular permutation is a power of some cycle.

5.3 Alternating Groups

The alternating groups are among the most important examples of groups. We study some of their properties in this section.

Theorem 5.39 *If a permutation σ can be expressed as a product of even number of transpositions, then every decomposition of σ into a product of transpositions must have an even number of transpositions. In symbols, if*

$$\sigma = \tau_1 \tau_2 \ldots \tau_k \text{ and } \sigma = \delta_1 \delta_2 \ldots \delta_m$$

where the τ's and the δ's are transpositions, then k and m are both even or both odd.

Proof We consider a polynomial p of n variable

$$
\begin{aligned}
p(a_1, \ldots, a_n) &= (a_1 - a_2)(a_1 - a_3) \ldots (a_1 - a_n) \\
&\quad (a_2 - a_3)(a_2 - a_4) \ldots (a_2 - a_n) \ldots (a_{n-1} - a_n) \\
&= \prod_{i<j} (a_i - a_j).
\end{aligned}
$$

If $\sigma \in S_n$, we define

$$\sigma^*\big(p(a_1, \ldots, a_n)\big) = \prod_{i<j}(a_{i\sigma} - a_{j\sigma}).$$

Suppose that $\tau = (r \ s)$ is a transposition with $r < s$. Then, we have

$$\tau^*\big(p(a_1, \ldots, a_n)\big) = \prod_{i<j}(a_{i\tau} - a_{j\tau}).$$

Note that $a_{r\tau} - a_{s\tau} = a_s - a_r = -(a_r - a_s)$, and if a_r and a_s do not exist in a factor, then this factor is fixed under τ. The other factors can be expressed in one of the following forms:

(1) $(a_s - a_k)(a_r - a_k)$, if $s < k$,
(2) $(a_k - a_s)(a_r - a_k)$, if $r < k < s$,
(3) $(a_k - a_s)(a_k - a_r)$, if $k < r$.

Therefore, we conclude that $\tau^*\big(p(a_1, \ldots, a_n)\big) = -f(a_1, \ldots, a_n)$.
   If $\sigma = \tau_1 \tau_2 \ldots \tau_k$, where $\tau_1, \tau_2, \ldots, \tau_k$ are transpositions, then

$$
\begin{aligned}
\sigma^*\big(p(a_1, \ldots, a_n)\big) &= \big(\tau_1 \tau_2 \ldots \tau_k\big)^*\big(p(a_1, \ldots, a_n)\big) \\
&= (-1)^k p(a_1, \ldots, a_n).
\end{aligned}
\tag{5.2}
$$

Similarly, if $\sigma = \delta_1 \delta_2 \ldots \delta_m$, where $\delta_1, \delta_2, \ldots, \delta_m$ are transpositions, then

$$\sigma^*\big(p(a_1, \ldots, a_n)\big) = \big(\delta_1\delta_2 \ldots \delta_m\big)^*\big(p(a_1, \ldots, a_n)\big) \tag{5.3}$$
$$= (-1)^m p(a_1, \ldots, a_n).$$

Comparing (5.2) and (5.3), we conclude that

$$(-1)^k = (-1)^m.$$

This implies that these two decompositions of σ as the product of transpositions are of the same parity. ∎

Therefore, any permutation is either the product of an odd number of transpositions or the product of an even number of transpositions, and no product of an even number of transpositions can be equal to a product of an odd number of transpositions.

Definition 5.40 A permutation which can be expressed as a product of an even number of transpositions is called an *even permutation*. A permutation which can be expressed as a product of an odd number of transpositions is called an *odd transposition*.

If we define the sign of a permutation σ as

$$\mathrm{sgn}(\sigma) = \frac{p(a_{1\sigma}, \ldots, a_{n\sigma})}{p(a_1, \ldots, a_n)},$$

then

$$\mathrm{sgn}(\sigma) = \begin{cases} 1 & \text{if } \sigma \text{ is even} \\ -1 & \text{if } \sigma \text{ is odd.} \end{cases}$$

If $\sigma, \delta \in S_n$, then

$$\mathrm{sgn}(\sigma\delta) = \frac{p(a_{1\sigma\delta}, \ldots, a_{n\sigma\delta})}{p(a_1, \ldots, a_n)}$$
$$= \frac{p(a_{1\sigma\delta}, \ldots, a_{n\sigma\delta})}{p(a_{1\sigma}, \ldots, a_{n\sigma})} \cdot \frac{p(a_{1\sigma}, \ldots, a_{n\sigma})}{p(a_1, \ldots, a_n)}$$
$$= \mathrm{sgn}(\sigma) \cdot \mathrm{sgn}(\delta).$$

To summarize in words, the sign of the product is the product of sign,

$$\text{even} \times \text{even} = \text{odd} \times \text{odd} = \text{even},$$
$$\text{even} \times \text{odd} = \text{odd} \times \text{even} = \text{odd}.$$

Theorem 5.41 *If A_n is the set of all even permutations, then A_n is a subgroup of S_n.*

Proof If $\sigma, \delta \in A_n$, then we have $\sigma\delta \in A_n$. Since A_n is a finite closed subset of the finite group S_n, it follows that A_n is a subgroup of S_n. ∎

A_n is called the *alternating group* of degree n.

Theorem 5.42 *If $n > 1$, the order of A_n is equal to $n!/2$.*

Proof For each even permutation σ, the permutation $(1\ 2)\sigma$ is odd, and if $\sigma \neq \delta$, then $(1\ 2)\sigma \neq (1\ 2)\delta$. Hence, there are at least as many odd permutations as there are even ones. On the other hand, for each odd permutation σ, the permutation $(1\ 2)\sigma$ is even, and if $\sigma \neq \delta$, then $(1\ 2)\sigma \neq (1\ 2)\delta$. Hence, there are at least as many even permutations as there are odd ones. This yields that there exist equal numbers of even and odd permutations. Since $|S_n| = n!$, it follows that $|A_n| = n!/2$.   ∎

Theorem 5.43 *For each $n \geq 3$, A_n is generated by cycles of length* 3.

Proof Suppose that $\sigma \in A_n$. Then, σ is a product of even number of transpositions. Let a, b, c, and d are four different numbers between 1 and n. Then, we have

$$(a\ b)(a\ c) = (a\ b\ c),$$
$$(a\ b)(c\ d) = (a\ c\ b)(c\ b\ d).$$

This completes the proof.   ∎

Theorem 5.44 *For each $n \geq 3$, A_n is generated by cycles of the form $(1\ a\ b)$, where $2 \leq a, b \leq n$, and $a \neq b$.*

Proof If $\sigma \in A_n$, then σ is a product of transpositions. Since $(a\ b) = (1\ a)(1\ b)(1\ a)$ for each $2 \leq a, b \leq n$, and $a \neq b$, it follows that σ is a product of transpositions of the form $(1\ a)$. Since σ is even, the number of transpositions of the form $(1\ a)$ in σ is even. But for each $a \neq b$, we have $(1\ a)(1\ b) = (1\ a\ b)$. This completes the proof.   ∎

Theorem 5.45 *For each $n \geq 3$, A_n is generated by cycles of the form $(1\ 2\ a)$, where $2 \leq a \leq n$.*

Proof If $n = 3$, then $A_3 = \{id, (1\ 2\ 3), (1\ 3\ 2)\}$ is generated by $(1\ 2\ 3)$. So, we assume that $n \geq 4$.

Each cycle of length 3 in A_n containing 1 and 2 is generated by the cycle of the form $(1\ 2\ a)$, because $(1\ a\ 2) = (1\ 2\ a)^{-1}$.

For each cycle of length 3 in A_n containing 1 but not 2, we have

$$(1\ a\ b) = (1\ 2\ b)(1\ 2\ a)(1\ 2\ b)(1\ 2\ b).$$

Now, by Theorem 5.44, the proof completes.   ∎

Theorem 5.46 *For each $n \geq 3$, A_n is generated by consecutive cycles of the form $(a\ a+1\ a+2)$, where $1 \leq a \leq n-2$.*

Proof If $n = 3$, then $A_3 = \langle (1\ 2\ 3) \rangle$. If $n = 4$, then by Theorem 5.45, $A_4 = \langle (1\ 2\ 3), (1\ 2\ 4) \rangle$. Since

$$(1\ 2\ 4) = (1\ 2\ 3)(2\ 3\ 4)(1\ 2\ 3)(1\ 2\ 3),$$

it follows that $A_4 = \langle (1\ 2\ 3), (2\ 3\ 4) \rangle$. Now, assume that $n \geq 5$. By Theorem 5.45, it suffices to show that $(1\ 2\ a)$ can be obtained from a product of consecutive cycles of length 3. We apply mathematical induction on a. Let $a \geq 5$ and $(1\ 2\ b)$ be a product of consecutive cycles of length 3, for $3 \leq b < a$. We have

$$(1\ 2\ a) = (1\ 2\ a - 1)(1\ 2\ a - 2)(a - 2\ a - 1\ a)(1\ 2\ a - 1)(1\ 2\ a - 2).$$

Now, the inductive assumption show that $(1\ 2\ a)$ is a product of consecutive cycles of length 3. ∎

Theorem 5.47 *For each $n \geq 3$, A_n is generated by*

(1) $(1\ 2\ 3)$ and $(1\ 2\ \ldots\ n)$ if n is odd;
(2) $(1\ 2\ 3)$ and $(2\ 3\ \ldots\ n)$ if n is even.

Proof Note that if $n = 3$, then we are done. So, we suppose that $n \geq 4$.
(1) Let n be odd and $\tau = (1\ 2\ \ldots\ n)$. Then, we conclude that $\tau \in A_n$. Moreover, for each $1 \leq a \leq n - 3$, by Lemma 5.32, we get

$$\tau^{-a}(1\ 2\ 3)\tau^a = (1\tau^a\ 2\tau^a\ 3\tau^a) = (a + 1\ a + 2\ a + 3) \in A_n.$$

Now, by Theorem 5.46, we are done.
(2) Let n is even and $\tau = (2\ 3\ \ldots\ n)$. Then, we have $\tau \in A_n$. Also, for each $1 \leq a \leq n - 3$, by Lemma 5.32, we obtain

$$\tau^{-a}(1\ 2\ 3)\tau^a = (1\tau^a\ 2\tau^a\ 3\tau^a) = (1\ a + 2\ a + 3) \in A_n.$$

Finally, since $(1\ a + 1\ a + 2)$ and $(1\ a\ a + 1)$ are in A_n, we can write

$$(1\ a + 1\ a + 2)(1\ a\ a + 1) = (a\ a + 1\ a + 2) \in A_n.$$

Now, by Theorem 5.46, the proof completes. ∎

Theorem 5.48 *If $n \geq 5$, then all cycles of length 3 are conjugate in A_n.*

Proof Suppose that σ and δ are two cycles of length 3 in A_n. By Theorem 5.34, there exists a permutation $\tau \in S_n$ such that $\tau^{-1}\sigma\tau = \delta$. If $\tau \in A_n$, then we are done. So, suppose that $\tau \notin A_n$. Let $\sigma = (a\ b\ c)$. Since $n \geq 5$, it follows that there exist x and y not in $\{a, b, c\}$. We set $\theta = (x\ y)$. Since $\theta^{-1}\sigma\theta = \sigma$, it follows that $(\theta\tau)^{-1}\sigma(\theta\tau) = \delta$, where $\theta\tau \in A_n$. ∎

Theorem 5.49 *For each $n \geq 4$, the center of A_n is*

$$Z(A_n) = \{id\}.$$

This means that A_n is centerless, for $n \geq 4$.

Proof We show that, for every non-identity permutation σ, there is a permutation in A_n that does not commute with σ.

Since σ is not the identity, it follows that σ maps an element a into b with $a \neq b$. Since $n \geq 4$, we can choose distinct c and d not equal to a and b. Now, we claim that the cycle $(b\ c\ d)$ does not commute with σ. Indeed, $\sigma(b\ c\ d)$ maps a into c, but $(b\ c\ d)\sigma$ maps a into b. Therefore, no σ other than the identity commutes with every element of A_n. In other words, no σ other than the identity is in the center of A_n. Thus, the only element of the center of A_n is the identity.   ∎

Exercises

1. Let σ and τ belong to S_n. Prove that $\sigma^{-1}\tau^{-1}\sigma\tau$ is an even permutation.
2. Prove that there is no permutation σ such that $\sigma^{-1}(1\ 3\ 4)\sigma = (1\ 2)(4\ 6\ 7)$.
3. Compute the order of each member of A_4. What arithmetic relationship do these orders have with the order of A_4?
4. Prove that A_5 has a subgroup of order 12.
5. Show that A_4 has no subgroup of order 6.
6. Show that the group A_5 contains no elements of order 4, and precisely 15 elements of order 2. How many elements of are there of orders 3, 6, respectively?
7. Show that A_8 contains an element of order 15.
8. Find a cyclic subgroup of A_8 that has order 4.
9. Find a non-cyclic subgroup of A_8 that has order 4.
10. Suppose that H is a subgroup of S_n of odd order. Prove that H is a subgroup of A_n.
11. Let n be an even positive integer. Prove that A_n has an element of order greater than n if and only if $n \geq 8$.
12. Let n be an odd positive integer. Prove that A_n has an element of order greater than $2n$ if and only if $n \geq 13$.
13. Let n be an even positive integer. Prove that A_n has an element of order greater than $2n$ if and only if $n \geq 14$.
14. Let $H = \{\sigma^2 \mid \sigma \in S_4\}$ and $K = \{\sigma^2 \mid \sigma \in S_5\}$. Prove that $H = A_4$ and $K = A_5$.
15. Let $H = \{\sigma^2 \mid \sigma \in S_6\}$. Prove that $H \neq A_6$.
16. Why does the fact that the orders of the elements of A_4 are 1, 2, and 3 imply that $|Z(A_4)| = 1$?
17. For $n > 1$, let H be the set of all permutations in S_n that can be expressed as a product of a multiple of four transpositions. Show that $H = A_5$.
18. Consider S_n for a fixed $n \geq 2$ and let σ be a fixed odd permutation. Show that every odd permutation in S_n is a product of σ and some permutation in A_n.
19. Show that if σ is a cycle of odd length, then σ^2 is a cycle.
20. Show that every permutation in A_n is a product of cycles of length n.

5.4   Worked-Out Problems

Problem 5.50 Show that if $n \geq m$, then the number of cycles of length m in S_n is given by

$$\frac{n(n-1)(n-2)\dots(n-m+1)}{m}. \tag{5.4}$$

Solution We count how many cycles of length m in S_n exist. We have to fill the boxes:

with the numbers $1, 2, \dots, n$ with no repetitions. We have n choice for the first box. Then, $n-1$ choice for the second box, $n-2$ choice for the third box, and so on. Finally, we have $n - m + 1$ choice for the last box. So, there are $n(n-1)(n-2)\dots(n-m+1)$ choices for a cycle of length m, but we emphasis that some them are the same. For example, the following cycles are the same:

- If $m = 2$, then $(a\ b) = (b\ a)$ (2 equivalent notations);
- If $m = 3$, then $(a\ b\ c) = (b\ c\ a) = (c\ a\ b)$ (3 equivalent notations);
- If $m = 4$, then $(a\ b\ c\ d) = (b\ c\ d\ a) = (c\ d\ a\ b) = (d\ a\ b\ c)$ (4 equivalent notations).

In general, by induction we deduce that for cycles of length m, there are m equivalent notations. Since we have $n(n-1)(n-2)\dots(n-m+1)$ choices to form a cycle of length m in which there are m equivalent notations, it follows that the number of cycles of length m in S_n is equal to (5.4). ∎

Problem 5.51 If n is at least 4, show that every element of S_n can be written as a product of two permutations, each of which has order 2. (Experiment first with cycles.)

Solution First, we begin with an example. Let $(a_1\ a_2\ \dots\ a_7)$ be a cycle. We consider

$$\alpha = (a_1\ a_7)(a_2\ a_6)(a_3\ a_5)$$
$$\beta = (a_2\ a_7)(a_3\ a_6)(a_4\ a_5).$$

Since α and β are products of disjoint transpositions, it follows that $o(\alpha) = o(\beta) = 2$. Moreover, it is easy to see that $\alpha\beta = (a_1\ a_2\ \dots\ a_7)$. Next, we generalize the above example to an arbitrary cycle $\sigma = (a_1\ a_2\ \dots\ a_n)$. We take

$$\alpha = (a_1\ a_n)(a_2\ a_{n-1})\dots(a_i\ a_{n-i+1})\dots(a_m\ a_{n-m+1}),$$
$$\beta = (a_2\ a_n)(a_3\ a_{n-1})\dots(a_{i+1}\ a_{n-i+1})\dots(a_{m+1}\ a_{n-m+1}),$$

where $m = [n/2]$. Again, since α and β are products of disjoint transpositions, it follows that $o(\alpha) = o(\beta) = 2$. Now, we claim that $\sigma = \alpha\beta$. Since α and β are products

Table 5.1 A short table of values $p(n)$

n	1	2	3	4	5	6	7	8	9	10
$p(n)$	1	2	3	5	7	11	15	22	30	42

of disjoint transpositions, what they do to any one a_i is determined just by the transposition containing that a_i. Hence, for $i \leq m$, the transposition $(a_i \, a_{n-i+1})$ in α sends a_i to a_{n-i+1} and then the transposition $(a_{i+1} \, a_{n-i+1})$ in β sends a_{n-i+1} to a_{i+1}. In view of this, if $i \leq m$, then $\alpha\beta$ sends a_i to a_{i+1}. Now, if $i > m$, then we take $j = n - i + 1$. We have $j \leq m$ and $i = n - j + 1$. So, the transposition $(a_j \, a_{n-j+1})$ is in α and it sends $a_i = a_{n-j+1}$ to a_j and the transposition $(a_{j-1+1} \, a_{n-(j-1)+1}) = (a_j \, a_{n-j+2})$ in β sends a_j to a_{n-j+2}. But since $j = n - i + 1$, it follows that $n - j + 2 = i + 1$. Therefore, we observe that $\alpha\beta$ sends a_i to a_{i+1} when $i > m$. Consequently, $\sigma = \alpha\beta$.

Finally, suppose that σ is an arbitrary permutation. We can write $\sigma = \sigma_1\sigma_2 \ldots \sigma_k$, where σ_is are disjoint cycles. According to the above argument, each of σ_i can be written as the product of two permutations α_i and β_i, where $o(\alpha_i) = o(\beta_i) = 2$, and α_i and β_i only permute the numbers appear in σ_i. Since σ_is are disjoint, if $j \neq i$, then α_i and β_i are disjoint from α_j and β_j. Consequently, α_i commutes with α_j and β_j, for all $i \neq j$. Therefore, we conclude that

$$\sigma = \sigma_1\sigma_2 \ldots \sigma_k = \alpha_1\beta_1\alpha_2\beta_2 \ldots \alpha_k\beta_k$$
$$= \alpha_1\alpha_2 \ldots \alpha_k\beta_1\beta_2 \ldots \beta_k.$$

Since a product of disjoint transpositions has order 2, it follows that

$$o(\alpha_1\alpha_2 \ldots \alpha_k) = o(\beta_1\beta_2 \ldots \beta_k).$$

This completes the proof.

∎

Problem 5.52 Let n be a positive integer. A sequence of positive integers $n_1, n_2, \ldots$, n_k such that $n_1 \geq n_2 \geq \cdots \geq n_k$ and $n = n_1 + n_2 + \cdots + n_k$, is called a *partition of* n. Let $p(n)$ denote the number of partitions of n. Table 5.1 is a short table of values $p(n)$: Show that the number of conjugate classes in the symmetric group S_n is $p(n)$.

Solution Let σ be a permutation in S_n. We can write σ as a product of distinct cycles as follows:

$$(a_1 \, a_2 \, \ldots \, a_{k_1})(b_1 \, b_2 \, \ldots \, b_{k_2}) \ldots (c_1 \, c_2 \, \ldots \, c_{k_j})$$

such that $k_1 \geq k_2 \geq \cdots \geq k_j$ and $k_1 + k_2 + \cdots + k_j = n$. This is a unique expression, and so for each permutation we obtain a unique partition. By Corollary 5.36, two permutations are conjugate if and only if they are similar; in other words, they give rise to same partition. Hence, corresponding to a conjugate class we get a unique partition of n.

Conversely, let $n_1 \geq n_2 \geq \cdots \geq n_r$ and $n = n_1 + n_2 + \cdots + n_r$ be a partition of n. Then, there is a permutation τ which has a cycle decomposition of the type

$$(x_1 \, x_2 \, \ldots \, x_{n_1})(y_1 \, y_2 \, \ldots \, y_{n_2}) \ldots (z_1 \, z_2 \, \ldots \, z_{n_r}).$$

Each $\delta \in S_n$ similar to τ is conjugate to τ, and every permutation in S_n conjugate to τ is similar to τ. In this way, for each permutation we can associate a unique conjugate class, namely, conjugate class of τ.

Therefore, there exists a one to one correspondence between conjugate classes in S_n and partitions of n. Consequently, the number of conjugate classes in S_n is equal to $p(n)$.

■

Problem 5.53 Let $s(n, k)$ denote the number of permutations in S_n which have exactly k cycles (including cycles of length 1). Show that

$$s(n, 1) = (n - 1)!$$

and for $k \geq 2$

$$s(n, k) = s(n - 1, k - 1) + (n - 1)s(n - 1, k).$$

Also, prove that

$$\sum_{k=1}^{n} s(n, k)x^k = x^{(n)} := x(x + 1) \ldots (x + n - 1).$$

The $s(n, k)$ are known as *Stirling numbers of the first kind*. The expression $x^{(n)}$ is known as the *nth upper factorial*.

Solution It is easy to see that $s(n, 1) = (n - 1)!$. In general, we sort the permutations in S_n with exactly k cycles into two parts, depending on whether the permutation contains the cycle (n) of length 1. There exist $s(n - 1, k - 1)$ permutations containing the cycle (n). The other permutations are formed by inserting n after any of the $n - 1$ elements in the $s(n - 1, k)$ permutations of $n - 1$ elements into k cycles. Consequently, for $k \geq 2$, we obtain $s(n, k) = s(n - 1, k - 1) + (n - 1)s(n - 1, k)$. Now, we can verify the formula for $s(n, k)$ by induction. It is easy to see that the formula holds for $n = 1$. Suppose the formula is true for $s(m, k)$, where $m < n$. Then, we have

$$\sum_{k=1}^{n} s(n, k)x^k$$

$$= (n-1)!x + \sum_{k=2}^{n} s(n-1, k-1)x^k + (n-1)\sum_{k=2}^{n} s(n-1, k)x^k$$

$$= (n-1)!x + x\sum_{k=1}^{n-1} s(n-1, k)x^k$$

$$+ (n-1)\left(\sum_{k=1}^{n} s(n-1, k)x^k - (n-2)!x\right)$$

$$= (n-1)!x + xx^{(n-1)} + (n-1)\left(x^{(n-1)} - (n-2)!x\right)$$
$$= (x+n-1)x^{(n-1)}$$
$$= x^{(n)},$$

as desired.   ∎

5.5 Supplementary Exercises

1. Let G be a group of order $2m$, let $g \in G$ have order 2, and let $\lambda_g : G \to G$ be defined by $g(x) = gx$. Show that λ_g is a product of m disjoint transpositions.
2. Show that the symmetry group of a rectangle which is not a square has order 4. By labeling the vertices, $1, 2, 3, 4$ represents the symmetry group as a group of permutations of the set $\{1, 2, 3, 4\}$.
3. Prove that S_X is abelian if and only if $|X| \leq 2$.
4. Let H be a subgroup of S_n. Show that either H is a subset of A_n or exactly half of the elements of H are even permutation.
5. List the elements of the following subgroup in S_4:

$$\langle(1\ 4)(2\ 3),\ (1\ 2)(3\ 4)\rangle.$$

6. Consider the permutation

$$\begin{pmatrix} 1 & 2 & 3 & 4 & 5 & 6 & 7 & 8 & 9 & 10 & 11 & 12 & 13 & 14 & 15 & 16 \\ 8 & 3 & 1 & 6 & 9 & 4 & 10 & 12 & 13 & 5 & 11 & 15 & 16 & 14 & 2 & 7 \end{pmatrix} \in S_{16}.$$

   (a) Find its sign and its order. Compute the centralizer σ and compute the number of elements in this centralizer;
   (b) Write down σ^{1000} as a product of disjoint cycles.

7. Let G be a non-abelian group of order $2p$ for some prime $p \neq 2$. Prove that G contains exactly $p - 1$ elements of order p and it contains exactly p elements of order 2.
8. If p is a prime number, show that in S_p there are $(p-1)! + 1$ elements x satisfying $x^p = e$.

9. In the symmetric group S_4 find two elements that neither commute with each other and are not conjugate to one another.

10. Let σ be the cycle $(1\ 2\ \ldots\ m)$. Show that σ^k is also a cycle of length m if and only if k is relatively prime to m.

11. Which permutation of the set $X = \{x_1,\ x_2,\ x_3,\ x_4,\ x_5\}$ leave the polynomial $x_1 + x_2 - x_3 - x_4$ invariant? Find a polynomial in these variables which is left invariant under all permutations in the group $\langle (x_1\ x_2\ x_3\ x_4),\ (x_2\ x_4) \rangle$ but not by all of S_X.

12. If $\sigma \in A_n$, prove that

$$C_{A_n}(\sigma) = C_{S_n}(\sigma) \quad \text{or} \quad |C_{A_n}(\sigma)| = \frac{1}{2}|C_{S_n}(\sigma)|.$$

13. If $\sigma = (1\ 2\ \ldots\ m) \in S_n$, show that $|C_{S_n}(\sigma)| = (n - m)!m$.

14. Let $a(n, m)$ denote the number of permutations $\sigma \in S_n$ such that $\sigma^m = Id$ (with $a(0, m) = 1$). Show that

$$\sum_{n=0}^{\infty} \frac{a(n, m)}{n!} x^n = \exp\left(\sum_{d \mid m} \frac{x^d}{d} \right).$$

15. Let $n \geq 2$ and let A be the set of all permutations in S_n of the form

$$\sigma_k = \prod_{1 \leq i \leq k/2} (i\ k - i),$$

for $k = 3,\ 4,\ \ldots,\ n + 1$. Show that A generates S_n and that each $\sigma \in S_n$ can be written as a product of $2n - 3$ or fewer elements from A.

16. Let p be a prime congruent to 1 (mod 4), and consider the set

$$X = \{(x, y, z) \in \mathbb{N}^3 \mid x^2 + 4yz = p\}.$$

Show that the function

$$(x, y, z) \mapsto \begin{cases} (x + 2z, z, y - x - z) & \text{if } x < y - z \\ (2y - x, y, x - y + z) & \text{if } y - z < x < 2y \\ (x - 2y, x - y + z, y) & \text{if } x > 2y \end{cases}$$

is a permutation of order 2 on X with exactly one fixed point. Conclude that the permutation $(x, y, z) \mapsto (x, z, y)$ must also have at least one fixed point, and so $x^2 + 4y^2 = p$ for some positive integers x and y.

17. Find the permutation representation of a cyclic group of order n.

18. **(Stirling Numbers of the Second Kind).** In Problem 5.53, we have seen that the Stirling numbers $s(n, k)$ of the first kind count the number of ways to partition a set of size n into k disjoint non-empty cycles. The Stirling numbers of the second

kind, denoted by $S(n, k)$, count the number of ways to partition a set of size n into k non-empty disjoint subsets. It is clear that $S(n, 1) = 1$.

(a) Find $S(n, 2)$;
(b) For $n > 0$, show that $S(n, k) = kS(n - 1, k) + S(n - 1, k - 1)$;
(c) Show that

$$x^n = \sum_{k=1}^{n} S(n, k)x^{(k)}.$$

19. In the symmetric group S_n, for each $k = 3, \ldots, n$, let

$$\sigma_k = \prod_{i=1}^{[\frac{k}{2}]} (i \, k - i),$$

for example

$$\sigma_3 = (1 \, 2),$$
$$\sigma_4 = (1 \, 3),$$
$$\sigma_5 = (1 \, 4)(2 \, 3),$$
$$\sigma_6 = (1 \, 5)(2 \, 4),$$
$$\sigma_7 = (1 \, 6)(2 \, 5)(3 \, 4),$$
$$\sigma_8 = (1 \, 7)(2 \, 6)(3 \, 5).$$

Show that permutations $\sigma_1, \sigma_2, \ldots, \sigma_n$ generate S_{n-1}.

20. An *affine geometry* comprises a set X whose elements are called *points* together with various subsets of X called *lines* such that

(a) Each pair of distinct points is contained in exactly one line;
(b) Each pair of distinct lines has at most one point in common;
(c) Given a line L and a point P not on it, there exists exactly one line L' which contains P and has no point in common with L;
(d) There are at least two lines. Figure 5.2 gives a pictorial representation of an affine geometry with 4 points and 6 lines.

Fig. 5.2 Affine geometry

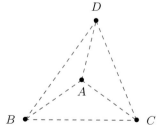

A *collineation* of an affine geometry is a permutation of the points of X which maps lines to lines. Show that the set of all collineations form a group under composition. What is the order of the collineation group of the above 4 element affine geometry?

Chapter 6
Group of Arithmetical Functions (Optional)

In this chapter, we show that the set of all arithmetical functions f with $f(1) \neq 0$ forms an abelian group with respect to a special operation. Then, we discuss an important subgroup of this group.

6.1 Arithmetical Functions

In this section we introduce several mathematical functions which play an important role in the study of divisibility properties of integers. We begin with the definition of arithmetical functions.

Definition 6.1 A real or complex-valued function defined on the positive integers is called an *arithmetical function*.

Example 6.2 Euler function φ defined in Definition 4.6 is a arithmetical function.

Theorem 6.3 (Gauss Theorem) *If n is a positive integer, then*

$$\sum_{d|n} \varphi(d) = n.$$

Proof Suppose that $X = \{1, 2, \ldots, n\}$ and let

$$S(d) = \{k \mid (k, n) = d, \ 1 \leq k \leq n\}.$$

Then, the sets $S(d)$ are disjoint and whose union is equal to X. If $f(d)$ is the number of integers in $S(d)$, then

$$\sum_{d|n} f(d) = n. \tag{6.1}$$

© The Author(s), under exclusive license to Springer Nature Singapore Pte Ltd. 2021
B. Davvaz, *A First Course in Group Theory*,
https://doi.org/10.1007/978-981-16-6365-9_6

Table 6.1 A short table of values $\mu(n)$

n	1	2	3	4	5	6	7	8	9	10	11	12	13	14	15
$\mu(n)$	1	−1	−1	0	−1	1	−1	0	0	1	−1	0	−1	1	1

Moreover, we have
$$(k, n) = d \Leftrightarrow (k/d, n/d) = 1,$$
$$0 < k \leq n \Leftrightarrow 0 < k/d \leq n/d.$$

So, if we take $q = k/d$, then there exists a one to one correspondence between the elements of $S(d)$ and those integers q satisfying $0 < k \leq n$ and $(q, n/d) = 1$. The number of such q is equal to $\varphi(n/d)$. Hence, we have $f(d) = \varphi(n/d)$. Now, by (6.1), we conclude that
$$\sum_{d \mid n} \varphi\left(\frac{n}{d}\right) = n.$$

This completes the proof, because when d runs through all divisors of n so does n/d. ∎

A positive integer n is *square-free* if $p^2 \nmid n$, for every prime p.

Definition 6.4 The Möbius function μ is defined as follows:

$$\mu(n) = \begin{cases} 1 & \text{if } n = 1 \\ 0 & \text{if } n \text{ is not square} - \text{free} \\ (-1)^k & \text{if } n \text{ is the product of } k \text{ distinct primes.} \end{cases}$$

Table 6.1 is a short table of values $\mu(n)$:

Theorem 6.5 *If n is a positive integer, then*

$$\sum_{d \mid n} \mu(d) = \left[\frac{1}{n}\right] = \begin{cases} 1 & \text{if } n = 1 \\ 0 & \text{if } n > 1. \end{cases}$$

Proof Clearly, we have $\mu(1) = 1$. Let $n > 1$ and $n = p_1^{\alpha_1} p_2^{\alpha_2} \ldots p_k^{\alpha_k}$. We know that $\mu(d)$ is non-zero if $d = 1$ or those divisors of n which are products of distinct primes. Consequently, we can write

$$\sum_{d \mid n} \mu(d) = \mu(1) + \mu(p_1) + \cdots + \mu(p_k) + \mu(p_1 p_2) + \cdots + \mu(p_{k-1} p_k)$$
$$+ \cdots + \mu(p_1 p_2 \ldots p_k)$$
$$= 1 + \binom{k}{1}(-1) + \binom{k}{2}(-1)^2 + \cdots + \binom{k}{k}(-1)^k$$
$$= (1 + (-1))^2 = 0.$$

This completes the proof. ∎

The Möbius function is related to the Euler function by the following theorem.

Theorem 6.6 *If n is a positive integer, then*

$$\varphi(n) = \prod_{d|n} \mu(d)\frac{n}{d}.$$

Proof We can write

$$\varphi(n) = \sum_{k=1}^{n} \left[\frac{1}{(n,k)}\right].$$

Then, by Theorem 6.5, we obtain

$$\varphi(n) = \sum_{k=1}^{n} \sum_{d|(n,k)} \mu(d) = \sum_{k=1}^{n} \sum_{\substack{d|n \\ d|k}} \mu(d). \tag{6.2}$$

For a fixed divisor d of n we must sum over all those $1 \le k \le n$ which are multiple of d. If $k = qd$, then

$$1 \le k \le n \Leftrightarrow 1 \le q \le \frac{n}{d}.$$

By using this fact and (6.2), we can write

$$\varphi(n) = \sum_{d|n} \sum_{q=1}^{n/d} \mu(d) = \sum_{d|n} \mu(d) \sum_{q=1}^{n/d} 1 = \sum_{d|n} \mu(d)\frac{n}{d}.$$

This completes our proof. ∎

Theorem 6.7 *If n is a positive integer, then*

$$\varphi(n) = n \prod_{p|n} \left(1 - \frac{1}{p}\right),$$

where p is prime.

Proof If $n = 1$, then there are no primes which divide 1. Hence, the product is empty, and so $\varphi(1) = 1$. Suppose that $n > 1$ and let $p_1, p_2, \ldots, p_m$ be the distinct prime divisors of n. The product can be written as

$$\prod_{p|n}\left(1-\frac{1}{p}\right) = \prod_{i=1}^{m}\left(1-\frac{1}{p_i}\right)$$

$$= \left(1-\frac{1}{p_1}\right)\left(1-\frac{1}{p_2}\right)\cdots\left(1-\frac{1}{p_m}\right)$$

$$= 1 - \sum\frac{1}{p_i} + \sum\frac{1}{p_i p_j} - \sum\frac{1}{p_i p_j p_k} + \cdots + \frac{(-1)^m}{p_1 p_2 \cdots p_m}.$$
$$(6.3)$$

On the last line of (6.3), in a term such as $\sum 1/p_i p_j p_k$, it means that we consider all possible products of $p_i p_j p_k$ of distinct prime factors of n taken three at a time. Moreover, each term on the last line of (6.3) is of the form $\pm 1/d$, where d is a divisor of n which is either 1 or a product of distinct primes. The numerator ± 1 is exactly $\mu(d)$. If d is divisible by the square of any prime p_i, then $\mu(d) = 0$. So, we observe that (6.3) is exactly the same as

$$\prod_{p|n}\left(1-\frac{1}{p}\right) = \sum_{d|n}\frac{\mu(d)}{d}.$$

Therefore, we conclude that

$$\varphi(n) = \sum_{d|n}\mu(d)\frac{n}{d} = n\sum_{d|n}\frac{\mu(d)}{d} = n\prod_{p|n}\left(1-\frac{1}{p}\right),$$

as desired.   ∎

Many properties of φ can be easily obtained from Theorem 6.7. Some of these are listed in the next theorem.

Theorem 6.8 *Euler function has the following properties:*

(1) $\varphi(p^\alpha) = p^\alpha - p^{\alpha-1}$, *for each prime number p and $\alpha \geq 1$;*
(2) $\varphi(mn) = \varphi(m)\varphi(n)\left(\dfrac{d}{\varphi(d)}\right)$, *where $d = (m, n)$;*
(3) $\varphi(mn) = \varphi(m)\varphi(n)$, *if $(m, n) = 1$;*
(4) $m|n$ *implies* $\varphi(m)|\varphi(n)$;
(5) $\varphi(n)$ *is even, for each $n \geq 3$. In addition, if n has k distinct odd prime factors, then $2^k|\varphi(n)$.*

Proof (1) It is enough if we consider $n = p^\alpha$ in Theorem 6.7.
    (2) By Theorem 6.7, for each positive integer n, we can write

$$\frac{\varphi(n)}{n} = \prod_{p|n}\left(1-\frac{1}{p}\right),$$

where p is prime. Each prime divisor of mn is either a prime divisor of m or of n, and those primes which divide both m and n also divide (m, n). Therefore, we have

$$\frac{\varphi(mn)}{mn} = \prod_{p|mn}\left(1 - \frac{1}{p}\right) = \frac{\prod_{p|m}\left(1 - \frac{1}{p}\right)\prod_{p|n}\left(1 - \frac{1}{p}\right)}{\prod_{p|(m,n)}\left(1 - \frac{1}{p}\right)} = \frac{\frac{\varphi(m)}{m}\frac{\varphi(n)}{n}}{\frac{\varphi(d)}{d}},$$

and consequently, we obtain (2).

(3) It is a special case of (2). In fact, it is enough to consider $d = 1$ in (2).

(4) Since $m|n$, it follows that there exists an integer $1 \le c \le n$ such that $n = cm$. If $c = n$, then $m = 1$ and (4) is obviously satisfied. So, let $c < n$. By using (2), we have

$$\varphi(n) = \varphi(mc) = \varphi(m)\varphi(c)\frac{d}{\varphi(d)} = d\varphi(m)\frac{\varphi(c)}{\varphi(d)}, \tag{6.4}$$

where $d = (m, c)$. Now, we use mathematical induction on n. If $n = 1$, then (4) is true obviously. Assume that (4) is true for each integer less than n. Since $c < n$ and $d|c$, by induction hypothesis, it follows that $\varphi(d)|\varphi(c)$. Therefore, we deduce that the right part of (6.4) is an integer multiple of $\varphi(m)$. This yields that $\varphi(m)|\varphi(n)$.

(5) If $n = 2^\alpha$ and $\alpha \ge 2$, then by part (1) we obtain $\varphi(n) = \varphi(2^\alpha) = 2^{\alpha-1}$. This means that $\varphi(n)$ is even. Now, we assume that n has at least one odd prime factor. Then, we have

$$\varphi(n) = n\prod_{p|n}\left(1 - \frac{1}{p}\right) = n\prod_{p|n}\left(\frac{p-1}{p}\right)$$

$$= n\frac{\prod_{p|n}(p-1)}{\prod_{p|n}p} = \frac{n}{\prod_{p|n}p}\prod_{p|n}(p-1).$$

Therefore, we conclude that

$$\varphi(n) = c(n)\prod_{p|n}(p-1).$$

where $c(n)$ is an integer. Since the product multiplying $c(n)$ is even, it follows that $\varphi(n)$ is even. In addition, each odd prime p gives a factor 2 to this product. Consequently, if n has k distinct odd prime factors, then $2^k|\varphi(n)$. ∎

Exercises

1. Show that if $n - 1$ and $n + 1$ are both primes, with $n > 4$, then

$$\varphi(n) \leq \frac{n}{3}.$$

2. Find all n for which $\varphi(n) \equiv 0 \pmod{4}$.
3. For each of the following statements either give a proof or exhibit a counter example.

   (a) If $(m, n) = 1$, then $(\varphi(m), \varphi(n)) = 1$;
   (b) If n is composite, then $(n, \varphi(n)) > 1$;
   (c) If the same primes divide m and n, then $n\varphi(m) = m\varphi(n)$.

4. Prove that $\varphi(n) > n/6$ for all n with at least 8 distinct prime factors.
5. Prove that

$$\sum_{d^2 \mid n} \mu(d) = \mu^2(n).$$

6.2 Dirichlet Product and Its Properties

The Dirichlet product is a binary operation defined on arithmetical functions. It is commutative, associative, and distributive over addition and has other important number-theoretical properties.

Definition 6.9 Let f and g be two arithmetical functions. We define the *Dirichlet product* of f and g by

$$(f * g)(n) = \sum_{d \mid n} f(d) g \left(\frac{n}{d} \right),$$

for all positive integer n.

Theorem 6.10 *Let AF be the set of arithmetical functions f such that $f(1) \neq 0$. Then, $(AF, *)$ forms an abelian group.*

Proof Since $(f * g)(1) = f(1)g(1)$, it follows that AF is closed under Dirichlet product. The commutative property is evident from noting that

$$\sum_{d \mid n} f(d) g \left(\frac{n}{d} \right) = \sum_{ab=n} f(a) g(b).$$

In order to prove the associative property, suppose that f, g, and h are any mathematical functions. Let $A = g * h$ and consider $f * A = f * (g * h)$. Then, we obtain

$$(f * A)(n) = \sum_{ad=n} f(a)A(d)$$

$$= \sum_{ad=n} f(a) \sum_{bc=d} g(b)h(c)$$

$$= \sum_{abc=n} f(a)g(b)h(c).$$

Similarly, if we assume that $B = f * g$ and consider $B * h = (f * g) * h$, then we have

$$(B * h)(n) = \sum_{ad=n} B(d)h(a)$$

$$= \sum_{ad=n} h(a)B(d)$$

$$= \sum_{ad=n} h(a) \sum_{bc=d} f(b)g(c)$$

$$= \sum_{abc=n} f(b)g(c)h(a).$$

So, we conclude that $f * (g * h) = (f * g) * h$, i.e., Dirichlet product is associative. Now, let

$$e(n) = \left[\frac{1}{n}\right] = \begin{cases} 1 & \text{if } n = 1 \\ 0 & \text{otherwise} \end{cases}$$

It is clear that $e \in AF$. Moreover, if $f \in AF$, then we have

$$(f * e)(n) = \sum_{d|n} f(d)e\left(\frac{n}{d}\right) = \sum_{d|n} f(d)\left[\frac{d}{n}\right] = f(n),$$

for each positive integer n. This means that e is the identity element of AF. Finally, we must show that for given $f \in AF$, there exists $f^{-1} \in AF$ such that $f * f^{-1} = e$.

Let f is given. We construct f^{-1} inductively. First, we need $(f * f^{-1})(1) = e(1)$, which occurs if and only if $f(1)f^{-1}(1) = 1$. Since $f(1) \neq 0$, it follows that $f^{-1}(1)$ is uniquely determined. Now, suppose that $n > 1$ and f^{-1} determined for each $k < n$. Then, we have to solve the equation $(f * f^{-1})(n) = e(n)$, or equivalently

$$\sum_{d|n} f\left(\frac{n}{d}\right) f^{-1}(d) = 0.$$

This can be written as

$$f(1)f^{-1}(n) \sum_{\substack{d|n \\ d<n}} f\left(\frac{n}{d}\right) f^{-1}(d) = 0,$$

and so

$$f^{-1}(n) = \frac{-1}{f(1)} \sum_{\substack{d|n \\ d<n}} f\left(\frac{n}{d}\right) f^{-1}(d).$$

Since the values $f^{-1}(d)$ are known for all divisors $d < n$, it follows that there is a uniquely determined value to $f^{-1}(n)$. Consequently, we may uniquely determine an f^{-1} for each $f \in AF$. ∎

Remark 6.11 If we consider the usual product of functions, i.e., $(f \cdot g)(n) = f(n)g(n)$, then the identity element is $I(n) = n$, for all positive integer n, because $I \cdot f = f$, for all function f. In the group AF, however, I is certainly not identity element. But it has the nice property of transferring each function f into its so-called sum-function S_f.

Definition 6.12 We define
$$S_f(n) = \sum_{d|n} f(d)$$

to be the *sum-function* of $f \in AF$.

Note that $S_f \in AF$ too. Moreover, it is easy to see that $I * f = f * I = S_f$, for all $f \in AF$.

Lemma 6.13 *The Dirichlet inverse of I is the Möbius function μ.*

Proof According to Theorem 6.5, we have

$$\sum_{d|n} \mu(d) = I(n).$$

In the notation of Dirichlet product this becomes $\mu * I = I * \mu = e$. Therefore, I and μ are inverses of each other. ∎

Theorem 6.14 *Each arithmetical function f can be expressed in terms of its sum-function S_f as*
$$f(n) = \sum_{d|n} \mu(d) S_f\left(\frac{n}{d}\right).$$

Proof We have

$$\mu * S_f = \mu * (I * f) = (\mu * I) * f = e * f = f,$$

and this completes the proof. ∎

Theorem 6.15 (Möbius Inversion Theorem) *For two arithmetic functions f and g we have the following equivalence:*

$$f(n) = \sum_{d|n} g(d) \iff g(n) = \sum_{d|n} f(d)\mu\left(\frac{n}{d}\right).$$

Proof If $f = g * I$, then we have

$$f * \mu = (g * I) * \mu = g * (I * \mu) = g * e = g.$$

On the other hand, if $f * \mu = g$, then we have

$$g * I = (f * \mu) * I = f * (\mu * I) = f * e = f.$$

This completes the proof. ∎

Exercises

1. Solve the following equations:

   (a) $\varphi(2^x 5^y) = 80$;
   (b) $\varphi(n) = 12$;
   (c) $\varphi(n) = 2n/3$;
   (d) $\varphi(n) = n/2$;
   (e) $\varphi(\varphi(n)) = 2^{13}3^3$.

2. If $(p, q) = 1$, prove that

$$\sum_{k=1}^{q-1}\left[\frac{kp}{q}\right] = \sum_{k=1}^{p-1}\left[\frac{kq}{p}\right].$$

   Hint: Use the identity

$$\left[\frac{kp}{q}\right] - \left[\frac{(q-k)p}{q}\right] = p - 1,$$

   for $k = 1, 2, \ldots, q - 1$, and show that both sums equal $(p - 1)(q - 1)/2$.

6.3 Multiplicative Functions

In this section, we study arithmetical functions called multiplicative functions. These functions have the property that their value at the product of two relatively prime integers is equal to the product of the value of the functions at these integers.

Definition 6.16 An arithmetical function f is called *multiplicative* if f is not identically zero and if $f(mn) = f(m)f(n)$, whenever $(m, n) = 1$. A multiplicative function f is called *completely multiplicative* if we also have $f(mn) = f(m)f(n)$, for all positive integers m and n.

Example 6.17 The Euler function φ is multiplicative. This is easily seen from part (3) of Theorem 6.8. But it is not completely multiplicative, since $\varphi(4) = 2$ and $\varphi(2)\varphi(2) = 1$.

Example 6.18 The Möbius function μ is multiplicative. This is easily seen from Definition 6.4. But it is not completely multiplicative, since $\mu(4) = 0$ and $\mu(2)\mu(2) = 1$.

Example 6.19 The identity element e of the group AF is completely multiplicative.

Example 6.20 Let f and g be two arithmetical functions. Then, the ordinary product fg and the quotient product f/g are defined by

$$(fg)(n) = f(n)g(n),$$
$$\left(\frac{f}{g}\right)(n) = \frac{f(n)}{g(n)}, \text{ whenever } g(n) \neq 0.$$

If f and g are multiplicative, so are fg and f/g.

Theorem 6.21 *If f is multiplicative, then $f(1) = 1$.*

Proof Since $(n, 1) = 1$, for all positive integer n, it follows that $f(n) = f(n)f(1)$. Since f is not identically zero, it follows that $f(n) \neq 0$, for some positive integer n. So, we get $f(1) = 1$. ∎

The following is a nice characterization of multiplicative functions.

Theorem 6.22 *Let f be an arithmetical function with $f(1) = 1$. Then,*

(1) f is multiplicative if and only if

$$f(p_1^{\alpha_1} \dots p_k^{\alpha_k}) = f(p_1^{\alpha_1}) \dots f(p_k^{\alpha_k}),$$

for all primes p_i and all positive integers α_i;

(2) If f is multiplicative, then f is completely multiplicative if and only if

$$f(p^\alpha) = f(p)^\alpha,$$

for all primes p and all positive integers α.

Proof The proof follows easily from the definition and is left as an exercise for the reader. ∎

The following are further examples of well-known arithmetical functions.

Example 6.23 Let n be a positive integer.

(1) The *divisor function* τ is defined to be the number of positive divisors of n, i.e.,

$$\tau(n) = |\{d \in \mathbb{N} \mid d|n\}|;$$

(2) The *sum of divisors function* σ is defined to be the sum of all positive divisors of n, i.e.,

$$\sigma(n) = \sum_{d|n} d.$$

Theorem 6.24 *The divisor function σ and the sum of divisors function σ are multiplicative. Their values at primes powers are given by*

$$\tau(p^\alpha) = \alpha + 1 \text{ and } \sigma(p^\alpha) = \frac{p^{\alpha+1} - 1}{p - 1}.$$

Proof In order to prove τ is multiplicative, assume that n and m are positive integers with $(m, n) = 1$. Note that if $d_1|m$ and $d_2|n$, then $d_1 d_2|mn$. Conversely, if $d|mn$, then $d = d_1 d_2$ such that $d_1|m$ and $d_2|n$. Therefore, there exists a one to one correspondence between the set of divisors of mn and the set

$$A = \{(d_1, d_2) \mid d_1|m \text{ and } d_2|n\}.$$

Since $\tau(mn)$ is the number of divisors of mn and $|A| = \tau(m)\tau(n)$, it follows that $\tau(mn) = \tau(m)\tau(n)$. Similarly, we can prove that σ is multiplicative. Indeed, we have

$$\sigma(mn) = \sum_{d|n} d = \sum_{\substack{d_1|m \\ d_2|n}} d_1 d_2 = \left(\sum_{d_1|m} d_1\right)\left(\sum_{d_2|n} d_2\right) = \sigma(m)\sigma(n).$$

Moreover, since the divisors of p^α are exactly $1, p, \ldots, p^\alpha$, it follows that $\tau(p^\alpha) = \alpha + 1$. Moreover, if we apply the geometric series formula, then we obtain

$$\sigma(p^\alpha) = 1 + p + \cdots + p^\alpha = \frac{p^{\alpha+1} - 1}{p - 1},$$

as desired. ∎

Theorem 6.25 *If f and g are multiplicative, so is their Dirichlet product $f * g$.*

Proof Let $h = f * g$ and $(m, n) = 1$. Then, we have

$$h(mn) = \sum_{d|mn} f(d)g\left(\frac{mn}{d}\right).$$

Suppose that $d = ab$ such that $a|m$ and $b|n$. Then, we obtain

$$h(mn) = \sum_{\substack{a|m \\ b|n}} f(ab)g\left(\frac{mn}{ab}\right) = \sum_{\substack{a|m \\ b|n}} f(a)f(b)g\left(\frac{m}{a}\right)g\left(\frac{n}{b}\right)$$

$$= \sum_{a|m} f(a)g\left(\frac{m}{a}\right)\sum_{b|n} f(b)g\left(\frac{n}{b}\right) = h(m)h(n).$$

Thus, h is also multiplicative.   ∎

Theorem 6.26 *If g and $f * g$ are multiplicative, then f is also multiplicative.*

Proof Assume that f is not multiplicative and let $h = f * g$. Then, there exist positive integers m and n with $(m, n) = 1$ such that $f(mn) \neq f(m)f(n)$. By the well-ordering principle, we choose such a pair m and n for which the product mn is as small as possible. We consider the following two cases:

   Case 1: If $mn = 1$, then $f(1) \neq f(1)f(1)$, and so $f(1) \neq 1$. Since $h(1) = f(1)g(1) = f(1) \neq 0$, it follows that h is not multiplicative, and this is a contradiction.

   Case 2: If $mn > 1$, then $f(ab) = f(a)f(b)$, for all positive integers a and b with $(a, b) = 1$ and $ab < mn$. Therefore, we have

$$h(mn) = \sum_{\substack{a|m \\ b|n}} f(ab)g\left(\frac{mn}{ab}\right)$$

$$= \sum_{\substack{a|m \\ b|n \\ ab<mn}} f(ab)g\left(\frac{mn}{ab}\right) + f(mn)g(1)$$

$$= \sum_{\substack{a|m \\ b|n \\ ab<mn}} f(a)f(b)g\left(\frac{m}{a}\right)g\left(\frac{n}{b}\right) + f(mn)$$

$$= \sum_{a|m} f(a)g\left(\frac{m}{a}\right)\sum_{b|n} f(b)g\left(\frac{n}{b}\right) - f(m)f(n) + f(mn)$$

$$= h(m)h(n) - f(m)f(n) + f(mn).$$

Since $f(mn) \neq f(m)f(n)$, it follows that $h(mn) \neq h(m)h(n)$. This shows that h is not multiplicative, and it is a contradiction.   ∎

Theorem 6.27 *If f is multiplicative, so is f^{-1}, its Dirichlet inverse.*

Proof The result can be easily obtained from Theorem 6.26. In fact, since both f and $f^{-1} * f = e$ are multiplicative, it follows that f^{-1} is multiplicative.   ∎

Corollary 6.28 *The set of multiplicative functions is a subgroup of AF.*

Proof It is clear by Theorems 6.25 and 6.27. ∎

Theorem 6.29 *Let f be multiplicative. Then, f is completely multiplicative if and only if $f^{-1}(n) = \mu(n)f(n)$, for each positive integer n.*

Proof Let $g(n) = \mu(n)f(n)$. If f is completely implicative, then

$$(g * f)(n) = \sum_{d|n} \mu(d)f(d)f\left(\frac{n}{d}\right) = f(n)\sum_{d|n}\mu(d) = f(n)e(n) = e(n),$$

because $f(1) = 1$ and $e(n) = 0$, for each $n > 1$. Hence, we have $g = f^{-1}$.

Conversely, suppose that $f^{-1}(n) = \mu(n)f(n)$. In order to prove that f is completely multiplicative, it is enough to show that $f(p^\alpha) = f(p)^\alpha$ for prime powers. Since $f * f^{-1} = f * \mu f = e$, it follows that $(f * \mu f)(n) = e(n)$. This yields that

$$\sum_{d|n} \mu(d)f(d)f\left(\frac{n}{d}\right) = 0,$$

for each integer n. Now, we take $n = p^\alpha$, then obtain

$$\mu(1)f(1)f(p^\alpha) + \mu(p)f(p)f(p^{\alpha-1}) = 0.$$

Therefore, we conclude that $f(p^\alpha) = f(p)f(p^{\alpha-1})$. This shows that $f(p^\alpha) = f(p)^\alpha$. ∎

Exercises

1. Prove that
$$\frac{\varphi(n)\sigma(n) + 1}{n}$$
   is an integer if n is prime, and it is not integer if n is divisible by square of a prime.
2. Prove that f is multiplicative if and only if its sum-function S_f is multiplicative.
3. Let $f(n) = [\sqrt{n}] - [\sqrt{n-1}]$. Prove that f is multiplicative but not completely multiplicative.
4. Solve the equation $\varphi(\sigma(2^n)) = 2^n$.

6.4 Worked-Out Problems

Problem 6.30 If $\varphi(m)|m - 1$, prove that there does not exist prime number p such that $p^2|m$.

Solution Suppose that there exists prime number p_j such that $p_j^2|m$. Let $m = p_1^{\alpha_1} \dots p_j^{\alpha_j} \dots p_k^{\alpha_k}$, where $\alpha_j \geq 2$. Then, we have

$$p_j \Big| p_j \left(p_j^{\alpha_j - 1} - p_j^{\alpha_j - 2} \right) = p_j^{\alpha_j} - p_j^{\alpha_j - 1} = \varphi(p_j^{\alpha_j}).$$

On the other hand, since $p_j^{\alpha_j}|m$, it follows that $\varphi(p_j^{\alpha_j})|\varphi(m)$. So, we conclude that $p_j|\varphi(m)$. Since $p_j \nmid m - 1$, it follows that $\varphi(m) \nmid m - 1$. This is a contradiction. ∎

Problem 6.31 If m is not prime and $\varphi(m)|m - 1$, prove that m has at least three distinct prime factors.

Solution Suppose that m has exactly two distinct prime factors. By Problem 6.30, since $\varphi(m)|m - 1$, we conclude that $m = pq$, where p and q are distinct primes. Then, we have

$$\begin{aligned}
\frac{m-1}{\varphi(m)} &= \frac{pq-1}{(p-1)(q-1)} \\
&= \frac{pq - p - q + 1 + p + q - 2}{(p-1)(q-1)} \\
&= \frac{(p-1)(q-1)}{(p-1)(q-1)} + \frac{p-1}{(p-1)(q-1)} + \frac{q-1}{(p-1)(q-1)} \\
&= 1 + \frac{1}{q-1} + \frac{1}{p-1}.
\end{aligned}$$

Since $\varphi(m)|m - 1$, it follows that $(m - 1)/\varphi(m)$ is an integer. Moreover, we have

$$1 < 1 + \frac{1}{q-1} + \frac{1}{p-1} \leq 3.$$

So, we must have

$$\frac{1}{q-1} + \frac{1}{p-1} = 1 \text{ or } 2.$$

This implies that $p = q = 2$ or $p = q = 3$. But this contradicts our assumption that p and q are distinct. Consequently, it is impossible for m to have only two distinct prime factors. ∎

Problem 6.32 Show that if $m \geq 2$, then the sum of all positive integers which are less than m and relatively prime to m is $(1/2)m\varphi(m)$.

Solution Suppose that $1 \leq k < m$ such that $(k, m) = 1$. If $(m - k, m) = d$, then $d|m$ and $d|m - k$. This implies that $d|k$, and so $d = 1$. Since $k < m$, it follows that $m - k \geq 1$. Also, we have $m - k < m$. Now, assume that

$$\begin{aligned}
A &= \{k \mid 1 \leq k \leq m, \ (k, m) = 1\}, \\
B &= \{m - k \mid 1 \leq k \leq m, \ (k, m) = 1\}.
\end{aligned}$$

There is a one to one correspondence between A and B given by $f(k) = m - k$. Consequently, we have $|A| = |B| = \varphi(m)$. Therefore, we get

$$\left(k_1 + (m - k_1)\right) + \cdots + \left(k_{\varphi(m)} + (m - k_{\varphi(m)})\right) = 2 \sum_{\substack{0 < k < m \\ (k,m)=1}} k.$$

This shows that

$$\underbrace{m + \cdots + m}_{\varphi(m) \text{ times}} = 2S,$$

where S is the required sum. Therefore, we deduce that $S = (1/2)m\varphi(m)$. ∎

Problem 6.33 Let $f : \mathbb{N} \to \mathbb{N}$ be multiplicative and strictly increasing. If $f(2) = 2$, prove that $f(n) = n$ for all n.

Solution We have $f(1) = 1$ and $f(2) = 2$. First, we show that $f(3) = 3$. Suppose that $f(3) = k + 3$, where k is a non-negative integer. We can write $f(6) = f(2)f(3) = 2(k + 3) = 2k + 6$. Since f is strictly increasing, it follows that $f(5) \leq 2k + 5$, and so $f(10) = f(2)f(5) \leq 4k + 10$. Hence, we must have $f(9) \leq 4k + 9$, which implies that $f(18) = f(2)f(9) \leq 8k + 18$. This gives

$$f(15) \leq 8k + 15. \tag{6.5}$$

On the other hand, we have

$$f(15) = f(3)f(5) \geq (k + 3)(k + 5) = k^2 + 8k + 15. \tag{6.6}$$

By (6.5) and (6.6) we conclude that $k = 0$. Hence, $f(3) = 3$, as desired. Now, we use mathematical induction. Suppose that

$$f(m) = m, \text{ for all } m = 1, 2, \ldots, 2k - 1,$$

where $k \geq 2$. Then, we have

$$f(4k - 2) = f\left(2(2k - 1)\right) = f(2)f(2k - 1)$$
$$= 2f(2k - 1) = 4k - 2.$$

Since f is strictly increasing, it follows that $f(m) = m$, for all $2k - 1 \leq m \leq 4k - 2$. In particular, we deduce that $f(2k) = 2k$ and $f(2k + 1) = 2k + 1$. This completes the proof.

Problem 6.34 Prove that

$$\frac{n}{\varphi(n)} = \sum_{d|n} \frac{\mu^2(d)}{\varphi(d)}.$$

Solution Since μ and φ are multiplicative and $\varphi(n) \neq 0$, for each positive integer n, it follows that μ^2/φ is multiplicative. So, if $G = S_{\mu^2/\varphi}$, sum-function, then

$$G(n) = \sum_{d\mid n} \frac{\mu^2(d)}{\varphi(d)}.$$

and G is multiplicative, since $(\mu^2/\varphi) * I = G$. Now, if $n = p_1^{\alpha_1} \ldots p_k^{\alpha_k}$, then

$$G(n) = G(p_1^{\alpha_1}) \ldots G(p_k^{\alpha_k})$$

$$= \left(1 + \frac{1}{\varphi(p_1)}\right) \ldots \left(1 + \frac{1}{\varphi(p_k)}\right)$$

$$= \frac{p_1}{p_1 - 1} \cdots \frac{p_k}{p_k - 1}$$

$$= \frac{n}{n\left(1 - \frac{1}{p_1}\right) \ldots \left(1 - \frac{1}{p_k}\right)}$$

$$= \frac{n}{\varphi(n)}.$$

This completes the proof.   ∎

Problem 6.35 Let G be a finite abelian group. We say that $f :: G \to \mathbb{C}$ is a *character* on G if (1) $f(x) \neq 0$, for all $x \in G$, and (2) $f(xy) = f(x)f(y)$, for all $x, y \in G$. Suppose that f is a character on the multiplicative group U_k. We may extend the domain of this character to the entire set of natural numbers in the following manner: First, let $\bar{n} \in U_k$ be the equivalence class modulo k containing n. We extend the domain of f to the entire set of natural numbers as follows:

$$\chi(n) = \begin{cases} f(\bar{n}) & \text{if } (n, k) = 1 \\ 0 & \text{otherwise.} \end{cases}$$

We call χ a *Dirichlet character modulo k*. It is easy to see that χ has the following two important properties:

(1) It is k-periodic, i.e., $\chi(a + k) = \chi(a)$, for all $a \in \mathbb{N}$;
(2) It is completely multiplicative.

Table 6.2 shows the Dirichlet characters modulo 5. It is not difficult to check that no more exist.

If χ is an arithmetical function that is both periodic and completely multiplicative, and it is not zero function, prove that it is a Dirichlet character.

Solution Let k be the minimal period of χ. Since χ is k-periodic, it is constant in each equivalence class modulo k. Moreover, since it is completely multiplicative, its

Table 6.2 Dirichlet characters modulo 5

x (mod 5)	0	1	2	3	4
χ_1	0	1	1	1	1
χ_2	0	1	−1	−1	1
χ_3	0	1	i	−i	−1
χ_4	0	1	−i	i	−1

values on U_k make it a character. Therefore, we need to show that $\chi(n) = 0$ if and only if $(n, k) \neq 1$. First, assume $(n, k) = 1$. Then we have $n^{\varphi(k)} \equiv 1 \pmod{k}$. Since χ is completely multiplicative and k-periodic, we have $\chi(n)^{\varphi(k)} = \chi\left(n^{\varphi(k)}\right) = \chi(1)$. Thus, if $\chi(n) = 0$, then $\chi(1) = 0$. However, if $\chi(1) = 0$, then we have $\chi(m) = \chi(1)\chi(m) = 0$, for all m. Hence, if $(n, k) = 1$ and $\chi(n) = 0$, then χ is zero function. On the other hand, we assume that there is n such that $(n, k) > 1$ and $\chi(n) \neq 0$. Then there is at least one prime p such that $p|k$ and $\chi(p) \neq 0$. Consider this p, and let m be any natural number. Because χ is k-periodic and completely multiplicative, we have

$$\chi(m)\chi(p) = \chi(mp) = \chi(mp + k) = \chi(p)\chi\left(m + \frac{k}{p}\right).$$

Since $\chi(p) \neq 0$, we must have $\chi(m) = \chi(m + k/p)$, for all m. But this means that χ is k/p-periodic. This violates the stipulation that k is the minimal period of χ. ∎

6.5 Supplementary Exercises

1. Let n be an integer with $n \geq 2$. Show that $\varphi(2^n - 1)$ is divisible by n.
2. If p is a prime and n is an integer such that $1 < n < p$, prove that

$$\varphi\left(\sum_{k=0}^{p-1} n^k\right) \equiv 0 \pmod{p}.$$

3. Let m and n be positive integers. Prove that, for some positive integer a, each of $\varphi(a), \varphi(a + 1), \ldots, \varphi(a + n)$ is a multiple of m.
4. If n is composite, prove that $\varphi(n) \leq n - \sqrt{n}$.
5. Find all solutions of $\varphi(n) = 4$, and prove that there are no more.
6. Find $m, n \in \mathbb{N}$ such that they have no prime divisors other than 2 and 3, $(m, n) = 18$, $\tau(m) = 21$, and $\tau(n) = 10$.
7. Determine an arithmetical function f such that

$$\frac{1}{\varphi(n)} = \sum_{d|n} \frac{1}{d} f\left(\frac{n}{d}\right) \quad (n \in \mathbb{N}).$$

8. An arithmetical function f is called *periodic* if there exists a positive integer k such that $f(n + k) = f(n)$ for each positive integer N; the integer k is called a period for f. Show that if f is completely multiplicative and periodic with period k, then the values of f are either 0 or roots of unity.

9. Let f be a multiplicative function satisfying $\lim_{p^m \to \infty} f(p^m) = 0$. Show that $\lim_{n \to \infty} f(n) = 0$.

10. Prove that the sum-function S_f of a multiplicative function f is given by

$$S_f(n) = \prod_{i=1}^{k} \left(1 + f(p_i) + f(p_i^2) + \cdots + f(p_i^{\alpha_i})\right),$$

whenever $n = p_1^{\alpha_1} \cdots p_k^{\alpha_k}$.

11. For any real number $x \geq 1$, prove that

$$\left| \sum_{n \leq x} \frac{\mu(n)}{n} \right| \leq 1.$$

12. Prove that a finite abelian group G of order n has exactly n distinct characters.

13. For two sequences of complex numbers $\{a_0, a_1, \ldots, a_n, \ldots\}$ and $\{b_0, b_1, \ldots, b_n, \ldots\}$ show that the following relations are equivalent:

$$a_n = \sum_{k=0}^{n} b_k \text{ for all } n \Leftrightarrow b_n = \sum_{k=0}^{n} (-1)^{k+n} a_k \text{ for all } n.$$

14. Prove that

$$\sum_{k=1}^{n} \tau(k) = \sum_{k=1}^{n} \left[\frac{n}{k}\right] \text{ and } \sum_{k=1}^{n} \sigma(k) = \sum_{k=1}^{n} k \left[\frac{n}{k}\right].$$

Chapter 7
Matrix Groups

Among the most important examples of groups are groups of matrices. In this chapter, matrix groups are defined and a number of standard examples are discussed.

7.1 Introduction to Matrix Groups

Let $A = (a_{ij})_{m \times n}$ and $B = (b_{ij})_{n \times p}$. We recall the *product* AB is the matrix $C = (a_{ij})_{m \times p}$, where

$$c_{ij} = \sum_{k=1}^{n} a_{ik} b_{kj}.$$

Theorem 7.1 *Matrix multiplication is associative.*

Proof Let $A = (a_{ij})_{m \times n}$, $B = (b_{ij})_{n \times p}$ and $C = (c_{ij})_{p \times q}$ be three arbitrary matrices. Suppose that $AB = D = (d_{ij})_{m \times p}$ and $BC = E = (e_{ij})_{n \times q}$. Moreover, suppose that

$$(AB)C = DC = X = (x_{ij})_{m \times q} \quad \text{and} \quad A(BC) = AE = Y = (y_{ij})_{m \times q}.$$

Then, we obtain

$$x_{ij} = \sum_{l=1}^{p} d_{il} c_{lj}$$

$$= \sum_{l=1}^{p} \left(\sum_{k=1}^{n} a_{ik} b_{kj} \right) c_{lj}$$

© The Author(s), under exclusive license to Springer Nature Singapore Pte Ltd. 2021
B. Davvaz, *A First Course in Group Theory*,
https://doi.org/10.1007/978-981-16-6365-9_7

$$= \sum_{l=1}^{p} (a_{i1}b_{1l} + a_{i2}b_{2l} + \cdots + a_{in}b_{nl})c_{lj}$$

$$= a_{i1} \sum_{l=1}^{p} b_{1l}c_{lj} + a_{i2} \sum_{l=1}^{p} b_{2l}c_{lj} + \cdots + a_{in} \sum_{l=1}^{p} b_{nl}c_{lj}$$

$$= \sum_{k=1}^{n} a_{ik} \left(\sum_{l=1}^{p} b_{il}c_{lj} \right)$$

$$= \sum_{k=1}^{n} a_{ik}e_{kj}$$

$$= y_{ij}.$$

Therefore, we deduce that $X = Y$. ∎

A matrix A with n rows and n columns is called a *square matrix* of order n. One example of a square matrix is the $n \times n$ *identity matrix* I_n. This is the matrix $I_n = (\delta_{ij})_{n \times n}$ defined by

$$\delta_{ij} = \begin{cases} 1 & \text{if } i = j \\ 0 & \text{if } i \neq j. \end{cases}$$

Definition 7.2 Let A be an $n \times n$ matrix over $\mathbb{F}$. An $n \times n$ matrix B such that $BA = I_n$ is called a *left inverse* of A. Similarly, an $n \times n$ matrix B such that $AB = I_n$ is called a *right inverse* of A. If $AB = BA = I_n$, then B is called an *inverse* of A and A is said to be *invertible*.

Lemma 7.3 *If a matrix A has a left inverse B and a right inverse C, then $B = C$.*

Proof Suppose that $BA = I_n$ and $AC = I_n$. Then, we obtain

$$B = BI_n = B(AC) = (BA)C = I_nC = C$$

as desired. ∎

Thus, if A has a left and a right inverse, then A is invertible and has a unique inverse, which we shall denote by A^{-1}.

Theorem 7.4 *Let A and B be two matrices of the same size over $\mathbb{F}$.*

(1) If A is invertible, so is A^{-1} and $(A^{-1})^{-1} = A$.
(2) If both A and B are invertible, so is AB and $(AB)^{-1} = B^{-1}A^{-1}$.

Proof (1) It is evident from the symmetry of the definition.
(2) We have

$$(AB)(B^{-1}A^{-1}) = A(BB^{-1})A^{-1} = AI_nA^{-1} = AA^{-1} = I_n.$$

Similarly, we obtain $(B^{-1}A^{-1})(AB) = I_n$. This completes our proof. ∎

Theorem 7.5 *If*
$$GL_n(\mathbb{F}) = \{A \in M_{n \times n} \mid A \text{ is invertible}\},$$

then $GL_n(\mathbb{F})$ is a group under the multiplication of matrices.

Proof The proof follows from Theorem 7.1, the definition of identity matrix and Theorem 7.4. ∎

$GL_n(\mathbb{F})$ is called the *general linear group*.

Definition 7.6 The following three operations on a matrix are called *elementary row operations*:

(1) Multiply a row through by a non-zero constant.
(2) Interchange two rows.
(3) Add a multiple of one row to another row.

Indeed, an elementary row operation is a special function f which associate with each matrix $A = (a_{ij})_{m \times n}$ a matrix $f(A) = (b_{ij})_{m \times n}$ such that

(1) $b_{ij} = a_{ij}$ if $i \neq k$, and $b_{kj} = ca_{kj}$,
(2) $b_{ij} = a_{ij}$ if i is different from both k and l, and $b_{kj} = a_{lj}$, $b_{lj} = a_{kj}$,
(3) $b_{ij} = a_{ij}$ if $i \neq k$, and $b_{kj} = a_{kj} + ca_{lj}$.

Lemma 7.7 *To each elementary row operation f there corresponds an elementary row operation g, of the same type as f, such that $f(g(A)) = g(f(A)) = A$, for each A.*

Proof (1) Suppose that f is the operation which multiplies the kth row of a matrix by the non-zero scalar c. Let g be the operation which multiplies row k by c^{-1}.

(2) Suppose that f is the operation which replaces row k by row k plus c times row l, where $k \neq l$. Let g be the operation which replaces row k by row k plus $-c$ times row l.

(3) If f interchanges rows k and l, let $g = f$.

In each of the above cases, we clearly have $f(g(A)) = g(f(A)) = A$, for each A. ∎

An $n \times n$ matrix is called an *elementary matrix* if it can be obtained from the identity matrix I_n by performing a single elementary row operation.

Lemma 7.8 *If the elementary matrix E results from performing a certain elementary row operation f on I_n and if A is an $m \times n$ matrix, then the product EA is the matrix that results this same row elementary operation is performed on A, i.e., $f(a) = EA$.*

Proof It is straightforward by considering the three types of elementary row operations.   ∎

Example 7.9 Consider the matrix

$$\begin{bmatrix} 1 & 5 & 4 & 1 \\ 3 & 2 & 3 & 0 \\ 1 & -1 & 3 & 1 \end{bmatrix}$$

and consider the elementary matrix

$$\begin{bmatrix} 1 & 0 & 0 \\ 0 & 1 & 0 \\ 4 & 0 & 1 \end{bmatrix}$$

which results from adding 4 times the first row of I_3 to the third row. The product EA is

$$\begin{bmatrix} 1 & 5 & 4 & 1 \\ 3 & 2 & 3 & 0 \\ 5 & 19 & 19 & 5 \end{bmatrix}.$$

Theorem 7.10 *Each elementary matrix belongs to $GL_n(\mathbb{F})$.*

Proof If A is an $n \times n$ elementary matrix, then A results from performing some row operation on I_n. Let B be the $n \times n$ matrix that results when the inverse operation is performed on I_n. Applying Lemma 7.7 and using the fact that inverse row operations cancel the effect of each other, it follows that $AB = I_n$ and $BA = I_n$. So, the elementary matrix B is the inverse of A. This yields that $A \in GL_n(\mathbb{F})$.   ∎

Definition 7.11 An $m \times n$ matrix R is called *row-reduced echelon matrix* if

(1) the first non-zero entry in each non-zero row of R is equal to 1,
(2) each column of R which contains the leading non-zero entry of some row has all its other entries 0,
(3) every row of R which has all its entries 0 occurs below every row which has a non-zero entry,
(4) if rows $1, \ldots, r$ are the non-zero rows of R, and if the leading non-zero entry of row i occurs in column k_i, $i = 1, \ldots, r$, then $k_1 < k_2 < \cdots < k_r$.

Example 7.12 (1) One example of row-reduced echelon matrix is I_n.
(2) The matrix

$$\begin{bmatrix} 0 & 1 & 2 & 0 & -3 \\ 0 & 0 & 0 & 1 & 2 \\ 0 & 0 & 0 & 0 & 0 \end{bmatrix}$$

a row-reduced echelon matrix, too.

Matrices that can be obtained from one another by a finite sequence of elementary row operations are said to be *row equivalent*. We left to the readers to show that each matrix is row equivalent to a unique row-reduced echelon matrix.

If one defines an *elementary column operation* and *column equivalent* in a manner analogous to that of elementary row operation and row equivalent, it is clear that each $m \times n$ matrix will be column equivalent to a column-reduced echelon matrix. Also, each elementary column operation will be of the form $A \rightarrow AE$, where E is an $n \times n$ elementary matrix, and so on.

Definition 7.13 A *diagonal matrix* is a square matrix in which the entries outside main diagonal are all zero. A diagonal matrix whose diagonal elements all contains the same scalar c is called a *scalar matrix*.

Corollary 7.14 *The set of all $n \times n$ invertible diagonal matrices is a subgroup of $GL_n(\mathbb{F})$.*

Definition 7.15 A matrix $A = \left(a_{ij}\right)_{n \times n}$ is called

(1) an *upper triangular* if $a_{ij} = 0$ for $i > j$.
(2) a *lower triangular* if $a_{ij} = 0$ for $i < j$.

Let $UT_n(\mathbb{F})$ be the set of upper triangular matrices such that all entries on the diagonal are non-zero.

Theorem 7.16 $UT_n(\mathbb{F})$ *is a subgroup of $GL_n(\mathbb{F})$.*

Proof Suppose that A and B are two arbitrary elements of $UT_n(\mathbb{F})$. Let $C = \left(c_{ij}\right)_{n \times n} = AB$. If $i > j$, then

$$c_{ij} = \sum_k a_{ik}b_{kj} = 0 \text{ and } c_{ii} = a_{ii}b_{ii} \neq 0,$$

which shows that $C \in UT_n(\mathbb{F})$.

Now, we notice that

(1) The elementary matrix corresponding to multiply the ith row by a non-zero constant c is the matrix with ones on the diagonal, expect in the ith row, which instead has a c, and zero everywhere else.
(2) If we consider a replacement that add a multiple of the ith row to jth row, where $i < j$, then this matrix has non-zero entries only along the diagonal and in the ijth entry, which is above the diagonal. Therefore, all entries bellow the diagonal are zero.

The above row operations are sufficient to row reduce A to I_n. Since A is upper triangular, it follows that there exists a sequence of row operations of the type described in the above that transforms A into I_n. Consequently, there is a sequence of upper triangular elementary matrices $E_1, E_2, \ldots, E_k$ such that $E_k \ldots E_2 E_1 A = I_n$, or equivalently

$$A^{-1} = E_k \ldots E_2 E_1. \tag{7.1}$$

The right side of (7.1) is a product of upper triangular matrices. Hence, the result of this product is also an upper triangular matrix. Hence, we conclude that the inverse of A and lies in $UT_n(\mathbb{F})$. This completes our proof. ∎

Let $LT_n(\mathbb{F})$ be the set of lower triangular matrices such that all entries on the diagonal are non-zero.

Theorem 7.17 $LT_n(\mathbb{F})$ *is a subgroup of* $GL_n(\mathbb{F})$.

Proof The proof is similar to the proof of Theorem 7.16. ∎

Theorem 7.18 *If A is n × n matrix, then the following are equivalent:*

(1) $A \in GL_n(\mathbb{F})$,
(2) *A is row equivalent to* I_n,
(3) *A is a product of elementary matrices.*

Proof Suppose that R is a row-reduced echelon matrix which is row equivalent to A. Then, by Lemma 7.8, there exist elementary matrices $E_1, \ldots, E_k$ such that

$$R = E_k \ldots E_1 A$$

Since each E_i is invertible, it follows that

$$A = E_1^{-1} \ldots E_k^{-1} R.$$

Since products of invertible matrices are invertible, we conclude that A is invertible if and only if R is invertible. Since R is a square row-reduced echelon matrix, it follows that R is invertible if and only if each row of R contains non-zero entry, that is, if and only if $R = I_n$. We have now shown that

$$A \text{ is invertible} \Leftrightarrow R = I,$$

and if $R = I$ then $A = E_1^{-1} \ldots E_k^{-1}$. This proves that (1), (2) and (3) are equivalent statements about A. ∎

Corollary 7.19 *The general linear group* $GL_n(\mathbb{F})$ *is generated by elementary matrices.*

Corollary 7.20 *Let A be an invertible matrix. If a sequence of elementary row operations reduces A to the identity, then that same sequence of operations when applied to* I_n *yields* A^{-1}.

Example 7.21 We want to find the inverse of

$$A = \begin{bmatrix} 1 & 1 & 2 \\ 3 & 4 & 5 \\ 2 & 6 & -1 \end{bmatrix},$$

if exists. If we are able to convert matrix A to the identity matrix by using elementary row operations, i.e.,

$$[A \mid I_n] \longrightarrow [I_n \mid B],$$

then A is invertible and $A^{-1} = B$. The computation can be carried out as follows:

$$\begin{bmatrix} 1 & 1 & 2 & 1 & 0 & 0 \\ 3 & 4 & 5 & 0 & 1 & 0 \\ 2 & 6 & -1 & 0 & 0 & 1 \end{bmatrix} \longrightarrow \begin{bmatrix} 1 & 1 & 2 & 1 & 0 & 0 \\ 0 & 1 & -1 & -3 & 1 & 0 \\ 2 & 6 & -1 & 0 & 0 & 1 \end{bmatrix}$$

$$\longrightarrow \begin{bmatrix} 1 & 1 & 2 & 1 & 0 & 0 \\ 0 & 1 & -1 & -3 & 1 & 0 \\ 0 & 4 & -5 & -2 & 0 & 1 \end{bmatrix} \longrightarrow \begin{bmatrix} 1 & 0 & 3 & 4 & -1 & 0 \\ 0 & 1 & -1 & -3 & 1 & 0 \\ 0 & 4 & -5 & -2 & 0 & 1 \end{bmatrix}$$

$$\longrightarrow \begin{bmatrix} 1 & 0 & 3 & 4 & -1 & 0 \\ 0 & 1 & -1 & -3 & 1 & 0 \\ 0 & 0 & -1 & 10 & -4 & 1 \end{bmatrix} \longrightarrow \begin{bmatrix} 1 & 0 & 0 & 34 & -13 & 3 \\ 0 & 1 & -1 & -3 & 1 & 0 \\ 0 & 0 & -1 & 10 & -4 & 1 \end{bmatrix}$$

$$\longrightarrow \begin{bmatrix} 1 & 0 & 0 & 34 & -13 & 3 \\ 0 & 1 & -1 & -3 & 1 & 0 \\ 0 & 0 & 1 & -10 & 4 & -1 \end{bmatrix} \longrightarrow \begin{bmatrix} 1 & 0 & 0 & 34 & -13 & 3 \\ 0 & 1 & 0 & -13 & 5 & -1 \\ 0 & 0 & 1 & -10 & 4 & -1 \end{bmatrix}.$$

Therefore, we get

$$A^{-1} = \begin{bmatrix} 34 & -13 & 3 \\ -13 & 5 & -1 \\ -10 & 4 & -1 \end{bmatrix}.$$

Definition 7.22 The *determinant* of a matrix A, written $\det(A)$, is a certain number associated to A. If $A = (a_{ij})_{n \times n}$, then determinant of A is defined by the following:

$$\det(A) = \sum_{\sigma \in S_n} \mathrm{sgn}(\sigma) \prod_{i=1}^{n} a_{i(i\sigma)}.$$

Theorem 7.23 *If $n = 1$, then $\det(A) = a_{11}$. If $n > 1$, then*

$$\det(A) = \sum_{j=1}^{n} a_{ij} A_{ij}, \quad i = 1, \ldots, n \tag{7.2}$$

and

$$\det(A) = \sum_{i=1}^{n} a_{ij} A_{ij}, \quad j = 1, \ldots, n, \tag{7.3}$$

such that A_{ij} is the ij-cofactor associated with A, i.e.,

$$A_{ij} = (-1)^{i+j} \det(M_{ij}),$$

where M_{ij} is $n-1) \times (n-1)$ matrix obtained from A by removing its ith row and jth column. The M_{ij}'s are called the minors of A.

Proof The proof follows from the expansion of the right sides of (7.2), (7.3) and using the definition of determinant.　　　　　　　　　　　　　　　　　　　　■

Formula (7.2) is called the *expansion of the determinant through the ith row* and Formula (7.3) is called the *expansion of the determinant through the jth column*. An alternative notation for the determinant of a matrix A is $|A|$ rather than $\det(A)$.

Example 7.24 Evaluate $\det(A)$, where

$$A = \begin{bmatrix} 2 & 3 & 3 \\ 4 & 5 & -1 \\ 1 & 2 & 3 \end{bmatrix}.$$

We have

$$\det(A) = 2 \begin{vmatrix} 5 & -1 \\ 2 & 3 \end{vmatrix} - 3 \begin{vmatrix} 4 & -1 \\ 1 & 3 \end{vmatrix} + 3 \begin{vmatrix} 4 & 5 \\ 1 & 2 \end{vmatrix}$$
$$= 2(5 \cdot 3 - (-1) \cdot 2) - 3(4 \cdot 3 - (-1) \cdot 1) + 3(4 \cdot 2 - 5 \cdot 1) = 4.$$

Lemma 7.25 *Let A be any $n \times n$ matrix.*

(1) If B is the matrix that results when a single row of A is multiplied by a constant c, then $\det(B) = c \det(A)$.
(2) If B is the matrix that results when two rows of A are interchanged, then $\det(B) = -\det(A)$.
(3) If B is the matrix that results when a multiple of one row of A is added to another row, then $\det(B) = \det(A)$.

Proof It is straightforward.　　　　　　　　　　　　　　　　　　　　　　■

Lemma 7.26 *If E is an $n \times n$ elementary matrix and A is any $n \times n$ matrix, then*

$$\det(EA) = \det(E) \det(A).$$

Proof We consider the three different types of elementary row operation.

(1) Let E be the elementary matrix obtained from I_n by multiplying the entries of some row of I_n by a non-zero scalar c. Then, we can row reduce E to I_n by multiplying the same row by $\frac{1}{c}$. Hence, we obtain $\det(I_n) = \frac{1}{c} \det(E)$, or equivalently $\det(E) = c$. Since EA is the matrix resulting from multiplying the entries of a row of A by c, it follows that $\det(EA) = c \det(A) = \det(E) \det(A)$.

(2) Let E be the elementary matrix obtained from I_n by interchanging two rows

of I_n. Then, $\det(E) = -\det(I_n) = -1$. Since EA is the matrix resulting from interchanging the corresponding two rows of A, it follows that $\det(EA) = (-1)\det(A)$, and so $\det(EA) = \det(E)\det(A)$.

(3) Let E be the elementary matrix obtained from I_n by adding a multiple of one row of I_n to another row of I_n. Since row operations of this type do not change the determinant, it follows that $\det(E) = \det(I_n) = 1$. Since EA is the result of adding a multiple of a row of A to another row of A, it follows that $\det(EA) = 1 \cdot \det(A) = \det(E)\det(A)$.

So, no matter what type of elementary matrix E is, we have shown that $\det(EA) = \det(E)\det(A)$. Now, we are done. ∎

Theorem 7.27 *A square matrix A is invertible if and only if $\det(A) \neq 0$.*

Proof Suppose that $E_1, \ldots, E_k$ are elementary matrices which place A in reduced row echelon matrix R. Then, $R = E_k \ldots E_1 A$. By Lemma 7.26, we have $\det(R) = \det(E_k) \ldots \det(E_1)\det(A)$. Since the determinant of an elementary matrix is non-zero, it follows that $\det(A)$ and $\det(R)$ are either both zero or both non-zero.

Now, if A is invertible, then the reduced row echelon form of A is I_n. In this case, $\det(R) = \det(I_n) = 1$ which implies that $\det(A) \neq 0$.

Conversely, if $\det(A) \neq 0$, then $\det(R) \neq 0$ which implies that R can not have a row of all zeros. This implies that $R = I_n$ and so A is invertible. ∎

Corollary 7.28 *We have*

$$GL_n(\mathbb{F}) = \{A \in M_{n \times n} \mid \det(A) \neq 0\}.$$

Theorem 7.29 *A square matrix $A = (a_{ij})_{n \times n}$ is invertible if and only if $AX = 0$, i.e.,*

$$\begin{bmatrix} a_{11} & a_{12} & \ldots & a_{1n} \\ a_{21} & a_{22} & \ldots & a_{2n} \\ \vdots & \vdots & \ddots & \vdots \\ a_{n1} & a_{n2} & \ldots & a_{nn} \end{bmatrix} \begin{bmatrix} x_1 \\ x_2 \\ \vdots \\ x_n \end{bmatrix} = \begin{bmatrix} 0 \\ 0 \\ \vdots \\ 0 \end{bmatrix}$$

has only the trivial solution.

Proof Suppose that A is invertible and X_0 be any solution to $AX = 0$. Thus, $AX_0 = 0$. Multiplying both sides of this equation by A^{-1} gives $A^{-1}(AX_0) = A^{-1}0$, or equivalently $(A^{-1}A)X_0 = 0$. This implies that $X_0 = 0$. Thus, $AX = 0$ has only the trivial solution.

Conversely, if $AX = 0$ has only the trivial solution, then A must be row equivalent to a reduced row echelon matrix R with n leading 1s. Hence, $R = I_n$. Now, by Theorem 7.18 our proof completes. ∎

Definition 7.30 Let A be a square matrix and $V_1, V_2, \ldots, V_n$ be the columns of A. We say that $V_1, V_2, \ldots, V_n$ are *linearly independent* if and only if the only solution for

$$x_1 V_1 + x_2 V_2 + \cdots + x_n V_n = 0 \quad (\text{with } x_i \in \mathbb{F})$$

is $x_1 = x_2 = \cdots = x_n = 0$. Note that the equation $x_1 V_1 + x_2 V_2 + \cdots + x_n V_n = 0$ can be rewrite as $AX = 0$.

Corollary 7.31 *A square matrix A is invertible if and only if the columns of A are linearly independent.*

Proof The proof results from Theorem 7.29 and Definition 7.30.   ∎

Theorem 7.32 *If $\mathbb{F}$ is a finite field with q elements, then*

$$|GL_n(\mathbb{F})| = \prod_{k=0}^{n-1} (q^n - q^k).$$

Proof We count $n \times n$ matrices whose columns are linearly independent. We can do this by building up a matrix from scratch. The first column can be anything other than the zero column, so there exist $q^n - 1$ possibilities. The second column must be linearly independent from the first column, which is to say that it must not be a multiple of the first column. Since there exist q multiples of the first column, it follows that there exist $q^n - q$ possibilities for the second column. In general, the ith column must be linearly independent from the first $i - 1$ columns, which means that it can not be a linear combination of the first $i - 1$ columns. Since there exist q^{i-1} linear combinations of the first $i - 1$ columns, it follows that there exist $q^n - q^{i-1}$ possibilities for the ith column. Once we build the entire matrix this way, we know that the column are all linearly independent by choice. Moreover, we can build any $n \times n$ matrix whose columns are linearly independent in this way. Consequently, there exist

$$(q^n - 1)(q^n - q) \ldots (q^n - q^{n-1})$$

matrices.   ∎

Theorem 7.33 *The center of $GL_n(\mathbb{F})$ is*

$$Z\big(GL_n(\mathbb{F})\big) = \{aI_n \mid a \in \mathbb{F}^*\}.$$

This means that the center of general linear group is the subgroup comprising scalar matrices.

Proof We suppose that $n > 1$. Each elements of $Z\big(GL_n(\mathbb{F})\big)$ must commute with every elements of $GL_n(\mathbb{F})$. In particular must commute with elementary matrices. We complete the our proof in four steps.

   Step 1: Each element of $Z\big(GL_n(\mathbb{F})\big)$ commutes with off-diagonal matrix units. Indeed, assume that $c \in \mathbb{F}^*$ and $e_{ij}(c)$ is a matrix with c in ijth entry and zeroes elsewhere. Specially, $e_{ij}(1)$ is called the ijth *matrix unit*. Now, we define $E_{ij}(c) = I_n + e_{ij}(c)$. Then, we have

(1) Since $E_{ij}(-c)$ is the inverse of $E_{ij}(c)$, it follows that $E_{ij}(c) \in GL_n(\mathbb{F})$.

(2) Any matrix that commutes with $E_{ij}(1)$ must commute with $e_{ij}(1)$

Step 2: If a matrix commutes with off-diagonal matrix units, then it is a diagonal matrix. Indeed, let A be a matrix such that $a_{ji} \neq 0$ for some $i \neq j$ and let $B = e_{ij}(1)$. Then, jjth entry of AB is non-zero, while the jjth entry of BA is zero. Consequently, any matrix commutes with all of off-diagonal matrix units $e_{ij}(1)$ can not have any off-diagonal entries.

Step 3: A *permutation matrix* is a matrix obtained by permuting the rows of I_n according to some permutation of the numbers 1 to n. A permutation matrix is invertible. Let A be a diagonal matrix with $a_{ii} \neq a_{jj}$ and let B be a permutation matrix obtained by permuting the ith and jth rows of I_n. Then, A does not commute with B. Therefore, we conclude that any diagonal matrix that commutes with all permutation matrices is scalar.

Step 4: Combining the first two steps yields that any matrix in $Z(GL_n(\mathbb{F}))$ must be diagonal, and the third step then yields that it must be scalar. ∎

Corollary 7.34 *For any $n \times n$ matrices A and B,*

$$AB = I_n \iff BA = I_n.$$

Proof It is enough to prove that $BA = I_n$ implies that $AB = I_n$. Let $BA = I_n$. If $AX = 0$, then $B(AX) = B0 = 0$. Hence, $(BA)X = 0$, or equivalently $I_n X = 0$. This implies that $X = 0$. Now, by Theorem 7.29, we conclude that A is invertible. Since $BA = I_n$, it follows that

$$A(BA)A^{-1} = AI_n A^{-1} = I_n.$$

This implies that $(AB)AA^{-1} = I_n$. Thus, we conclude that $AB = I_n$. ∎

Corollary 7.35 *Let A and B be two $n \times n$ matrices. If AB is invertible, then A and B are invertible.*

Proof Suppose that $C = B(AB)^{-1}$ and $D = (AB)^{-1}A$. Then, we obtain

$$AC = A(B(AB)^{-1}) = (AB)(AB)^{-1} = I_n,$$
$$DB = ((AB)^{-1}A)B = (AB)^{-1}(AB) = I_n.$$

Now, by Corollary 7.34, we deduce that $C = A^{-1}$ and $D = B^{-1}$. ∎

The result in Lemma 7.26 can be generalized to any two $n \times n$ matrices.

Theorem 7.36 *If A and B are two $n \times n$ matrices, then*

$$\det(AB) = \det(A)\det(B).$$

Proof We consider two cases:

(1) If A is invertible, then there exist elementary matrices $E_1, \ldots, E_k$ such that $A = E_1 \ldots E_k$. Then, we have

$$
\begin{aligned}
\det(AB) &= \det(E_1 \ldots E_k B) \\
&= \det(E_1) \det(E_2 \ldots E_k B) \\
&\vdots \\
&= \det(E_1) \ldots \det(E_k) \det(B) \\
&= \det(E_1 E_2) \det(E_3) \ldots \det(E_k) \det(B) \\
&\vdots \\
&= \det(E_1 \ldots E_k) \det(B) \\
&= \det(A) \det(B).
\end{aligned}
$$

(2) If A is not invertible, by Corollary 7.35, AB is not invertible. This yield that $\det(AB) = 0 = \det(A) \det(B)$, and so the theorem holds.   ■

Corollary 7.37 *If A is $n \times n$ invertible matrix, then*

$$
\det(A^{-1}) = \frac{1}{\det(A)}.
$$

Proof Since $AA^{-1} = I_n$, it follows that $\det(AA^{-1}) = \det(I_n)$ and so $\det(A^{-1})$ $\det(A) = 1$. Since $\det(A) \neq 0$, the proof can be completed by dividing through by $\det(A)$.   ■

Theorem 7.38 *If*

$$
SL_n(\mathbb{F}) = \{A \in GL_n(\mathbb{F}) \mid \det(A) = 1\},
$$

then $SL_n(\mathbb{F})$ is a subgroup of $GL_n(\mathbb{F})$.

Proof Let $A, B \in SL_n(\mathbb{F})$ be arbitrary. Since $\det(A) = \det(B) = 1$, by Theorem 7.36 we conclude that $\det(AB) = 1$. This implies that $AB \in SL_n(\mathbb{F})$. Moreover, by Corollary 7.37, $\det(A^{-1}) = 1$, and hence $A^{-1} \in SL_n(\mathbb{F})$.   ■

$SL_n(\mathbb{F})$ is called the *special linear group*.

Theorem 7.39 *The center of $SL_n(\mathbb{F})$ is*

$$
Z\big(SL_n(\mathbb{F})\big) = \{aI_n \mid a^n = 1\}.
$$

Proof The proof results from the definition of $SL_n(\mathbb{F})$ and Theorem 7.33.   ■

Theorem 7.40 *The special linear group $SL_2(\mathbb{F})$ is generated by matrices of the form*

$$\begin{bmatrix} 1 & 0 \\ t & 1 \end{bmatrix} \text{ and } \begin{bmatrix} 1 & r \\ 0 & 1 \end{bmatrix}$$

for all $r, s \in \mathbb{F}^*$.

Proof Suppose that

$$\begin{bmatrix} a & b \\ c & d \end{bmatrix}$$

is an arbitrary element of $SL_2(\mathbb{F})$. We consider the following three cases:

Case 1: If $b \neq 0$, then

$$\begin{bmatrix} a & b \\ c & d \end{bmatrix} = \begin{bmatrix} 1 & 0 \\ (d-1)b^{-1} & 1 \end{bmatrix} \begin{bmatrix} 1 & b \\ 0 & 1 \end{bmatrix} \begin{bmatrix} 1 & 0 \\ (a-1)b^{-1} & 1 \end{bmatrix}.$$

Case 2: If $c \neq 0$, then

$$\begin{bmatrix} a & b \\ c & d \end{bmatrix} = \begin{bmatrix} 1 & (a-1)c^{-1} \\ 0 & 1 \end{bmatrix} \begin{bmatrix} 1 & 0 \\ c & 1 \end{bmatrix} \begin{bmatrix} 1 & (d-1)c^{-1} \\ 0 & 1 \end{bmatrix}.$$

Case 3: If $b = c = 0$, then

$$\begin{bmatrix} a & 0 \\ 0 & a^{-1} \end{bmatrix} = \begin{bmatrix} 1 & 0 \\ (1-a)a^{-1} & 1 \end{bmatrix} \begin{bmatrix} 1 & 1 \\ 0 & 1 \end{bmatrix} \begin{bmatrix} 1 & 0 \\ a-1 & 1 \end{bmatrix} \begin{bmatrix} 1 & -a^{-1} \\ 0 & 1 \end{bmatrix},$$

and we are done. ∎

Let $e_{ij}(\lambda)$ is a matrix with λ in ijth entry and zeroes elsewhere, and let $E_{ij}(\lambda) = I_n + e_{ij}(\lambda)$.

Theorem 7.41 *The special linear group $SL_n(\mathbb{F})$ is generated by matrices of the form $E_{ij}(\lambda)$ with $i \neq j$ and $\lambda \in \mathbb{F}^*$.*

Proof We apply mathematical induction. If $n = 2$, then by Theorem 7.40, we are done. Suppose that the statement is true for each $(n-1) \times (n-1)$ matrix. Let A be any arbitrary matrix in $SL_n(\mathbb{F})$. Multiplying A by $E_{ij}(\lambda)$ on the left or right is an elementary row or column operation:

$$E_{ij}(\lambda)A = \begin{bmatrix} a_{11} & \cdots & a_{1n} \\ \vdots & \vdots & \vdots \\ a_{i1} + \lambda a_{j1} & \cdots & a_{in} + \lambda a_{jn} \\ \vdots & \vdots & \vdots \\ a_{n1} & \cdots & a_{nn} \end{bmatrix}$$

and

$$AE_{ij}(\lambda) = \begin{bmatrix} a_{11} & \cdots & a_{1j} + \lambda a_{1i} & \cdots & a_{1n} \\ \vdots & \vdots & \vdots & & \vdots \\ a_{n1} & \cdots & a_{nj} + \lambda a_{ni} & \cdots & a_{nn} \end{bmatrix}.$$

These are the only operations we may use. Since $\det(A) = 1$, it follows that some entry of the first column of A is not 0. If $a_{k1} \neq 0$ with $k > 1$, then

$$E_{1k}\left(\frac{1 - a_{11}}{a_{k1}}\right) A = \begin{bmatrix} 1 & \cdots \\ \vdots & \ddots \end{bmatrix}. \tag{7.4}$$

If $a_{21}, \ldots, a_{n1}$ are all 0, then $a_{11} \neq 0$ and

$$E_{21}\left(\frac{1}{a_{11}}\right) A = \begin{bmatrix} a_{11} & \cdots \\ 1 & \cdots \\ \vdots & \ddots \end{bmatrix}.$$

Hence, by (7.4) with $k = 2$, we obtain

$$E_{12}(1 - a_{11}) E_{21}\left(\frac{1}{a_{11}}\right) A = \begin{bmatrix} 1 & \cdots \\ \vdots & \ddots \end{bmatrix}.$$

When we have a matrix with upper left entry 1, multiplying it on the left by $E_{i1}(\lambda)$ for $i \neq 1$ will all λ to the $1i$th entry, hence with a suitable λ, we can make the $i1$th entry of the matrix 0. Consequently, multiplication on the left by suitable matrices of the form $E_{ij}(\lambda)$ produces a block matrix

$$\begin{bmatrix} 1 & * \\ 0 & B \end{bmatrix}$$

whose first column is all 0's except for the upper left entry, which is 1. Multiplying this matrix on the right by $E_{1j}(\lambda)$ for $j \neq 1$ adds λ to the $1j$th entry without changing a column other than the jth column. With a suitable choice of λ we can make the $1j$th entry equal to 0, and carrying this out for $j = 2, \ldots, n$ leads to a block matrix

$$\begin{bmatrix} 1 & 0 \\ 0 & A' \end{bmatrix}. \tag{7.5}$$

Since this matrix is in $SL_n(\mathbb{F})$, it follows that $\det(A') = 1$. This implies that $A' \in SL_{n-1}(\mathbb{F})$. By induction hypothesis, A' is a product of elementary matrices $E_{ij}(\lambda)_{(n-1) \times (n-1)}$. We say

$$A' = E_1 E_2 \ldots E_r.$$

So, we have

$$\begin{bmatrix} 1 & 0 \\ 0 & A' \end{bmatrix} = \begin{bmatrix} 1 & 0 \\ 0 & E_1 E_2 \dots E_r \end{bmatrix} = \begin{bmatrix} 1 & 0 \\ 0 & E_1 \end{bmatrix}\begin{bmatrix} 1 & 0 \\ 0 & E_2 \end{bmatrix} \cdots \begin{bmatrix} 1 & 0 \\ 0 & E_r \end{bmatrix}.$$

This yields that A is a product of $n \times n$ matrices $E_{ij}(\lambda)$.   ∎

Definition 7.42 Let A be an $m \times n$ matrix. Then, A^t, the *transpose* of A, is the matrix obtained by interchanging the rows and columns of A. Geometrically, A^t is obtained from A by reflecting across the diagonal of A.

We say A is *symmetric* if $A^t = A$ and A is *skew-symmetric* if $A^t = -A$.

Assuming that the sizes of the matrices are such that the operations can be performed, the transpose operation has the following properties:

(1) $(A^t)^t = A$,
(2) $(A + B)^t = A^t + B^t$,
(3) $(cA)^t = cA^t$, where $c \in \mathbb{F}$,
(4) $(AB)^t = B^t A^t$.

Corollary 7.43 *If A is any square matrix, then $\det(A) = \det(A^t)$.*

Corollary 7.44 *Any orthogonal matrix has determinant equal to 1 or -1.*

Theorem 7.45 *If*
$$O_n(\mathbb{F}) = \{A \in GL_n(\mathbb{F}) \mid A^t A = I_n\},$$

then $O_n(\mathbb{F})$ is a subgroup of $GL_n(\mathbb{F})$.

Proof Let A and B be two arbitrary elements of $O_n(\mathbb{F})$. Then, we have $A^t A = I_n$ and $B^t B = I_n$. This implies that

$$(AB)^t(AB) = B^t A^t A B = B^t I_n B = B^t B = I_n.$$

So, we conclude that $AB \in O_n(\mathbb{F})$. Moreover, we have $A^t = A^{-1}$, and so

$$(A^{-1})^t A^{-1} = A A^{-1} = I_n.$$

This implies that $A^{-1} \in O_n(\mathbb{F})$.   ∎

$O_n(\mathbb{F})$ is called the *orthogonal group*. We can define

$$SO_n(\mathbb{F}) = SL_n(\mathbb{F}) \cap O_n(\mathbb{F}).$$

$SO_n(\mathbb{F})$ is called the *special orthogonal group*.
One can gets a whole of examples by fixing $Q \in M_n(\mathbb{F})$ and defining

$$G_Q(\mathbb{F}) = \{A \in GL_n(\mathbb{F}) \mid A^t Q A = Q\}.$$

Theorem 7.46 $G_Q(\mathbb{F})$ *is a subgroup of* $GL_n(\mathbb{F})$.

Proof Suppose that A and B are two arbitrary elements of $G_Q(\mathbb{F})$. Then we have $AB \in G_Q(\mathbb{F})$. Indeed,

$$(AB)^t Q(AB) = B^t A^t Q A B = B^t Q B = Q.$$

In addition, we have $A^{-1} \in G_Q(\mathbb{F})$. Indeed, when multiplying the equality $Q = A^t Q A$ to the right by A^{-1} and to the left by $(A^{-1})^t$, then we obtain $(A^{-1})^t Q A^{-1} = Q$. ∎

One important case is when $n = 2m$ and we take Q to be the matrix

$$J_m = \begin{bmatrix} 0 & -I_m \\ I_m & 0 \end{bmatrix}.$$

Then, $G_Q(\mathbb{F})$ is denoted by $SP_{2m}(\mathbb{F})$ and is called *symplectic group*

Exercises

1. If A is a symmetric matrix, is the matrix A^{-1} symmetric?
2. Show that the non-zero elements of $Mat_{n \times n}(\mathbb{C})$ is not group under multiplication.
3. For what values of a and b the following matrix belong to $GL_4(\mathbb{R})$:

$$\begin{bmatrix} b & 0 & a & 0 \\ 0 & 0 & b & a \\ 0 & a & 0 & b \\ b & a & 0 & 0 \end{bmatrix}.$$

4. The *trace* of a square matrix $A = (a_{ij})_{n \times n}$ is

$$tr(A) = \sum_{i=1}^{n} a_{ii}.$$

Let A and B be two square matrices.

(a) Prove that $tr(AB) = tr(BA)$;
(b) If B is invertible, show that the formula $tr(B^{-1}AB) = tr(A)$.

5. Let A be a 2×2 matrix over a field $\mathbb{F}$. Prove that $\det(I_2 + A) = 1 + \det(A)$ if and only if $tr(A) = 0$.
6. Let A be an $n \times n$ matrix over a field $\mathbb{F}$. Prove that there are at most n distinct scalars $c \in \mathbb{F}$ such that $\det(cI_2 - A) == 0$.

7. If A and B are invertible matrices and the involved partitioned products are defined, show that

$$\begin{bmatrix} A & B \\ C & 0 \end{bmatrix}^{-1} = \begin{bmatrix} 0 & C^{-1} \\ B^{-1} & -B^{-1}AC^{-1} \end{bmatrix}.$$

8. Show that the non-zero elements of $Mat_{n \times n}(\mathbb{C})$ is not group under multiplication.
9. Let $A = (a_{ij})_{n \times n}$ be a square matrix with $a_{ij} = \max\{i, j\}$. Compute $\det(A)$.
10. Let $A = (a_{ij})_{n \times n}$ be a square matrix with $a_{ij} = 1/(i + j)$. Show that A is invertible.
11. Show that the set

$$G = \left\{ \begin{bmatrix} \cosh x & \sinh x \\ \sinh x & \cosh x \end{bmatrix} \mid x \in \mathbb{R} \right\},$$

in which sinh and cosh are hyperbolic functions, is a group under multiplication of matrices.
12. Show by example that if $AC = BC$, then it does not follow that $A = B$. However, show that if C is invertible the conclusion $A = B$ is valid.
13. Show that $SL_2(\mathbb{C}) = SP_2(\mathbb{C})$.
14. Does the set $\{A \in GL_2(\mathbb{R}) \mid A^2 = I_2\}$ form a subgroup of $GL_2(\mathbb{R})$?
15. Let A be a 2×2 matrix over a field $\mathbb{F}$, and suppose that $A^2 = 0$. Show that for each scalar c that $\det(cI_2 - A) = c^2$.
16. Determine the *one-dimensional Lorentz group*. That is, all 2×2 matrices A such that

$$A^t \begin{bmatrix} 1 & 0 \\ 0 & -1 \end{bmatrix} A = \begin{bmatrix} 1 & 0 \\ 0 & -1 \end{bmatrix}.$$

17. Determine $GL_2(\mathbb{Z}_2)$, $SL_2(\mathbb{Z}_2)$, $O_2(\mathbb{Z}_2)$, $SO_2(\mathbb{Z}_2)$, and $SP_2(\mathbb{Z}_2)$.
18. (a) Let G be the group of all 2×2 matrices $\begin{bmatrix} a & b \\ c & d \end{bmatrix}$, where $ad - bc \neq 0$ and a, b, c, d are integers modulo 3. Show that $|G| = 48$;
    (b) If we modify the example of G in part (a) by insisting that $ad - bc = 1$, then what is $|G|$?
19. (a) Let G be the group of all 2×2 matrices $\begin{bmatrix} a & b \\ c & d \end{bmatrix}$, where a, b, c, d are integers modulo p, p is a prime number, such that $ad - bc \neq 0$; indeed, $G = GL_2(\mathbb{Z}_p)$. What is $|G|$?
    (b) Let H be the subgroup of G of part (a) defined by

$$\left\{ \begin{bmatrix} a & b \\ c & d \end{bmatrix} \mid ad - bc = 1 \right\};$$

indeed, $H = SL_2(\mathbb{Z}_p)$. What is $|H|$?
20. Let

$$H = \left\{ \begin{bmatrix} 1 & a & b \\ 1 & 1 & c \\ 0 & 0 & 1 \end{bmatrix} \mid a, b, c \in \mathbb{Z}_3 \right\}.$$

(a) Show that H is a subgroup of $SL_3(\mathbb{Z}_3)$;
(b) How many elements does H have?
(c) Find three subgroups of H of order 9. Are these subgroups of order 9 abelian? Is H abelian?

21. Find the centralizer of $\begin{bmatrix} 1 & 0 \\ 1 & 1 \end{bmatrix}$ in $GL_2(\mathbb{R})$.

22. Let $A = \begin{bmatrix} 0 & -1 \\ 1 & 0 \end{bmatrix}$ and $B = \begin{bmatrix} 0 & 1 \\ -1 & -1 \end{bmatrix}$ be two elements of $SL_2(\mathbb{Q})$. Show that $o(A) = 4$, $o(B) = 3$, but AB has infinite order.

23. Let $A \in Mat_{2 \times 2}(\mathbb{C})$. Show that there is no matrix solution $B \in Mat_{2 \times 2}(\mathbb{C})$ to $AB - BA = I_2$. What can you say about the same problem with $A, B \in Mat_{n \times n}(\mathbb{C})$?

24. Let $\lambda \in \mathbb{F}$. Prove that the matrix $E_{ij}(\lambda)$ is of order p if and only if the characteristic of $\mathbb{F}$ is p.

7.2   More About Vectors in $\mathbb{R}^n$

Vectors are the way we represent points in two, three, or n dimensional space. A *vector* can be considered as an object which has a length and a direction, or it can be thought of as a point in space, with coordinates representing that point. In terms of coordinates, we use the notation

$$X = \begin{bmatrix} x \\ y \end{bmatrix}$$

for a column vector. This vector denotes the point in the xy-plane which is x units in the horizontal direction and y units in the vertical direction from the origin. In n dimensional space, we represent a vector by

$$X = \begin{bmatrix} x_1 \\ x_2 \\ \vdots \\ x_n \end{bmatrix}.$$

Let $X = [x_1 \ x_2 \ \ldots \ x_n]^t$ and $Y = [y_1 \ y_2 \ \ldots \ y_n]^t$ be two vectors in $\mathbb{R}^n$. We define the *inner product* $\langle X, Y \rangle$ by

$$\langle X, Y \rangle = x_1 y_1 + x_2 y_2 + \cdots x_n y_n.$$

The *length* of X is given by

$$\|X\| = \sqrt{X \cdot X} = \sqrt{x_1^2 + x_2^2 + \cdots x_n^2}.$$

The vector $U = \frac{X}{\|X\|}$ is called the *unit vector* in the X-direction. The angle θ between two vectors X and Y is calculated by

$$\langle X, Y \rangle = \|X\| \|Y\| \cos \theta,$$

where $0 \le \theta \le \pi$. Two vector X and Y are said to be *orthogonal* if $\langle X, Y \rangle = 0$.

Definition 7.47 Non-zero vectors $X_1, X_2, \ldots, X_k$ in $\mathbb{R}^n$ form an *orthogonal set* if they are orthogonal to each other, i.e., $\langle X_i, X_j \rangle = 0$ for all $i \ne j$. An *orthonormal set* is an orthogonal set with the additional property that all vectors are unit, i.e., $\|X_i\| = 1$ for all $i = 1, \ldots, k$.

Lemma 7.48 *Any orthogonal set S is linearly independent.*

Proof Let $X_1, X_2, \ldots, X_k$ be distinct vectors in S and

$$Y = c_1 X_1 + c_2 X_2 + \cdots + c_k X_k.$$

Then, we have

$$\langle Y, X_j \rangle = \langle \sum_{i=1}^{k} c_i X_j, X_i \rangle = c_j \langle X_j, X_j \rangle = c_j \|X_j\|^2.$$

Since $\langle X_j, X_j \rangle \ne 0$, it follows that

$$c_j = \frac{\langle Y, X_j \rangle}{\|X_j\|^2}, \quad 1 \le j \le k.$$

Thus, when $Y = 0$, each $c_j = 0$. ∎

Gram-Schmidt orthogonalization process: Let $X_1, X_2, \ldots, X_k$ be any independent vectors. Then one may construct orthogonal vectors $V_1, V_2, \ldots, V_k$ as follows:

$$V_1 = X_1,$$

$$V_2 = X_2 - \frac{\langle X_2, V_1 \rangle}{\|V_1\|^2} V_1,$$

$$\vdots$$

$$V_j = X_j - \sum_{i=1}^{j-1} \frac{\langle X_j, V_i \rangle}{\|V_i\|^2} V_i,$$

$$\vdots$$

Fig. 7.1 The projection of
U onto V

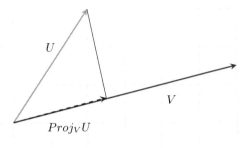

Then, $V_1, V_2, \ldots, V_k$ are orthogonal vectors.

The *vector projection* of a vector U on a non-zero vector V is the orthogonal projection of U on a straight line parallel to V, see Fig. 7.1.

Lemma 7.49 *The vector projection of a vector U on a non-zero vector V is equal to*

$$Proj_V U = \frac{V V^t}{\|V\|^2} U.$$

Proof The projection vector of U onto V is a scalar multiple of V, i.e., $Proj_V U = cV$. Since the vector $U - cV$ is perpendicular to V, it follows that $\langle V, U - cV \rangle = 0$. Hence, we have $\langle V, U \rangle - c\langle V, V \rangle = 0$. This implies that

$$c = \frac{\langle V, U \rangle}{\langle V, V \rangle}.$$

Thus, we conclude that

$$Proj_V U = \frac{\langle V, U \rangle}{\langle V, V \rangle} V. \tag{7.6}$$

When we multiply a vector by a scalar, it does not matter whether we put the scalar before or after the vector. So, by (7.6), we get

$$Proj_V U = V \frac{\langle V, U \rangle}{\langle V, V \rangle}.$$

Since $\langle V, U \rangle = V^t U$ and $\langle V, V \rangle = \|V\|^2$, it follows that

$$Proj_V U = V \frac{V^t U}{\|V\|^2} = \frac{V V^t}{\|V\|^2} U,$$

as desired.   ∎

If we define

$$P = \frac{VV^t}{\|V\|^2},$$

then the projection formula becomes $Proj_V U = PU$. The matrix P is called the *projection matrix*. So, we can project any vector onto the vector V by multiplying by the matrix P.

Exercises

1. (a) Suppose that $\theta \in \mathbb{R}$. Show that

$$(\cos\theta, \sin\theta), \quad (-\sin\theta, \cos\theta)$$

and

$$(\cos\theta, \sin\theta), \quad (\sin\theta, -\cos\theta)$$

are orthonormal bases of $\mathbb{R}^2$;
   (b) Show that each orthonormal basis of $\mathbb{R}^2$ is of the form given by one of the two possibilities of part (a).
2. Suppose that $X, Y \in \mathbb{R}^n$. Prove that $\langle X, Y \rangle = 0$ if and only if

$$\|X\| \le \|X + aY\|,$$

for all $a \in \mathbb{R}$.
3. Find vectors $X, Y \in \mathbb{R}^2$ such that X is a scalar multiple of $\begin{bmatrix} 1 \\ 3 \end{bmatrix}$, Y is orthogonal to $\begin{bmatrix} 1 \\ 3 \end{bmatrix}$, and $\begin{bmatrix} 1 \\ 2 \end{bmatrix} = X + Y$.
4. Apply Gram-Schmidt orthogonalization process to the following vectors in $\mathbb{R}^3$:

$$\begin{bmatrix} 1 \\ 2 \\ 0 \end{bmatrix}, \begin{bmatrix} 8 \\ 1 \\ -6 \end{bmatrix} \text{ and } \begin{bmatrix} 0 \\ 0 \\ 1 \end{bmatrix}.$$

5. Find the angle between a diagonal of a cube and one of its edges.
6. If $C = \|A\|B + \|B\|A$, where A, B, and C are all non-zero vectors, show that C bisects the angle between A and B.

7.3 Rotation Groups

A rotation matrix describes the rotation of an object. In this section, we derive the matrix for a general rotation.

Theorem 7.50 *Any* 2×2 *orthogonal matrix with entries in* $\mathbb{R}$ *and determinant* 1 *has the form*

$$\text{Rot}(\theta) = \begin{bmatrix} \cos\theta & -\sin\theta \\ \sin\theta & \cos\theta \end{bmatrix},$$

and represents a rotation in $\mathbb{R}^2$ *by an angle* θ *counterclockwise, with center the origin.*

Proof Suppose that

$$\begin{bmatrix} a & b \\ c & d \end{bmatrix}$$

be an orthogonal matrix with entries in $\mathbb{R}$. Then, we have

$$A^t A = \begin{bmatrix} a & c \\ b & d \end{bmatrix} \begin{bmatrix} a & b \\ c & d \end{bmatrix} = \begin{bmatrix} 1 & 0 \\ 0 & 1 \end{bmatrix}.$$

So, both columns of A are mutually orthogonal unit vectors. The only two unit vectors orthogonal to $\begin{bmatrix} a \\ c \end{bmatrix}$ are $\begin{bmatrix} -c \\ a \end{bmatrix}$ and $\begin{bmatrix} c \\ -a \end{bmatrix}$. Hence, one of these must be $\begin{bmatrix} b \\ d \end{bmatrix}$. Consequently, there exist two possibilities for A:

$$\begin{bmatrix} a & -c \\ c & a \end{bmatrix} \quad \text{or} \quad \begin{bmatrix} a & c \\ c & -a \end{bmatrix}.$$

Since $a^2 + b^2 = 1$, it follows that the first matrix has determinant 1 and the second matrix has determinant -1. If we consider the first matrix, then there exists a unique angle θ such that $\cos\theta = a$ and $\sin\theta = b$. Hence, we denote the resulting matrix A by

$$\text{Rot}(\theta) = \begin{bmatrix} \cos\theta & -\sin\theta \\ \sin\theta & \cos\theta \end{bmatrix},$$

to emphasis that it is a function of θ. Now, we obtain

$$\text{Rot}(\theta) \begin{bmatrix} x \\ y \end{bmatrix} = \begin{bmatrix} x\cos\theta - y\sin\theta \\ x\sin\theta + y\cos\theta \end{bmatrix}.$$

The right side is the image of the vector $\begin{bmatrix} x \\ y \end{bmatrix}$ under a rotation about the origin by angle θ.   ∎

Fig. 7.2 The x- and y-axes and the resulting x'- and y'-axes formed by a rotation through an angle θ

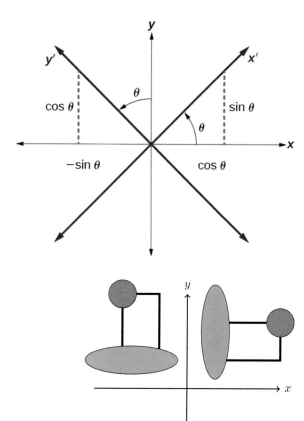

Fig. 7.3 Rotation with $\theta = \pi/2$

Indeed, according to Theorem 7.50, if the x and y axes are rotated through an angle θ, then every point on the plane may be thought of as having two representations:

(1) (x, y) on the xy plane with the original x-axis and y-axis.
(2) (x', y') on the new plane defined by the new, rotated axes, called the x'-axis and y'-axis, see Fig. 7.2.

Example 7.51 In particular, the matrix

$$\begin{bmatrix} 0 & -1 \\ 1 & 0 \end{bmatrix}$$

is used for $\theta = \pi/2$, see Fig. 7.3.

The matrix

Fig. 7.4 Rotation with
$\theta = \pi$

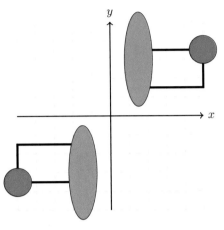

Fig. 7.5 Rotation with
$\theta = 3\pi/2$

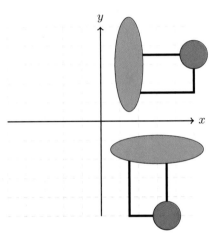

$$\begin{bmatrix} -1 & 0 \\ 0 & -1 \end{bmatrix}$$

is used for $\theta = \pi$, see Fig. 7.4.

The matrix

$$\begin{bmatrix} 0 & 1 \\ -1 & 0 \end{bmatrix}$$

is used for $\theta = 3\pi/2$, see Fig. 7.5.

Lemma 7.52 *If A is an n × n orthogonal matrix with* $\det(A) = 1$, *then*

$$\det(A - I_n) = (-1)^n \det(A - I_n).$$

Fig. 7.6 A rotation in $\mathbb{R}^3$

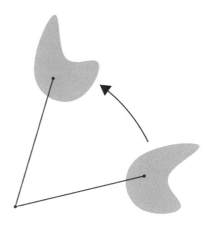

Proof We have

$$
\begin{aligned}
\det(A - I_n) &= \det(A^t)\det(A - I_n) \\
&= \det\left(A^t(A - I_n)\right) \\
&= \det(I_n - A^t) \\
&= \det\left((I_n - A)^t\right) \\
&= \det(I_n - A) \\
&= (-1)^n \det(A - I_n),
\end{aligned}
$$

as desired. ∎

Corollary 7.53 *If A is a 3×3 orthogonal matrix with $\det(A) = 1$, then there exists a non-zero column vector $X \in \mathbb{R}^3$ such that $AX = X$.*

Proof In Lemma 7.52, let $n = 3$. Then, we deduce that $\det(A - I_n) = 0$. This yields that $A - I_n$ is not invertible. So, $(A - I_n)X = 0$ has a non-zero solution. Consequently, there exists a non-zero column vector $X \in \mathbb{R}^3$ such that $AX = X$. ∎

Definition 7.54 A rotation in $\mathbb{R}^3$ is a matrix $R(e_r, \theta)$ determined by a unit vector $e_R \in \mathbb{R}^3$ and an angle θ. More precisely, $R(e_R, \theta)$ has the line through e_R as the axis and the plane perpendicular to the line is rotated by the angle θ in the counterclockwise direction (see Fig. 7.6).

Corollary 7.53 can be used to show that any 3×3 orthogonal matrix with determinant 1 represents a rotation in $\mathbb{R}^3$ about an axis passing through the origin.

Theorem 7.55 *Each rotation matrix in $\mathbb{R}^3$ lies in $SO_3(\mathbb{R})$.*

Proof Suppose that $R(e_R, \theta)$ is a rotation in $\mathbb{R}^3$, where e_R is the unit vector in the direction of the axis of rotation. By Gram-Schmidt orthogonalization process, we can find two more vectors. So, we can consider an orthonormal set $\{V_1, V_2, V_3\}$ of $\mathbb{R}^3$ such that $V_3 = e_R$. Therefore, the matrix $[V_1 \ V_2 \ V_3]$ is orthogonal. Now, by referring to Theorem 7.50, this rotation described by

$$R = R(e_R, \theta) = \begin{bmatrix} \cos\theta & -\sin\theta & 0 \\ \sin\theta & \cos\theta & 0 \\ 0 & 0 & 1 \end{bmatrix}.$$

Since $R^t R = I_3$ and $\det(R) = 1$, we conclude that $R \in SO_3(\mathbb{R})$.   ∎

Theorem 7.56 *Each matrix in $SO_3(\mathbb{R})$ is a rotation.*

Proof Suppose that $R \in SO_3(\mathbb{R})$ is an arbitrary element. By Lemma 7.53, there exists a column unit vector V_3 such that $R V_3 = V_3$. By Gram-Schmidt orthogonalization process, we can extend this to an orthonormal set $\{V_1, V_2, V_3\}$. Then, the matrix $A = [V_1\ V_2\ V_3] \in SO_3(\mathbb{R})$. Hence, we conclude that $RA \in SO_3(\mathbb{R})$. We can write

$$\begin{aligned} RV_1 &= aV_1 + bV_2, \\ RV_2 &= cV_1 + dV_2, \\ RV_3 &= V_3. \end{aligned}$$

Since

$$A^{-1}RA = \begin{bmatrix} a & b & 0 \\ c & d & 0 \\ 0 & 0 & 1 \end{bmatrix} \in SO_3(\mathbb{R}),$$

it follows that

$$A^{-1}RA = \begin{bmatrix} a & b \\ c & d \end{bmatrix} \in SO_2(\mathbb{R}),$$

which means that it is a planer rotation matrix $\mathrm{Rot}(\theta)$. Consequently, we have $R = R(V_3, \theta)$.   ∎

Corollary 7.57 *The set of rotation in $\mathbb{R}^3$ can be identified with $SO_3(\mathbb{R})$.*

Exercises

1. For any rotation matrix R, show that $R^t = R^{-1}$.
2. Use standard trigonometric identities to verify that

$$\mathrm{Rot}(\theta)\mathrm{Rot}(\phi) = \mathrm{Rot}(\theta + \phi).$$

   Deduce that $\mathrm{Rot}(2\theta) = \mathrm{Rot}(\theta)^2$ and $\mathrm{Rot}(-\theta) = \mathrm{Rot}(\theta)^{-1}$.

3. Use rotation matrix to find the image of the vector $\begin{bmatrix} -2 \\ 1 \\ 2 \end{bmatrix}$ if it is rotated

   (a) $30°$ about the x-axis;
   (b) $-30°$ about the x-axis;

(c) $45°$ about the y-axis;
(d) $-45°$ about the y-axis;
(e) $90°$ about the z-axis;
(f) $-90°$ about the z-axis.

7.4   Reflections in $\mathbb{R}^2$ and $\mathbb{R}^3$

Reflection matrix is the matrix which can be used to make reflection transformation of a figure.

Theorem 7.58 *A reflection across the line $y = mx$ is performed by matrix*

$$\text{Ref}(\theta) = \begin{bmatrix} \cos 2\theta & \sin 2\theta \\ \sin 2\theta & -\cos 2\theta \end{bmatrix},$$

in terms of an angle θ that mirror makes with the positive x-axis (Fig. 7.7).

Proof The formula for the distance of $P(x, y)$ from the line $mx - y = 0$ is

$$d = \frac{mx - y}{\sqrt{1 + m^2}},$$

where $m = \tan \theta$. Hence, we have

$$d = \frac{x \tan \theta - y}{\sqrt{1 + \tan^2 \theta}} = \frac{x \tan \theta - y}{\sec \theta} = x \sin \theta - y \cos \theta.$$

So, we conclude that

Fig. 7.7 A reflection across the line $y = mx$

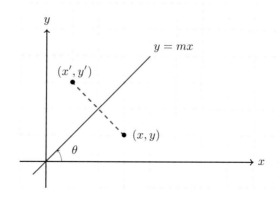

Fig. 7.8 Reflection with $\theta = \pi/2$

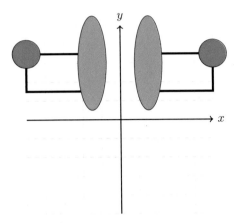

$$x' = x - 2d \sin\theta = x - 2\sin\theta(x\sin\theta - y\cos\theta)$$
$$= x(1 - 2\sin^2\theta) + 2y\sin\theta\cos\theta = x\cos 2\theta + y\sin 2\theta$$

and

$$y' = y + 2d\cos\theta = y + 2\cos\theta(x\sin\theta - y\cos\theta)$$
$$= 2x\sin\theta\cos\theta + y(1 - 2\cos^2\theta) = x\sin 2\theta - y\cos 2\theta.$$

Consequently, we obtain

$$\begin{bmatrix} x' \\ y' \end{bmatrix} = \begin{bmatrix} \cos 2\theta & \sin 2\theta \\ \sin 2\theta & -\cos 2\theta \end{bmatrix} \begin{bmatrix} x \\ y \end{bmatrix},$$

and we are done.   ∎

Example 7.59 In particular, the matrix

$$\begin{bmatrix} 0 & 1 \\ 1 & 0 \end{bmatrix}$$

is used for $\theta = \pi/2$, see Fig. 7.8. The matrix

$$\begin{bmatrix} -1 & 0 \\ 0 & -1 \end{bmatrix}$$

is used for $\theta = \pi$, see Fig. 7.9.

Corollary 7.60 *The set of reflections in $\mathbb{R}^2$ is exactly the set of elements in $O_2(\mathbb{R})$ that do not belong to $SO_2(\mathbb{R})$.*

Definition 7.61 A *reflection* in $O_3(\mathbb{R})$ is a matrix A whose effect is to send every column vector of $\mathbb{R}^3$ to its mirror image with respect to a plane S containing the

Fig. 7.9 Reflection with $\theta = \pi$

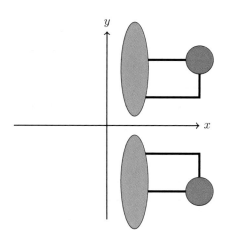

origin. More precisely, suppose that

$$S = \{ax + by + cz = 0 \mid a, b, c \in \mathbb{R}\}$$

is a plane through the origin. We say a matrix A makes a *reflection across S* if

(1) $AX = X$ for all vectors X in S,
(2) $AX = -X$ for all vectors X perpendicular to S.

If N is chosen to be a unit column vector perpendicular to S, then

$$AX = X - 2\langle X, N \rangle N, \tag{7.7}$$

for all $X \in \mathbb{R}^3$.

According to (7.7), we obtain

$$
\begin{aligned}
AX &= X - 2\langle X, N \rangle N \\
   &= X - 2N\langle X, N \rangle \\
   &= X - 2N(N^t X) \\
   &= X - 2(N N^t)X \\
   &= (I_n - 2N N^t)X.
\end{aligned}
$$

The matrix $(I_n - 2N N^t)$ is called *reflection matrix* for the plane S, and is also sometimes called a *Householder matrix*.

Example 7.62 Let $X^t = [x_1 \ x_2 \ x_3]$. Table 7.1 describes some reflections of the column vector X^t across some planes.

Example 7.63 We want to compute the reflection of the column vector

Table 7.1 Some reflections of the column vector X^t across some planes

Operator	Equations defining the image	Reflection matrix
Reflection across the xy plane	$x_1' = x + 0y + 0z$ $x_2' = 0x + y + 0z$ $x_3' = 0x + 0y - z$	$\begin{bmatrix} 1 & 0 & 0 \\ 0 & 1 & 0 \\ 0 & 0 & -1 \end{bmatrix}$
Reflection across the xz plane	$x_1' = x + 0y + 0z$ $x_2' = 0x - y + 0z$ $x_3' = 0x + 0y + z$	$\begin{bmatrix} 1 & 0 & 0 \\ 0 & -1 & 0 \\ 0 & 0 & -1 \end{bmatrix}$
Reflection across the yz plane	$x_1' = -x + 0y + 0z$ $x_2' = 0x + y + 0z$ $x_3' = 0x + 0y + z$	$\begin{bmatrix} -1 & 0 & 0 \\ 0 & 1 & 0 \\ 0 & 0 & 1 \end{bmatrix}$

$$X = \begin{bmatrix} -1 \\ 2 \\ -2 \end{bmatrix}$$

across the plane $2x - y + 3z = 0$. The column vector

$$V = \begin{bmatrix} 2 \\ -1 \\ 3 \end{bmatrix}$$

is normal to the plane and $\|V\|^2 = \langle V, V \rangle = 2^2 + (-1)^2 + 3^2 = \sqrt{14}$. So, a unit normal vector is

$$N = \frac{V}{\|V\|} = \frac{1}{\sqrt{14}} \begin{bmatrix} 2 \\ -1 \\ 3 \end{bmatrix}.$$

Consequently, the reflection matrix is equal to

$$I_n - 2NN^t = I_n - \frac{1}{7} V V^t$$

$$= I_n - \frac{1}{7} \begin{bmatrix} 2 \\ -1 \\ 3 \end{bmatrix} [2 \quad -1 \quad 3]$$

$$= \begin{bmatrix} \dfrac{3}{7} & \dfrac{2}{7} & \dfrac{-6}{7} \\ \dfrac{2}{7} & \dfrac{6}{7} & \dfrac{3}{7} \\ \dfrac{-6}{7} & \dfrac{3}{7} & \dfrac{-2}{7} \end{bmatrix}.$$

Theorem 7.64 *A reflection across a plane S can be performed by the matrix*

$$\text{Ref}(\theta) = \begin{bmatrix} \cos 2\theta & \sin 2\theta & 0 \\ \sin 2\theta & -\cos 2\theta & 0 \\ 0 & 0 & 0 \end{bmatrix}.$$

Proof Let V_1 be a unit vector in $\mathbb{R}^3$, V_2 be a unit vector orthogonal to V_1, and let $V_3 = U \times V$. Then, (V_1, V_2, V_3) is a orthogonal triple. Suppose that $A = \text{Ref}(\theta)$ is a reflection in $\mathbb{R}^3$ across the plane S through the origin whose unit normal column vector is N, and orthogonal to V_3. We can find an angle θ such that

$$N = -\sin\theta V_1 + \cos\theta V_2.$$

Using Eq. (7.7), we obtain

$$\begin{aligned} AV_1 &= (1 - 2\sin^2\theta)V_1 + 2\sin\theta\cos\theta V_2, \\ AV_2 &= 2\sin\theta\cos\theta V_1 + (1 - 2\cos^2\theta)V_2, \\ AV_3 &= V_3. \end{aligned}$$

That is,

$$\begin{aligned} AV_1 &= \cos 2\theta V_1 + \sin 2\theta V_2, \\ AV_2 &= \sin 2\theta V_1 - \cos 2\theta V_2, \\ AV_3 &= V_3. \end{aligned}$$

Therefore, we obtain

$$A = \text{Ref}(\theta) = \begin{bmatrix} \cos 2\theta & \sin 2\theta & 0 \\ \sin 2\theta & -\cos 2\theta & 0 \\ 0 & 0 & 0 \end{bmatrix},$$

as required. ∎

Exercises

1. Use standard trigonometric identities to verify that
   (a) $\text{Ref}(\theta)\text{Ref}(\phi) = \text{Rot}(2(\theta - \phi))$,
   (b) $\text{Rot}(\theta)\text{Ref}(\phi) = \text{Ref}(\phi + \frac{1}{2}\theta)$,
   (c) $\text{Ref}(\phi)\text{Rot}(\theta) = \text{Ref}(\phi - \frac{1}{2}\theta)$.

2. Find a matrix which represents a reflection in the line $y = 2x$.
3. Prove that each element of $O_3(\mathbb{R})$ can be expressed as a product of at most three reflections.

Fig. 7.10 A translation in $\mathbb{R}^2$

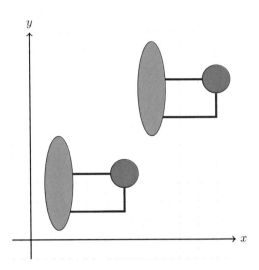

4. Find the matrix which induces reflection with respect to given vector $[a\ b\ c]^t$ in $\mathbb{R}^3$. Check your result for the case $[1\ 0\ 0]^t$.

7.5   Translation and Scaling Matrices

Suppose that $\mathcal{T}_n$ is the set of all translations of a fixed point $(x_1, x_2, \ldots, x_n)$ in $\mathbb{R}^n$. If

$$A = \begin{bmatrix} a_1 \\ a_2 \\ \vdots \\ a_n \end{bmatrix} \in \mathcal{T}_n.$$

then

$$\begin{bmatrix} x_1' \\ x_2' \\ \vdots \\ x_n' \end{bmatrix} = \begin{bmatrix} x_1 \\ x_2 \\ \vdots \\ x_n \end{bmatrix} + \begin{bmatrix} a_1 \\ a_2 \\ \vdots \\ a_n \end{bmatrix}.$$

$\mathcal{T}_n$ with addition of columns matrices as the composition rule is a group. This group is called the *translation group*.

Example 7.65 A translation is shown in Fig. 7.10 by $A = \begin{bmatrix} 5 \\ 5 \end{bmatrix} \in \mathcal{T}_2$.

Example 7.66 The set of all translations parallel to the x-axis, i.e., all translations of the form

$$\left\{ \begin{bmatrix} a \\ 0 \end{bmatrix} \mid a \in \mathbb{R} \right\}$$

is a subgroup of $\mathcal{T}_2$.

Example 7.67 The set of all translations parallel to the y axis, i.e., all translations of the form

$$\left\{ \begin{bmatrix} 0 \\ b \end{bmatrix} \mid b \in \mathbb{R} \right\}$$

is a subgroup of $\mathcal{T}_2$.

A scaling can be represent by a *scaling matrix*. To scale an object by a vector $V' = [c_1 \, c_2 \, \ldots \, c_n]$ $(c_1 = \cdots = c_n \neq 0)$, each point X would need to be multiplied by the following diagonal matrix:

$$A = \begin{bmatrix} c_1 & 0 & \ldots & 0 \\ 0 & c_2 & \ldots & 0 \\ \vdots & \vdots & \ddots & \vdots \\ 0 & 0 & \ldots & c_n \end{bmatrix}$$

As shown below, the multiplication give the expected result:

$$AX = \begin{bmatrix} c_1 & 0 & \ldots & 0 \\ 0 & c_2 & \ldots & 0 \\ \vdots & \vdots & \ddots & \vdots \\ 0 & 0 & \ldots & c_n \end{bmatrix} \begin{bmatrix} x_1 \\ x_2 \\ \vdots \\ x_n \end{bmatrix} = \begin{bmatrix} c_1 x_1 \\ c_2 x_2 \\ \vdots \\ c_n x_n \end{bmatrix}.$$

Remark 7.68 The set of all diagonal matrices with non-zero determinants in $GL_n(\mathbb{R})$ forms a subgroup.

Such a scaling changes the diameter of an object by a factor between the scale factors, the area by a factor between the smallest and the largest product of two scale factors, and the volume by the product of all three.

The scaling is uniform if and only if the scaling factors are equal ($v_1 = \cdots = v_n$). If all except one of the scale factors are equal to 1, we have directional scaling.

In the case where $v_1 = \cdots = v_n = c$, scaling changes the area of any surface by a factor of c^2 (see Fig. 7.11) and the volume of any solid object by a factor of c^3 (see Fig. 7.12).

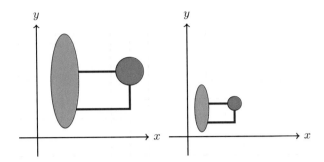

Fig. 7.11 Scaling with $c = 0.5$ in $\mathbb{R}^2$

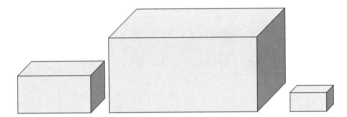

Fig. 7.12 Scaling with $c = 2$ and $c = 0.5$ in $\mathbb{R}^3$

Exercises

1. Prove that every positive rigid motion in $\mathbb{R}^3$ can be obtained by a rotation about a l-axis, followed by a translation along l (this type of motion is called *screw*).
2. Consider the graph $y = e^x$. Suppose that the graph is dilated from the y-axis by a factor of 3, reflected in the x-axis, and then translated 1 unit to the left and 2 units down. What is the equation of the resulting graph?
3. Consider the graph $y = x^2$. Suppose the graph is dilated from the x axis by a factor of 2, and then translated 3 units to the right. What is the equation of the resulting graph?
4. Suppose the graph of $y = f(x)$ is transformed by a dilation of factor k from the x axis, factor $h \neq 0$ from the y axis, and then translated c units to the right and d units up. Show that the resulting graph has equation

$$y = kf\left(\frac{1}{h}x - c\right) + d.$$

5. Two geometric shapes A and B are shown in Fig. 7.13. Shape A has one point at $(3, 4)$ and shape B has one point at $(0, -2)$. Calculate a chain of matrices that, when post-multiplied by the vertices of shape A, will transform all the vertices of shape A into the vertices of shape B, i.e., translate and rotate the shape point

Fig. 7.13 Two geometric
shapes A and B

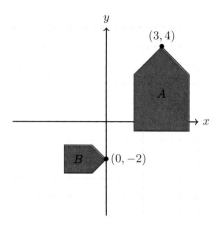

(3, 4) to (0, −2). The transformation must also scale the size of shape A by half
to shape B.

7.6 Dihedral Groups

The dihedral groups form an important infinite family of examples of finite groups.
They arise as groups of symmetries of the regular n-gons, and they play an important
role in group theory, geometry, and chemistry.

Definition 7.69 The *dihedral group* D_n is the group of symmetries of a regular
polygon with n vertices.

We may consider this polygon as having vertices on the unit circle, with vertices
labelled $1, 2, \ldots, n - 1$ starting at $(1, 0)$ and proceeding counterclockwise at angles
in multiples of $2\pi/n$ radians. There are two types of symmetries of the regular n-gon,
each one giving rise to n elements in the group D_n:

(1) Rotations $R_0, R_1, \ldots, R_{n-1}$, where R_k is rotation of angle $2k\pi/n$.
(2) Reflections $S_0, S_1, \ldots, S_{n-1}$, where S_k is reflection about the line through the
    origin and making an angle of $k\pi/n$ with the horizontal axis.

The group operation is given by composition of symmetries. Similar to permutations, if x and y are two elements in D_n, then $xy = y \circ x$. That means that xy is the
symmetry obtained by applying first x, followed by y.

So, according to our discussion in Sects. 7.3 and 7.4, the elements of D_n can be
thought 2×2 matrices, with group operation corresponding to matrix multiplication.
Indeed, we have

$$R_k = \begin{bmatrix} \cos\left(\dfrac{2k\pi}{n}\right) & -\sin\left(\dfrac{2k\pi}{n}\right) \\[3mm] \sin\left(\dfrac{2k\pi}{n}\right) & \cos\left(\dfrac{2k\pi}{n}\right) \end{bmatrix},$$

$$S_k = \begin{bmatrix} \cos\left(\dfrac{2k\pi}{n}\right) & \sin\left(\dfrac{2k\pi}{n}\right) \\[3mm] \sin\left(\dfrac{2k\pi}{n}\right) & -\cos\left(\dfrac{2k\pi}{n}\right) \end{bmatrix}.$$

Now, it is not difficult to see that the following relations hold in D_n:

$$\begin{aligned} R_i R_j &= R_{i+j}, \\ R_i S_j &= S_{i+j}, \\ S_i R_j &= S_{i-j}, \\ S_i S_j &= R_{i-j}, \end{aligned}$$

where $0 \le i, j \le n - 1$, and both $i + j$ and $i - j$ are computed modulo n. The Cayley table for D_n can be readily computed from the above relations. In particular, we see that R_0 is the identity, $R_i^{-1} = R_{n-i}$ and $S_i^{-1} = S_i$.

Example 7.70 In Example 3.39, we investigated D_3, the symmetry group of the equilateral triangle. The matrix representation of D_3 is given by

$$R_0 = \begin{bmatrix} 1 & 0 \\ 0 & 1 \end{bmatrix}, \quad R_1 = \begin{bmatrix} -\dfrac{1}{2} & -\dfrac{\sqrt{3}}{2} \\[3mm] \dfrac{\sqrt{3}}{2} & \dfrac{1}{2} \end{bmatrix}, \quad R_2 = \begin{bmatrix} -\dfrac{1}{2} & \dfrac{\sqrt{3}}{2} \\[3mm] -\dfrac{\sqrt{3}}{2} & -\dfrac{1}{2} \end{bmatrix},$$

$$S_0 = \begin{bmatrix} 1 & 0 \\ 0 & -1 \end{bmatrix}, \quad S_1 = \begin{bmatrix} -\dfrac{1}{2} & \dfrac{\sqrt{3}}{2} \\[3mm] \dfrac{\sqrt{3}}{2} & \dfrac{1}{2} \end{bmatrix}, \quad S_2 = \begin{bmatrix} -\dfrac{1}{2} & -\dfrac{\sqrt{3}}{2} \\[3mm] -\dfrac{\sqrt{3}}{2} & \dfrac{1}{2} \end{bmatrix}.$$

Example 7.71 In Example 3.40, we investigated D_4, the symmetry group of a square. The matrix representation of D_4 is given by

$$R_0 = \begin{bmatrix} 1 & 0 \\ 0 & 1 \end{bmatrix}, \quad R_1 = \begin{bmatrix} 0 & -1 \\ 1 & 0 \end{bmatrix}, \quad R_2 = \begin{bmatrix} -1 & 0 \\ 0 & -1 \end{bmatrix}, \quad R_3 = \begin{bmatrix} 0 & 1 \\ -1 & 0 \end{bmatrix},$$

$$S_0 = \begin{bmatrix} 1 & 0 \\ 0 & -1 \end{bmatrix}, \quad S_1 = \begin{bmatrix} 0 & 1 \\ 1 & 0 \end{bmatrix}, \quad S_2 = \begin{bmatrix} -1 & 0 \\ 0 & 1 \end{bmatrix}, \quad S_3 = \begin{bmatrix} 0 & -1 \\ -1 & 0 \end{bmatrix}.$$

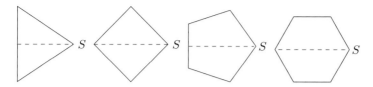

Fig. 7.14 Examples of polygons

Theorem 7.72 *Let R be a counterclockwise rotation of the n-gon by $2\pi/n$ radians and let S be a reflection across a line through a vertex. See examples in the polygons in Fig. 7.14. Then*

(1) The n rotations in D_n are I, R, R^2, ..., R^{n-1};
(2) $o(S) = 2$;
(3) The n reflections in D_n are S, RS, R^2S, ..., $R^{n-1}S$.

Proof (1) The rotations I, R, R^2, ..., R^{n-1} are distinct since R has order n.

(2) Simply consider what applying s twice to each vertex will do to it.

(3) Since I, R, R^2, ..., R^{n-1} are distinct, it follows that S, RS, R^2S, ..., $R^{n-1}S$ are distinct. In addition, for each j, R^jS is not a rotation because if $R^jS = R^i$, then $S = R^{i-j}$, while S is not a rotation. ∎

Corollary 7.73 *The dihedral group D_n has 2n elements and*

$$D_n = \{I, \ R, \ R^2, \ \ldots, \ R^{n-1}, \ S, \ RS, \ R^2S, \ \ldots, \ R^{n-1}S\}.$$

In particular, all elements of D_n with order greater than 2 are powers of R.

Corollary 7.74 *The dihedral group D_n is generated by R and S.*

Proof Since every element of D_n is a product of R and S, by Theorem 4.40 it follows that $\{R, \ S\}$ generate D_n. ∎

Theorem 7.75 *In the dihedral group D_n, we have $RS = SR^{-1}$.*

Proof Since RS is a reflection, it follows that $(RS)^2 = I$, or equivalently $RSRS = I$. This implies that $RS = S^{-1}R^{-1}$. Since $o(S) = 2$, it follows that $S = S^{-1}$ and so we obtain $RS = SR^{-1}$. ∎

Corollary 7.76 *For each $1 \leq j \leq n$, we have $R^kS = SR^{-k}$.*

Proof It is straightforward by mathematical induction. ∎

Corollary 7.77 *The dihedral group D_n is not abelian if $n \geq 3$.*

Theorem 7.78 *For each $n \geq 3$, the center of the dihedral group D_n is*

(1) $Z(D_n) = \{I\}$ if n is odd;

(2) $Z(D_n) = \{I, \ R^{n/2}\}$ *if n is even.*

Proof In order to determine the center of D_n it suffices to determine those elements which commute with the generators R and S. Since $n \geq 3$, it follows that $R^{-1} \neq R$.

If $R^k S \in Z(D_n)$ for some $1 \leq k \leq n$, then $R(R^k S) = (R^k S)R$. Then, we must have

$$R^{k+1} S = R(R^k S) = (R^k S)R = R^k(SR) = R^k(R^{-1}S) = R^{k-1}S.$$

This implies that $R^2 = I$ a contradiction. Thus, we conclude that no reflections are in the center of D_n since reflections do not commute with R.

Similarly, if for $1 \leq j \leq n$, $R^j S = SR^j$, then $R^j S = R^{-j}S$. Hence, $R^{2j} = I$. Since R has order n, it follows that

$$R^{2j} = I \ \Leftrightarrow \ n | 2j.$$

If n is odd, then

$$R^{2j} = I \ \Leftrightarrow \ n | j \ \Leftrightarrow \ j \text{ is multiple of } n \ \Leftrightarrow \ R^j = I.$$

So, if n is odd, then the only rotation that could be in $Z(D_n)$ is I.

If n is even, then

$$R^{2j} = I \ \Leftrightarrow \ n | 2j \ \Leftrightarrow \ (n/2) | j.$$

Hence, the only choice for j are $j = 0$ and $j = (n/2)$. So, we have

$$R^j = R^0 = I \text{ or } R^j = R^{n/2}.$$

To show that $R^{n/2}$ is in $Z(D_n)$, we check it commutes with every rotations and reflections in D_n. Clearly, $R^{n/2}$ commutes with all rotations, since all rotations are powers of R. Hence, we check $R^{n/2}$ commutes with each reflections. Indeed, we have

$$R^{n/2}(R^i S) = R^{(n/2)+i} S \qquad (7.8)$$

and

$$\begin{aligned} (R^i S)R^{n/2} &= R^i(SR^{n/2}) = R^i R^{-n/2} S \\ &= R^i R^{n/2} S = R^{i+(n/2)} S = R^{(n/2)+i} S. \end{aligned} \qquad (7.9)$$

From (7.8) and (7.9), we obtain $R^{n/2}(R^i S) = (R^i S)R^{n/2}$.

Therefore, we conclude that

$$Z(D_n) = \begin{cases} \{I\} & \text{if } n \text{ is odd} \\ \{I, R^{n/2}\} & \text{if } n \text{ is even.} \end{cases}$$

∎

Theorem 7.79 *The conjugacy classes of the dihedral group D_n are as follows:*

(1) If n is odd,

- *the identity element: $\{I\}$,*
- *$(n-1)/2$ conjugacy classes of size 2: $\{R, R^{-1}\}, \{R^2, R^{-2}\}, \ldots, \{R^{(n-1)/2}, R^{-(n-1)/2}\}$,*
- *all the reflections: $\{R, RS, R^2 S, \ldots, R^{n-1} S\}$.*

(2) If n is even,

- *two conjugacy classes of size 1: $\{I\}, \{R^{n/2}\}$,*
- *$(n/2) - 1$ conjugacy classes of size 2: $\{R, R^{-1}\}, \{R^2, R^{-2}\}, \ldots, \{R^{(n/2)-1}, R^{-((n/2)-1)}\}$,*
- *the reflection divided into two conjugacy classes:*
  $\{R^{2k} S \mid 0 \le k \le (n/2) - 1\}$ and $\{R^{2k+1} S \mid 0 \le k \le (n/2) - 1\}$.

Proof We know that every element of D_n is of the form R^i or $R^i S$ for some integer i. Thus, in order to determine the conjugacy class of an element X, we compute

$$R^{-i} X R \text{ and } (R^i S)^{-1} X (R^i S).$$

Since $R^{-i} R^j R^i = R^j$ and $(R^i S)^{-1} R^j (R^i S) = R^{-j}$, it follows that the only conjugates of R^j in D_n are R^j and R^{-j}. It is necessary the more computation to be sure nothing more is conjugate as well. Now, we try to find the conjugacy class of S. We have

$$R^{-i} S R^i = R^{n-2i} S,$$
$$(R^i S)^{-1} S (R^i S) = R^{2i} S.$$

Since $1 \le i \le n$, it follows that R^{2i} and R^{n-2i} run through powers of R divisible by 2.

Let n be odd. Since 2 is invertible modulo n, it follows that we can solve the linear congruence equation $2i \equiv k \pmod{n}$ for each i. This yields that

$$\{R^{2i} S \mid i \in \mathbb{Z}\} = \{R^k S \mid k \in \mathbb{Z}\}.$$

Hence, every reflections in D_n is conjugate to S.

Now, let n be even. We only obtain half the reflections as conjugates of S. The other half are conjugate to RS. Indeed, we have

$$R^{-i} (RS) R^i = R^{-2i+1} S,$$
$$(R^i S)^{-1} (RS)(R^i S) = R^{2i-1} S.$$

Since $1 \le i \le n$, this gives us $\{RS, R^3 S, \ldots, R^{n-1} S\}$. ∎

Example 7.80 *An application of D_4.* One application of D_4 is in the design of a letter-facing machine. Imagine letters entering a conveyor belt to be postmarked.

They are placed on the conveyor belt at random so that two sides are parallel to the belt. Suppose that a postmarker can recognize a stamp in the top right corner of the envelope, on the side facing up. In Fig. 7.15, a sequence of machines is shown that will recognize a stamp on any letter, no matter what position in which the letter starts. The letter P stands for a postmarker. The letters R and S stand for rotating and flipping machines that perform the motions of R and S. The arrows pointing up indicate that if a letter is postmarked, it is taken off the conveyor belt for delivery. If a letter reaches the end, it must not have a stamp. Letter-facing machines like this have been designed. One economic consideration is that R-machines tend to cost more than S-machines. R-machines also tend to damage more letters. Taking these facts into consideration, the reader is invited to design a better letter-facing machine. Assume that R-machines cost \$1000 and S-machines cost \$750. Be sure that all corners of incoming letters will be examined as they go down the conveyor belt.

Exercises

1. Show that D_4 is non-abelian group of order 8, where each element of D_4 is of the form $a^i b^j$, $0 \le i \le 3$ and $0 \le j \le 1$.
2. Draw the Hasse diagram for subgroups of the dihedral group D_4.
3. In the dihedral group D_n, suppose that R is a rotation and that S is a reflection. Use the fact that RS is also a reflection, together with the fact that reflections have order 2, to show that RSR is the inverse of S.
4. For each of the snowflakes in Fig. 7.16 find the symmetry group and locate the axes of the reflective symmetry.
5. Find the symmetry group of the Iranian architecture in Fig. 7.17.
6. Design a better letter-facing machine. How can you verify that a letter facing machine does indeed check every corner of a letter? Can it be done on paper without actually sending letters through it?
7. Let S be a reflection in the dihedral group D_n and R be a rotation in D_n. Determine

   (a) $C_{D_n}(S)$ when n is odd;
   (b) $C_{D_n}(S)$ when n is even;
   (c) $C_{D_n}(R)$.

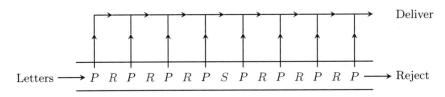

Fig. 7.15 A letter facer

Fig. 7.16 Snowflakes

7.7 Quaternion Group

Consider the case $\mathbb{F} = \mathbb{C}$, the complex numbers, and the set of eight elements

$$Q_8 = \left\{ \pm \begin{bmatrix} 1 & 0 \\ 0 & 1 \end{bmatrix}, \pm \begin{bmatrix} i & 0 \\ 0 & -i \end{bmatrix}, \pm \begin{bmatrix} 0 & 1 \\ -1 & 0 \end{bmatrix}, \pm \begin{bmatrix} 0 & i \\ i & 0 \end{bmatrix} \right\}.$$

One can use the notation

$$I = \begin{bmatrix} i & 0 \\ 0 & -i \end{bmatrix}, \ J = \begin{bmatrix} 0 & 1 \\ -1 & 0 \end{bmatrix} \text{ and } K = \begin{bmatrix} 0 & i \\ i & 0 \end{bmatrix}.$$

To avoid confusion, we may show the identity matrix by 1 instead of I_2. Then, we obtain

Fig. 7.17 Iranian architecture

Fig. 7.18 The rules
involving I, J, and K in the
quaternion group

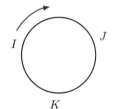

$$I^2 = J^2 = K^2 = -1,$$
$$IJ = -JI = K,$$
$$JK = -JK = I,$$
$$KI = -IK = J.$$

So, $Q_8 = \{\pm 1, \ \pm I, \ \pm J, \ \pm K\}$.

The rules involving I, J and K can be remembered by using Fig. 7.18.

Going clockwise, the product of two consecutive elements is the third one. The same is true for going counterclockwise, except that we obtain the negative of the third element. This group was invented by William Hamilton in 1834. The quaternions are used to describe rotations in three-dimensional space, and they are used in physics. The quaternions can be used to extend the complex numbers in a natural way.

Theorem 7.81 Q_8 *is a subgroup of* $GL_2(\mathbb{C})$ *and so in particular* Q_8 *is a group of order* 8.

Proof Clearly, Q_8 is non-empty. By the above relations, it is closed under multiplication. In addition, ± 1 are their own inverse, and all the other elements have an inverse which is minus themselves (since $I(-I) = 1$, etc.). Therefore, Q_8 is closed under inverse. ∎

Q_8 is called *quaternion group*. The quaternion group is not abelian. Its Cayley table is shown below.

·	1	−1	I	−I	J	−J	K	−K
1	1	−1	I	−I	J	−J	K	−K
−1	−1	1	−I	I	−J	J	−K	K
I	I	I	−1	1	K	−K	−J	J
−I	−I	I	1	−1	−K	K	J	−J
J	J	−J	−K	K	−1	1	I	−I
−J	−J	J	K	−K	1	−1	−I	I
K	K	−K	J	−J	−I	I	−1	1
−K	−K	K	−J	J	I	I	1	−1

Theorem 7.82 *The quaternion group Q_8 is generated by J and K.*

Proof We observe that each element of Q_8 is of the form $J^r K^s$ for some integers r and s. ∎

Remark 7.83 Note that I, J and K have order 4 and that any two of them generate the entire group.

The proper subgroups of Q_8 are $\langle I \rangle$, $\langle J \rangle$, $\langle K \rangle$ and $\langle -1 \rangle$. We have

$$\langle -1 \rangle = \langle I \rangle \cap \langle J \rangle \cap \langle K \rangle$$

and the center of Q_8 is $\langle -1 \rangle$.

Exercises

1. Show that Q_8 has exactly one element of order 2 and six elements of order 4.
2. Draw the Hasse diagram for subgroups of the quaternion group Q_8.

7.8 Worked-Out Problems

Problem 7.84 Let n be a positive integer and

$$A = \begin{bmatrix} 0 & 0 & -1 \\ 0 & 1 & 0 \\ 1 & 0 & 0 \end{bmatrix}.$$

Prove that $A^n = I_3$ if and only if $4 | n$.

Solution First, we compute A^2, A^3, A^4, etc. We find that

$$A^2 = \begin{bmatrix} 0 & 0 & -1 \\ 0 & 1 & 0 \\ 1 & 0 & 0 \end{bmatrix} \begin{bmatrix} 0 & 0 & -1 \\ 0 & 1 & 0 \\ 1 & 0 & 0 \end{bmatrix} = \begin{bmatrix} -1 & 0 & -1 \\ 0 & 1 & 0 \\ 0 & 0 & -1 \end{bmatrix},$$

$$A^3 = \begin{bmatrix} -1 & 0 & 0 \\ 0 & 1 & 0 \\ 0 & 0 & -1 \end{bmatrix} \begin{bmatrix} 0 & 0 & -1 \\ 0 & 1 & 0 \\ 1 & 0 & 0 \end{bmatrix} = \begin{bmatrix} 0 & 0 & 1 \\ 0 & 1 & 0 \\ -1 & 0 & 0 \end{bmatrix},$$

$$A^4 = \begin{bmatrix} 0 & 0 & 1 \\ 0 & 1 & 0 \\ -1 & 0 & 0 \end{bmatrix} \begin{bmatrix} 0 & 0 & -1 \\ 0 & 1 & 0 \\ 1 & 0 & 0 \end{bmatrix} = \begin{bmatrix} 1 & 0 & 0 \\ 0 & 1 & 0 \\ 0 & 0 & 1 \end{bmatrix} = I_3.$$

Now, if $4|n$, then we can write $n = 4q$. Hence, $A^n = A^{4q} = (A^4)^q = I_3^q = I_3$.

Conversely, if $A^n = I_3$, then by the division algorithm, there exist integers q and r such that $n = 4q + r$ with $0 \le r < 4$. Next, we obtain

$$A^r = A^{n-4q} = A^n (A^{-4})^q = I_3 I_3^q = I_3.$$

Since A, A^2 and A^3 are not equal to I_3, it follows that $r = 0$. Consequently, we obtain $4|n$. ∎

Problem 7.85 Let n be a positive integer and

$$A = \begin{bmatrix} 1 & 1 & 0 \\ 0 & 1 & 1 \\ 0 & 0 & 1 \end{bmatrix}.$$

Compute A^n.

Solution Suppose that

$$B = \begin{bmatrix} 0 & 1 & 0 \\ 0 & 0 & 1 \\ 0 & 0 & 0 \end{bmatrix}.$$

Then, we observe that $B^n = 0$, for each $n \ge 3$. Note that $A = I_3 + B$. Since I_3 and B commute, we can use the binomial theorem. So, we can write

$$A^n = (I_3 + B)^n = \sum_{i=0}^{n} \binom{n}{i} I_3^{n-i} B^i.$$

Therefore, we obtain

$$A^n = I_3 + nB + \frac{n(n-1)}{2}B^2 = \begin{bmatrix} 0 & n & \frac{n(n-1)}{2} \\ 0 & 1 & n \\ 0 & 0 & 1 \end{bmatrix},$$

and we are done.   ∎

7.9 Supplementary Exercises

1. Prove that the elements of $GL_n(\mathbb{R})$ which have integer entries and determinant equal to 1 or -1 form a subgroup of $GL_n(\mathbb{R})$.
2. Let $A = \begin{bmatrix} 0 & -1 \\ 1 & 0 \end{bmatrix}$, $B = \begin{bmatrix} 0 & -1 \\ 1 & -1 \end{bmatrix} \in GL_2(\mathbb{Z})$. Show that A has order 4, B has order 3 and AB has infinite order.
3. The matrix

$$\begin{bmatrix} 1 & t_0 & t_0^2 & \cdots & t_0^n \\ 1 & t_1 & t_1^2 & \cdots & t_1^n \\ \vdots & \vdots & \vdots & \vdots & \vdots \\ 1 & t_n & t_n^2 & \cdots & t_n^n \end{bmatrix}$$

   is called a *Vandermonde matrix*. Show that such a matrix is invertible, when t_0, $t_1, \ldots, t_n$ are $n+1$ distinct elements of $\mathbb{C}$.
4. Show by example that there are matrices A and B for which $\lim_{n \to \infty} A^n$ and $\lim_{n \to \infty} B^n$ both exist, but for which $\lim_{n \to \infty} (AB)^n$ does not exist.
5. Let $A = \begin{bmatrix} a & b \\ c & d \end{bmatrix}$ be a matrix over a field $\mathbb{F}$. Then, the set of all matrices of the form $p(A)$, where p is a polynomial over $\mathbb{F}$, is a commutative ring R with identity. If B is a 2×2 matrix over R, the determinant of B is then a 2×2 matrix over $\mathbb{F}$, of the form $p(A)$. Suppose that B is the 2×2 matrix over R as follows:

$$\begin{bmatrix} A - aI_2 & -bI_2 \\ -cI_2 & A - dI_2 \end{bmatrix}.$$

   Show that $\det(B) = p(A)$, where

$$p(x) = x^2 - (a+d)x + \det(A), \tag{7.10}$$

   and also that $p(A) = 0$. The polynomial p in (7.10) is the characteristic polynomial of A.
6. Given a matrix $A = \begin{bmatrix} a & b \\ c & d \end{bmatrix}$ in $GL_2(\mathbb{Z}_p)$. Consider its characteristic polynomial.

(a) If the roots λ_1 and λ_2 are distinct in $\mathbb{Z}_p$, show that A is conjugate to both
$$\begin{bmatrix} \lambda_1 & 0 \\ 0 & \lambda_2 \end{bmatrix} \text{ and } \begin{bmatrix} \lambda_2 & 0 \\ 0 & \lambda_1 \end{bmatrix}.$$

(b) If $\lambda_1 = \lambda_2$, show that A is $\begin{bmatrix} \lambda_1 & 0 \\ 0 & \lambda_1 \end{bmatrix}$ or it is conjugate to $\begin{bmatrix} \lambda_1 & 1 \\ 0 & \lambda_1 \end{bmatrix}$ in $GL_2(\mathbb{Z}_p)$.

7. Let
$$R = \begin{bmatrix} r_{11} & r_{12} & r_{13} \\ r_{21} & r_{22} & r_{23} \\ r_{31} & r_{32} & r_{33} \end{bmatrix}$$

be a rotation matrix in $\mathbb{R}^3$. Show that the rotation axis is the vector

$$V = \frac{1}{\sin \theta} \begin{bmatrix} r_{32} - r_{23} \\ r_{13} - r_{31} \\ r_{21} - r_{12} \end{bmatrix}.$$

If the angle of rotation θ is different from π and 0, then

$$\theta = \cos^{-1} \left(\frac{r_{11} + r_{22} + r_{33} - 1}{2} \right).$$

What happens if the rotation angle is very small? Describe a robust method to find the rotation axis (your method should include the cases for $\theta = 0$ and $\theta = \pi$).

Chapter 8
Cosets of Subgroups and Lagrange's Theorem

In group theory, the result known as Lagrange's theorem states that for a finite group G the order of any subgroup divides the order of G. Lagrange's theorem is one of the central theorems of group theory. In order to prove this theorem we introduce the notion of cosets of a subgroup.

8.1 Cosets and Their Properties

To understand Lagrange's theorem, one has to look at cosets of a subgroup. This notion was invented by Galois in 1830, although the term was coined by G.A. Miller in 1910.

Definition 8.1 Let G be a group and H be a subgroup of G. For any $a \in G$, the set $aH = \{ah \mid h \in H\}$ is called the *left coset of H in G containing a*. Analogously, $Ha = \{ha \mid h \in H\}$ is called the *right coset of H in G containing a*.

In other words, a coset is what we get when we take a subgroup and shift it (either on the left or on the right). The best way to think about cosets is that they are shifted subgroups, or translated subgroups. Note that a lies in both aH and Ha, since $a = ae = ea$. If the left and right cosets coincide or if it is clear from the context to which type of coset that we are referring to, we will use the word coset without specifying left or right. In additive notation, we get $a + H$ (which usually implies that we deal with a commutative group where we do not need to distinguish left and right cosets). Any element of a coset is called a representative of that coset.

Example 8.2 Let $G = \mathbb{Z}$, the additive group of integers, and $H = \langle m \rangle = m\mathbb{Z}$, the subgroup generated by m, for some $m \in \mathbb{Z}$. The set of left cosets is $\{H, \ 1 + H, \ 2 + H, \ \ldots, \ m - 1 + H\}$. Since $a + H = H + a$, left cosets coincide with right cosets.

B. Davvaz, *A First Course in Group Theory*,
https://doi.org/10.1007/978-981-16-6365-9_8

Example 8.3 Let $G = S_3$, the symmetric group of an equilateral triangle. Take the subgroup $H = \{id, (1\ 3)\}$. Then, the left cosets are

$$(1\ 2)H = \{(1\ 2),\ (1\ 2)(1\ 3)\} = \{(1\ 2),\ (1\ 2\ 3)\} = (1\ 2\ 3)H,$$
$$(1\ 3)H = \{(1\ 3),\ (1\ 3)(1\ 3)\} = \{(1\ 3),\ id\} = H,$$
$$(2\ 3)H = \{(2\ 3),\ (2\ 3)(1\ 3)\} = \{(2\ 3),\ (1\ 3\ 2)\} = (1\ 3\ 2)H.$$

Example 8.4 If $G = GL_2(\mathbb{R})$ and $H = SL_2(\mathbb{R})$, then for any matrix $A \in G$, the coset AH is the set of all 2×2 matrices with the same determinant as A. Thus, the coset $\begin{bmatrix} 3 & 0 \\ 2 & 1 \end{bmatrix} H$ is the set of all 2×2 matrices of determinant 3.

Theorem 8.5 (Properties of Cosets) *Let G be a group and H be a subgroup of G. Then,*

(1) $a \in aH$;
(2) $aH = H$ if and only if $a \in H$;
(3) $(ab)H = a(bH)$ and $H(ab) = (Ha)b$;
(4) $aH = bH$ if and only if $a \in bH$;
(5) $aH = bH$ or $aH \cap bH = \emptyset$;
(6) $aH = bH$ if and only if $a^{-1}b \in H$;
(7) $|aH| = |bH|$;
(8) aH is a subgroup of G if and only if $a \in H$.

Proof (1) We have $a = ae \in \{ah \mid h \in H\} = aH$.

(2) Suppose that $aH = H$. Then, $a = ae \in aH = H$. Conversely, let $a \in H$. We show that $aH \subseteq H$ and $H \subseteq aH$. The first inclusion follows immediately from the closure of H. In order to show that $H \subseteq aH$, assume that $h \in H$ is an arbitrary element. Since $a \in H$ and $h \in H$, it follows that $a^{-1}h \in H$. Consequently, we get $h = eh = (aa^{-1})h = a(a^{-1}h) \in H$.

(3) This follows directly from $(ab)h = a(bh)$ and $h(ab) = (ha)b$, for all $h \in H$.

(4) If $aH = bH$, then $a = ae \in aH = bH$. Conversely, if $a \in bH$, then there exists $h \in H$ such that $a = bh$. So, we deduce that $aH = (bh)H = b(hH) = bH$.

(5) This property follows directly from (4). Because if $aH \cap bH \neq \emptyset$, then there exists $c \in aH \cap bH$. This implies that $cH = aH$ and $cH = bH$, and so $aH = bH$.

(6) We observe that $aH = bH$ if and only if $H = a^{-1}bH$. Now, the result follows from (2).

(7) In order to show that $|aH| = |bH|$, it is enough to define a one to one function from aH onto bH. For this, we define $f : aH \to bH$ by $f(ah) = bH$. Obviously, f is onto. Moreover, from the cancellation law, f is one to one. This shows that f is a one to one correspondence.

(8) Assume that aH is a subgroup of G. Then, it contains the identity e. So, we get $aH \cap eH \neq \emptyset$. Then, from (5) we have $aH = eH = H$. Now, from (2), we conclude that $a \in H$. Conversely, if $a \in H$, then again by (2), we obtain $aH = H$. ∎

Remark 8.6 In Theorem 8.5, analogous results hold for right cosets.

Theorem 8.7 *If H and K are two subgroups of a group G, then for any $a, b \in G$ either $aH \cap bK = \emptyset$ or $aH \cap bK = c(H \cap K)$, for some $c \in G$.*

Proof Suppose that $aH \cap bK \neq \emptyset$, and let $c \in aH \cap bK$. Since $c \in cH$ and $c \in cK$, it follows that $c \in aH \cap cH$ and $c \in bK \cap cK$. Consequently, we obtain $aH = cH$ and $bK = cK$, and so $aH \cap bK = cH \cap cK$. Moreover, it is not difficult to see that $c(H \cap K) \subseteq cH \cap cK$. On the other hand, if $x \in cH \cap cK$, then $x = ch = ck$, for some $h \in H$ and $k \in K$. This implies that $h = c^{-1}x = k \in H \cap K$. So, $x \in c(H \cap K)$. Therefore, we conclude that $aH \cap bK = cH \cap cK = c(H \cap K)$. ■

Theorem 8.8 *Let G be a group and H be a subgroup of G. If $\mathcal{R}$ is the set of distinct right cosets of H in G and $\mathcal{L}$ is the set of distinct left cosets of H in G, then $|\mathcal{R}| = |\mathcal{L}|$.*

Proof Consider the mapping $f : \mathcal{R} \to \mathcal{L}$ defined by $f(Ha) = a^{-1}H$. First, we need to show that f is well defined. Namely, if we take two different representations of the same right coset we must show our mapping sends them to the same image. Suppose that $Ha = Hb$. Then, we know that $Hab^{-1} = H$ or $ab^{-1} \in H$. Let $ab^{-1} = h$, where $h \in H$. Then, $a^{-1}h = b^{-1}$ and so $a^{-1}hH = b^{-1}H$, or equivalently $a^{-1}H = b^{-1}H$. Consequently, we have $f(Ha) = f(Hb)$ as desired. Therefore, f is well defined.

Now, assume that $f(Ha) = f(Hb)$. Then, $a^{-1}H = b^{-1}H$. This yields that $ba^{-1}H = H$, or equivalently $ba^{-1} \in H$. If $ba^{-1} = h$, where $h \in H$, then $b = ha$. Consequently, $Hb = Hha$ and so $Ha = Hb$. Thus, f is one to one. In addition, if $aH \in \mathcal{L}$ is an arbitrary element, then $f(Ha^{-1}) = (a^{-1})^{-1}H = aH$. This shows that f is onto. Therefore, f is a one to one correspondence between $\mathcal{L}$ and $\mathcal{R}$. ■

Exercises

1. List the left and right cosets of the subgroup $\{id, (1\,2)(3\,4), (1\,3)(2\,4), (1\,4)(2\,3)\}$ in A_4.

2. If G is a group and $\{H_i \mid i \in I\}$ is a family of subgroups of G, show that
$$\left(\bigcap_{i \in I} H_i\right)a = \bigcap_{i \in I} H_i a.$$

3. Suppose that $G = Q_8$ and let $H = \langle -1 \rangle$ and $K = \langle I \rangle$ be subgroups of G. Find

   (a) the left cosets of H and K in G;
   (b) the right cosets of H and K in G.

4. Suppose that H and K are subgroups of G and there are elements a and b in G such that $aH \subseteq bK$. Prove that $H \subseteq K$.

5. Let G be a group and $a \in G$. If $o(a) = 30$, how many cosets of $\langle a^4 \rangle$ in $\langle a \rangle$ are there? List them.

6. Give an example of a group G having a subgroup H and two elements a, b such that $Ha = Hb$ but $aH \neq bH$.

7. Find an example of a subgroup H of a group G and elements a and b in G such that $aH \neq Hb$ and $aH \cap Hb \neq \emptyset$.

8. Suppose that H and K are subgroups of a group G with $H \leq K$. Show that for each $a \in G$, either $aH \subseteq K$ or $aH \cap K = \emptyset$.

8.2   Geometric Examples of Cosets

When a group is defined in terms of vectors and matrices, we can often get a picture of the group and its cosets.

Example 8.9 Let $G = \mathbb{R}^2 = \{xi + yj \mid x, y \in \mathbb{R}\}$, the additive group of vectors in a plane, where $i = (1, 0)$ and $j = (0, 1)$. Suppose that $H = \{xi \mid x \in \mathbb{R}\}$ and $v = ai + bj$ is a vector in G. Then, the left coset of H in G containing v is

$$v + H = \{(a + x)i + bj \mid x \in \mathbb{R}\}.$$

This is the line parallel to H that passes through the endpoint of v. The left cosets of H in G, in general, are the lines parallel to H. Two parallel lines are either equal or disjoint, so any two left cosets of H in G are equal or disjoint. In Fig. 8.1, the cosets H in G containing v and v' are equal while those of u, v, and w are disjoint.

Example 8.10 Let $\mathbb{C}^*$ be the group of non-zero complex numbers (see Example 3.36), and let $H = \{a + bi \mid a^2 + b^2 = 1, \ a, b \in \mathbb{R}\}$ be the subgroup of $\mathbb{C}^*$, which is defined in Example 3.66. First, notice that geometrically H is a circle of radius 1 with center $(0, 0)$. Let $4 + 3i$ be an element of $\mathbb{C}^*$. We want to study the coset of H in $\mathbb{C}^*$ containing $4 + 3i$. Of course, $\mathbb{C}^*$ is abelian, so there is no distinction between the left and the right cosets, and they are the same. The coset of H in $\mathbb{C}^*$ containing $4 + 3i$ is of the form $(4 + 3i)H$, and it is a circle with center $(0, 0)$ too. However, $4 + 3i$ scales the equation so that 1 becomes $4^2 + 3^2 = 25$. In order to see this,

Fig. 8.1 The cosets of $v + H$ in $\mathbb{R}^2$

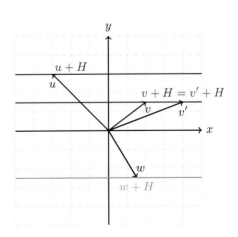

Fig. 8.2 H and its cosets $2H$ and $4H$ in $\mathbb{C}^*$

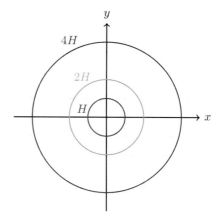

we have $(4 + 3i)(a + bi) = (4a - 3b) + (3a + 4b)i$. So, the multiplication $4 + 3i$ sends the point $(a, b) \in H$ to the point $(4a - 3b, \ 3a + 4b)$. Now, this gives us the equation

$$(4a - 3b)^2 + (3a + 4b)^2 = 16a^2 + 9b^2 - 24ab + 9a^2 + 16b^2 - 24ab$$
$$= 25a^2 + 25b^2 = 25(a^2 + b^2) = 25.$$

Now, we consider the general case. If $z \in \mathbb{C}^*$ be an arbitrary number, we can write $z = re^{i\theta}$, where r is the absolute value of z and θ is the argument of z. Consider $e^{i\theta} H$. Multiplying any complex number by $e^{i\theta}$ simply rotates anticlockwise through angle θ about the origin. Hence, we deduce that $e^{i\theta} H = H$. So, we obtain $zH = rH$. What does multiplying by r do? It scales the circle H by a factor of r. Two different positive real numbers $r \neq r'$ give different cosets $rH \neq r'H$, since the first has radius r and the second has radius r'. For illustration, see Fig. 8.2. Therefore, the cosets of H in $\mathbb{C}^*$ are the circles centered at the origin of positive radius.

Example 8.11 Let $G = \{(a, b) \mid a, b \in \mathbb{R} \text{ and } a > 0\}$ be the group defined in Example 3.42 with the following binary operation:

$$(a, b) \star (c, d) = (ac, bc + d),$$

for all $(a, b), \ (c, d) \in G$. Suppose that

$$H = \{(1, y) \mid y \in \mathbb{R}\},$$
$$K = \{(x, 0) \mid x \in \mathbb{R} \text{ and } x > 0\}.$$

It is easy to check that H and K are subgroups of G. These subgroups are indicated in Fig. 8.3. Now, let $(a, b) \in G$. What are the left and right cosets of H in G containing (a, b)? We observe that

Fig. 8.3 Subgroups H
and K

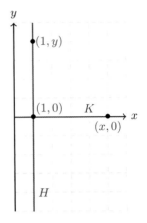

Fig. 8.4 Left and right
cosets of H in G containing
(a, b)

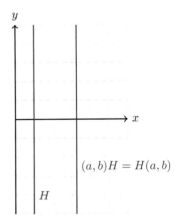

$$(a, b)(1, y) = (a, b + y) \text{ and } (1, y)(a, b) = (a, ay + b),$$

for all $(1, y) \in H$. Clearly, the numbers $b + y$ and $ay + b$ run over $\mathbb{R}$. This yields
that the left and the right cosets of H in G containing (a, b) are the same. In other
words, we have

$$(a, b)H = H(a, b) = \{(a, z) \mid z \in \mathbb{R}\},$$

and this is the vertical line parallel to H passing through the point (a, b), see Fig. 8.4.
Now, we determine the right coset of K in G containing (a, b). We have

$$(x, 0)(a, b) = (xa, b),$$

for all $(x, 0) \in K$. Since $x > 0$, it follows that (xa, b) runs through all points of the
form (z, b) with $z > 0$. Consequently, we can write

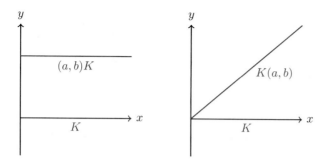

Fig. 8.5 Left and right cosets of K in G containing (a, b)

$$K(a, b) = \{(z, b) \mid z \in \mathbb{R} \text{ and } z > 0\}.$$

In view of this, the right coset of K in G containing (a, b) is determined by b alone and it is independent of the choice of a. So, it is horizontal line through (a, b). See the left hand picture in Fig. 8.5. Finally, we try to find the left coset of K in G containing (a, b). We have

$$(a, b)(x, 0) = (ax, bx),$$

for all $(x, 0) \in K$. If we take $ax = r$, then $bx = br/a$. So, we can write

$$(a, b)K = \left\{ (r, s) \mid r, s \in \mathbb{R}, \ r > 0, \text{ and } s = \frac{b}{a} r \right\}.$$

This is a half line out of the origin with slope b/a. See the right hand picture in Fig. 8.5. We observe that the left and right cosets of K in G containing (a, b) are not the same if $b \neq 0$.

Example 8.12 We know that $G = \mathbb{R}^3$ forms a group under the following binary operation:

$$(x_1, y_1, z_1) + (x_2, y_2, z_2) = (x_1 + x_2, y_1 + y_2, z_1 + z_2).$$

If we consider the subset

$$H = \{(x, y, z) \mid 2x + 7y - 3z = 0\},$$

then we observe that H is a subgroup of G. In geometrical terms, H is a plane through the origin in a three-dimensional space. The left coset $(a, b, c) + H$ is, in general terms, nothing more than the plane P passing through the point (a, b, c) and parallel to H in $\mathbb{R}^3$. For illustration, in Fig. 8.6 we consider $(a, b, c) = (0, 2, 0)$.

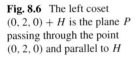

Fig. 8.6 The left coset
$(0, 2, 0) + H$ is the plane P
passing through the point
$(0, 2, 0)$ and parallel to H

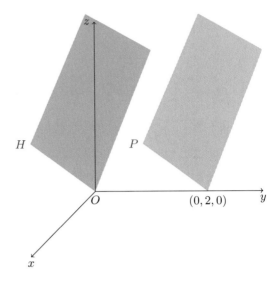

Exercises

1. In $\mathbb{R}^2$ under component addition, let $H = \{(x, 5x) \mid x \in \mathbb{R}\}$, the subgroup of all points on the line $y = 5x$. Show that the left coset $(3, 2) + H$ is the straight line passing through the point $(3, 2)$ and parallel to the line $y = 5x$.
2. Consider the additive group $\mathbb{R}$ and its subgroup $\mathbb{Z}$. Describe a coset $t + \mathbb{Z}$ geometrically. Show that the set of all cosets of $\mathbb{Z}$ in $\mathbb{R}$ is $\{t + \mathbb{Z} \mid 0 \le t < 1\}$. What are the analogous results for $\mathbb{Z}^2 \subseteq \mathbb{R}^2$?
3. Show that the function sine assigns the same value to each element of any fixed left coset of the subgroup $\langle 2\pi \rangle$ of the additive group $\mathbb{R}$ of real numbers. Thus sine induces a well-defined function on the set of cosets; the value of the function on a coset is obtained when we choose an element x of the coset and compute $\sin x$.

8.3   Lagrange's Theorem

Let H be a subgroup of a group G, which may be of finite or infinite order. We reconsider some parts of Theorem 8.5 in a different view. We exhibit two partitions of G by defining two equivalence relations, $\equiv_L$ and $\equiv_R$ on G.

Theorem 8.13 *Let H be a subgroup of G. Let the relation $\equiv_L$ be defined on G by*

$$a \equiv_L b (\mathrm{mod}\ H) \iff a^{-1}b \in H.$$

Also, let the relation $\equiv_R$ be defined on G by

$$a \equiv_R b(\mathrm{mod}\ H) \ \Leftrightarrow \ ab^{-1} \in H.$$

Then, $\equiv_L$ *and* $\equiv_R$ *are both equivalence relations on* G.

Proof We show that $\equiv_L$ is an equivalence relation, and leave the proof for $\equiv_R$ as exercise. We must verify the following conditions, for all $a, b, c \in G$,

(1) $a \equiv_L a(\mathrm{mod}\ H)$;
(2) $a \equiv_L b(\mathrm{mod}\ H)$ implies $b \equiv_L a(\mathrm{mod}\ H)$;
(3) $a \equiv_L b(\mathrm{mod}\ H)$ and $b \equiv_L c(\mathrm{mod}\ H)$ imply $a \equiv_L c(\mathrm{mod}\ H)$.

Now, we prove the above items.
(1) Since H is a subgroup, it follows that $a^{-1}a = e$ and $e \in H$, which is what we were required to demonstrate.
(2) Suppose that $a \equiv_L b(\mathrm{mod}\ H)$. Then $a^{-1}b \in H$. Since H is a subgroup, it follows that $(a^{-1}b)^{-1} \in H$. But $(a^{-1}b)^{-1} = b^{-1}(a^{-1})^{-1} = b^{-1}a$. Hence, $b^{-1}a \in H$ and $b \equiv_L a(\mathrm{mod}\ H)$.
(3) Finally, we require that $a \equiv_L b(\mathrm{mod}\ H)$ and $b \equiv_L c(\mathrm{mod}\ H)$ force $a \equiv_L c(\mathrm{mod}\ H)$. Indeed, we have $a^{-1}b \in H$ and $b^{-1}c \in H$. Again, since H is a subgroup, we conclude that $(a^{-1}b)(b^{-1}a) = a^{-1}c \in H$, from which it follows that $a \equiv_L c(\mathrm{mod}\ H)$. ∎

If $[a]_L$ and $[a]_R$ are the equivalence classes containing a related to $\equiv_L$ and $\equiv_R$, respectively, then we have $[a]_L = aH$ and $[a]_R = Ha$.

The left cosets of H in G define a partition of G, i.e.,

(1) For each $a \in G$, $aH \neq \emptyset$;
(2) For any $a, b \in G$, $aH = bH$ or $aH \cap bH = \emptyset$;
(c) $G = \bigcup_{a \in G} aH$.

We can now prove the theorem of Lagrange.

Theorem 8.14 (Lagrange's Theorem) *Let* H *be a subgroup of a finite group* G. *Then the order of* H *is a divisor of the order of* G.

Proof Let us start by recalling that the left cosets of H forms a partition of G. Since G is finite, it follows that there exists a finite number of disjoint left cosets, namely $a_1 H, a_2 H, \ldots, a_n H$. So, we have

$$G = a_1 H \cup a_2 H \cup \cdots \cup a_n H \text{ and } a_i H \cap a_j H = \emptyset, \text{ for all } i \neq j.$$

Let us look at the cardinality of G. We can write

$$|G| = |a_1 H \cup a_2 H \cup \cdots \cup a_n H| = \sum_{i=1}^{n} |a_i H|.$$

By Theorem 8.5 (7), the cosets of H in G all have some size of H, i.e., $|a_i H| = |H|$, for all $1 \leq i \leq n$. This yields that

$$|G| = \underbrace{|H| + |H| + \cdots + |H|}_{n \text{ times}} = n|H|.$$

This shows that the order of H is a divisor of the order of G. ■

There are several corollaries of Lagrange's theorem.

Corollary 8.15 *If G is a finite group and $a \in G$, then $o(a)||G|$.*

Proof Since $\langle a \rangle$ is a subgroup of G, then the order of $\langle a \rangle$ is a divisor of order G. By Corollary 4.27, we know that $o(a) = |\langle a \rangle|$. This shows that $o(a)||G|$. ■

Corollary 8.16 *If G is a finite group and $a \in G$, then $a^{|G|} = e$.*

Proof By Corollary 8.15, we have $o(a)||G|$. Thus, there is a positive integer q such that $|G| = o(a)q$. Consequently, we have

$$a^{|G|} = a^{o(a)q} = (a^{o(a)})^q = e^q = e,$$

as desired. ■

Corollary 8.17 (Euler's Theorem) *If a and n are positive integers such that $(a, n) = 1$, then $a^{\varphi(n)} \equiv 1 (\bmod\ n)$.*

Proof Let U_n be the group of units of $\mathbb{Z}_n$ under multiplication modulo n. By Corollary 4.7, we have $|U_n| = \varphi(n)$. So, by Corollary 8.16, we have $\overline{a}^{\varphi(n)} = \overline{1}$, for all $\overline{a} \in U_n$, which in turn translates into $n|a^{\varphi(n)} - 1$, or equivalently $a^{\varphi(n)} \equiv 1 (\bmod\ n)$. ■

Remark 8.18 In Problem 1.74, we realized a proof for Fermat's little theorem. Since $\varphi(p) = p - 1$, for any prime p, we can conclude Fermat's little theorem from Euler's theorem too.

Corollary 8.19 *Every group of prime order is cyclic.*

Proof Assume that G is of prime order p, and let $a \in G$ be an element different from the identity. Then, the cyclic subgroup $\langle a \rangle$ of G generated by a has at least two elements, a and e. But by Lagrange's theorem, the order $m \geq 2$ of $\langle a \rangle$ must divide the prime p. Consequently, we must have $m = p$ and $\langle a \rangle = G$. This shows that G is cyclic. ■

Corollary 8.20 *Let G be a group and p be a prime number. If G has exactly r subgroups of order p, then it has $r(p - 1)$ elements of order p.*

Proof In each subgroup of order p, all non-identity elements have order p. In addition, an element of order p generates a subgroup of order p. By Lagrange's theorem, the intersection of distinct subgroups of order p is trivial subgroup, so their non-identity elements are disjoint from each other. Consequently, each subgroup of order p has its own $p - 1$ elements of order p, not shared by the other subgroups of order p. Therefore, the number of elements of order p is $r(p - 1)$. ■

The following natural question arises: Is the converse of Lagrange's theorem true? That is, if d is a divisor of the order of G, then does G necessarily have a subgroup of order d. The standard example of the alternating group A_4, which has order 12 but has no subgroup of order 6, shows that the converse of Lagrange's theorem is not true.

Theorem 8.21 *The group A_4 of order 12 has no subgroups of order 6.*

Proof To verify this, recall that A_4 has eight elements of order 3 and suppose that H is a subgroup of order 6. Let σ be any element of order 3 in A_4. If σ is not in H, then $A_4 = H \cup \sigma H$. But then $\sigma^2 \in H$ or $\sigma^2 \in \sigma H$. If $\sigma^2 \in H$, then so is $(\sigma^2)^2 = \sigma^4 = \sigma$, so this case is ruled out. If $\sigma^2 \in \sigma H$, then $\sigma^2 = \sigma h$, for some $h \in H$, but this also implies that $\sigma \in H$. This argument shows that any subgroup of A_4 of order 6 must contain all eight elements of A_4 of order 3, which is absurd. ∎

Theorem 8.22 *If $n > 4$, then A_n has no subgroup of order $n!/4$.*

Proof If H is a subgroup of order $n!/4$, then H has only two left cosets in A_n. So, if σ is a cycle of length 3 in A_n, then the cosets H, σH, and $\sigma^2 H$ cannot be all distinct. Equality of any two of the above cosets implies either $\sigma \in H$ or $\sigma^2 \in H$. Now, $\sigma^2 \in H$ implies $\sigma = \sigma^4 \in H$. Thus, H contains all cycles of length 3. Now, since A_n is generated by cycles of length 3, it follows that $H = A_n$, a contradiction. ∎

Remark 8.23 Theorem 4.31 expresses that the converse of Lagrange's theorem holds for any finite cyclic group.

Theorem 8.24 *Let G be a group of order n. If G has at most one cyclic subgroup of order d where $d|n$, then G is cyclic.*

Proof For each divisor d of n, let $\theta_G(d)$ denote the number of elements of G of order d. For a given positive integer n such that $d|n$, let $\theta_G(d) \neq 0$. Then, there exists $a \in G$ of order d which generates a cyclic group $\langle a \rangle$ of order d of G. We prove that all elements of G of order d belong to $\langle a \rangle$. If $x \in G$ is an element such that $o(x) = d$ and $x \notin \langle a \rangle$, then $\langle x \rangle$ is another subgroup of order d such that $\langle a \rangle \neq \langle x \rangle$. This contradicts the hypothesis. Consequently, if $\theta_G(d) \neq 0$, then $\theta_G(d) = \varphi(d)$, for all positive integer $d|n$. In general, we can write $\theta_G(d) \leq \varphi(d)$, for all positive integer $d|n$. So, we obtain

$$n = \sum_{d|n} \theta_G(d) \leq \sum_{d|n} \varphi(d) = n.$$

This yields that $\theta_G(d) = \varphi(d)$, for all $d|n$. In particular, we have $\theta_G(n) = \varphi(n) \geq 1$. Hence, there exists at least one element of G of order n. This shows that G is cyclic. ∎

Exercises

1. Show that a group with at least two elements but with no proper non-trivial subgroups must be finite and of prime order.

2. Suppose that H and K are unequal subgroups of a group G, each of order 16. Prove that $24 \leq |H \cup K| \leq 31$.

3. For a group G and a subgroup H of G, let $Ha_1 \cup Ha_2 \cup \cdots \cup Ha_n$ be a decomposition of G into disjoint right cosets of H in G. Show that $a_1^{-1}H \cup a_2^{-1}H \cup \cdots \cup a_n^{-1}H$ is a decomposition of G into left cosets of H in G.

4. Let G be a finite group and let H and K be subgroups with relatively prime order. Prove that $H \cap K = \{e\}$.

5. Show that if H and K are subgroups of a group G, and have orders 56 and 63, respectively, then the subgroup $H \cap K$ must be cyclic.

6. Suppose that K is a proper subgroup of H and H is a proper subgroup of G. If $|K| = 42$ and $|G| = 420$, what are the possible orders of H?

7. Use Fermat's little theorem to show that if $p = 4n + 3$ is prime, there is no solution to the equation $x^2 \equiv 1 \pmod{p}$.

8. If G is a finite group with fewer than 100 elements and G has subgroups of orders 10 and 25, what is the order of G?

9. Let G be a group and $a, b \in G$. If $a^5 = e$ and $aba^{-1} = b^2$, find $o(b)$ if $b \neq e$.

10. Does A_5 contain a subgroup of order m for each factor m of 60?

11. Let G be an abelian group of order $2n$ with n odd. Prove that G has precisely one element of order 2.

12. Let G be a finite group. Show that the following conditions are equivalent:

    (1) G is cyclic;
    (2) For each positive integer d, the number of $a \in G$ such that $a^d = e$ is less than or equal to d;
    (3) For each positive integer d, G has at most one subgroup of order d;
    (4) For each positive integer d, G has at most $\varphi(d)$ elements of order d.

13. Let G be a finite group and H be a subgroup of G. Let $l(a)$ be the smallest positive integer m such that $a^m \in H$. Prove that $l(a) | o(a)$.

14. Let n be a positive integer and let m be a factor of $2n$. Show that D_n contains a subgroup of order m.

15. Using Lagrange's theorem, show that the binomial coefficient

$$\binom{n}{m} = \frac{n!}{m!(n-m)!},$$

for every integers m and n with $n \geq 1$ and $0 \leq m \leq n$, is an integer. *Hint:* Consider a subgroup of S_n of order $m!(n-m)!$.

8.4 Index of Subgroups

A special name and notation have been adopted for the number of left (or right) cosets of a subgroup in a group.

Definition 8.25 Let H be a subgroup of a group G. The number of distinct left (or right) cosets of H in G is the *index* of H in G. The index of H in G is denoted by $[G : H]$.

Note that if G is finite, then the index $[G : H]$ divides $|G|$.

Example 8.26 Consider Example 8.3, and let $H = \{id, \ (1\ 3)\}$ be the subgroup of S_3. Then, we observe that $[S_3 : H] = 3$.

Example 8.27 A left coset of $SL_2(\mathbb{F})$ in $GL_2(\mathbb{F})$ has the form $XSL_2(\mathbb{F}) = \{XA \mid A \in SL_2(\mathbb{F})\}$, where $X \in GL_2(\mathbb{F})$. If $\det(X) = c$, then $\det(XA) = c$. Therefore, all matrices in $XSL_2(\mathbb{F})$ have the determinant equal to c. Conversely, let $B \in GL_2(\mathbb{F})$ such that $\det(B) = c$. Take $A = X^{-1}B$, then $B = XA$. On the other hand, $\det(A) = \det(X^{-1}B) = \det(X^{-1})\det(B) = c^{-1}c = 1$. This means that $A \in SL_2(\mathbb{F})$. So, we deduce that $B \in XSL_2(\mathbb{F})$. Consequently, we have $XSL_2(\mathbb{F}) = \{A \mid A \in GL_2(\mathbb{F})$ and $\det(A) = c\}$. Hence, each left coset of $SL_2(\mathbb{F})$ in $GL_2(\mathbb{F})$ has the above description, for some non-zero element $c \in \mathbb{F}$. This shows that $[GL_2(\mathbb{F}) : SL_2(\mathbb{F})]$ is infinite.

Theorem 8.28 *If H and K are subgroups of a group G such that $K \leq H$, then* $[G : K] = [G : H][H : K]$.

Proof Suppose that $G = \bigcup_{i \in I} a_i H$, the union of distinct left cosets of H in G, where $|I| = [G : H]$, and let $H = \bigcup_{j \in J} b_j K$, the union of distinct left cosets of K in H, where $|J| = [H : K]$. Then, we can write

$$G = \bigcup_{i \in I} a_i H = \bigcup_{i \in I} a_i \left(\bigcup_{j \in J} b_j K \right) = \bigcup_{(i,j) \in I \times J} a_i b_j K.$$

Now, we claim that $a_i b_j K$'s are distinct. Assume that $a_i b_j K = a_k b_l K$, for some $i, k \in I$ and $j, l \in J$. Since $b_j K \subseteq H$ and $b_l K \subseteq H$, it follows that $a_i H \cap a_k H \neq \emptyset$. This implies that $a_i H = a_k H$, and so $i = k$. Then, we obtain $b_j K = b_l K$, and hence $j = l$. Therefore, we deduce that $a_i b_j K$'s are distinct left cosets of K in G. This yields that $[G : K] = [G : H][H : K]$. ∎

Theorem 8.29 (Poincaré Lemma) *If H and K are two subgroups of finite index in a group G such that $[G : H] = m$ and $[G : K] = n$, then $H \cap K$ is also of finite index in G and $[m, n] \leq [G : H \cap K] \leq mn$. In particular, if m and n are relatively prime, then $[G : H \cap K] = mn$. Moreover, if G is finite, then $G = HK$.*

Proof Let $a_1 H, a_2 H, \ldots, a_m H$ and $b_1 K, b_2 K, \ldots, b_n K$ be distinct left cosets of H and K in G, respectively. Now, by Theorem 8.7, for every $1 \le i \le m$ and $1 \le j \le n$ either $a_i H \cap b_j K = \emptyset$ or $a_i H \cap b_j K = c_{ij}(H \cap K)$, for some $c_{ij} \in G$. On the other hand, $x(H \cap K) = xH \cap xK$. Therefore, each left coset of $H \cap K$ is determined by intersection of a left coset of H and a left coset of K in G. Consequently, distinct number of left cosets of $H \cap K$ is at most equal to mn.

Now, by Theorem 8.28, we can write

$$[G : H \cap K] = [G : H][H : H \cap K] = m[H : H \cap K],$$
$$[G : H \cap K] = [G : K][K : H \cap K] = n[K : H \cap K],$$

and hence $m|[G : H \cap K]$ and $n|[G : H \cap K]$. Therefore, we conclude that the least common multiple of m and n divides $[G : H \cap K]$.   ∎

Corollary 8.30 *In Theorem 8.29, if m and n are relatively prime, then* $[G : H \cap K] = mn$.

Proof It is straightforward.   ∎

Theorem 8.31 *If H and K are subgroups of a group G, then*

$$[H : H \cap K] \le [G : K].$$

Proof Suppose that $\mathcal{A}$ is the set of all disjoint left cosets of $H \cap K$ in H and $\mathcal{B}$ is the set of all disjoint left cosets of K in G. We define $f : \mathcal{A} \to \mathcal{B}$ by $f\big(h(H \cap K)\big) = hK$, for all $h \in H$. If $h(H \cap K) = h'(H \cap K)$, for some $h, h' \in H$, then $h^{-1}h' \in H \cap K \subseteq K$. This implies that $hK = h'K$. So, we conclude that f is well defined. Moreover, if $hK = h'K$, for some $h, h' \in H$, then $h^{-1}h' \in K$. Since H is a subgroup and $h, h' \in H$, it follows that $h^{-1}h' \in H$. Hence, we have $h^{-1}h' \in H \cap K$, which implies that $h(H \cap K) = h'(H \cap K)$. Therefore, f is one to one. This yields this $|\mathcal{A}| \le |\mathcal{B}|$, or equivalently $[H : H \cap K] \le [G : K]$.   ∎

Exercises

1. If G is an infinite cyclic group and $\{e\} \ne H \le G$, prove that $[G : H]$ is finite.
2. Show that the integers have infinite index in the additive group of rational numbers.
3. Suppose that G is a finite group and H, K are subgroups of G such that $K \subset H$ and $[G : K]$ is prime. Prove that $H = G$.
4. Determine the index of the following subgroups in the corresponding groups:

    (a) $\{\bar{0}, \bar{3}, \bar{6}, \bar{9}\}$ in $\mathbb{Z}_{12}$;
    (b) $\mathbb{R}$ in $(\mathbb{C}, +)$;
    (c) $3\mathbb{Z}$ in $\mathbb{Z}$.

5. Show that if H is a subgroup of index 2 in a finite group G, then every left coset of H is also a right coset of H.
6. Two subgroups H and K of a group G are said to be *commensurable* if $H \cap K$ is of finite index in both H and K. Show that commensurability is an equivalence relation on the subgroups of G.

8.5 A Counting Principle and Double Cosets

Let A and B be two subsets of a group G. The set $AB = \{ab \mid a \in A, \ b \in B\}$ consisting of the products of elements $a \in A$ and $b \in B$ is said to be the *product* of A and B. The associative law of multiplication gives us $(AB)C = A(BC)$ for any three subsets A, B, and C. The product of two subgroups is not necessarily a subgroup. We have the following theorem.

Theorem 8.32 *Let H and K be two subgroups of a group G. Then, the following two conditions are equivalent:*

(1) The product HK is a subgroup of G;
(2) $HK = KH$.

Proof Suppose that HK is a subgroup of G. Then, for any $h \in H$ and $k \in K$, we have $h^{-1}k^{-1} \in HK$ and so $kh = (h^{-1}k^{-1})^{-1} \in HK$. Thus, $KH \subseteq HK$. Now, if x is any element of HK, then $x^{-1} = hk \in HK$, for some $h \in H$ and $k \in K$. So, we obtain $x = (x^{-1})^{-1} = (hk)^{-1} = k^{-1}h^{-1} \in KH$. This means that $HK \subseteq KH$. Thus, we conclude that $HK = KH$.

On the other hand, suppose that $HK = KH$, i.e., if $h \in H$ and $k \in K$, then $hk = k_1 h_1$ for some $h_1 \in H$ and $k_1 \in K$. In order to prove that HK is a subgroup of G, we must verify that it is closed and every element in HK has its inverse in HK. Suppose that $x = hk \in HK$ and $y = h'k' \in HK$, where $h, h' \in H$ and $k, k' \in K$. Then, we have $xy = hkh'k'$, but since $kh' \in KH = HK$, it follows that $kh' = h_2 k_2$ with $h_2 \in H$ and $k_2 \in K$. Hence, $xy = h(h_2 k_2)k' = (hh_2)(k_2 k') \in HK$. Clearly, $x^{-1} = k^{-1}h^{-1} \in KH = HK$. Consequently, HK is a subgroup of G. ∎

Corollary 8.33 *If H and K are subgroups of the abelian group G, then HK is a subgroup of G.*

Theorem 8.34 *If H and K are finite subgroups of a group G, then*

$$|HK| = \frac{|H||K|}{|H \cap K|}.$$

Proof Suppose that $A = H \cap K$, and let the index of A in H is n. Then, we can write $H = h_1 A \cup h_2 A \cup \cdots \cup h_n A$, in which $h_i A \cap h_j A = \emptyset$, for any $i \neq j$. On the other hand, we have $HK = h_1 AK \cup h_2 AK \cup \cdots \cup h_n AK$. Since $AK = K$, it follows that $HK = h_1 K \cup h_2 K \cup \cdots \cup h_n K$. We claim that $h_i K \cap h_j K = \emptyset$, for any $i \neq j$. Indeed, if $h_i K = h_j K \neq \emptyset$, for some $i \neq j$, then there exist $k_1, k_2 \in K$ such that $h_i k_1 = h_j k_2$, and then $h_i h_j^{-1} = k_2 k_1^{-1} \in A$. This implies that $h_i A = h_j A$, a contradiction. Therefore, we obtain $|HK| = n|K|$. Since $n = |H|/|A|$, it follows that $|HK| = |H||K|/|H \cap K|$.   ∎

Corollary 8.35 *If H and K are finite subgroups of a group G such that $\min\{|H|, |K|\} > \sqrt{|G|}$, then $H \cap K \neq \{e\}$.*

Proof Since $HK \subseteq G$, it follows that $|HK| \leq G$. On the other hand, by Theorem 8.34, we have

$$|G| \geq |HK| = \frac{|H||K|}{|H \cap K|} > \frac{\sqrt{|G|}\sqrt{|G|}}{|H \cap K|} = \frac{|G|}{|H \cap K|}.$$

So, we conclude that $|H \cap K| > 1$. This yields that $H \cap K \neq \{e\}$.   ∎

Definition 8.36 Let G be a group and H, K be two subgroups of G and let $x \in G$. Then, the set $HxK = \{hxk \mid h \in H \text{ and } k \in K\}$ is called a *double coset* of H and K in G.

Lemma 8.37 *For any $x, y \in G$, two double cosets HxK and HyK are either disjoint or identical.*

Proof Suppose that $a \in HxK \cap HyK$. Then, $a = hxk = h'yk'$, for some $h, h' \in H$ and $k, k' \in K$. This implies that $x = h^{-1}h'yk'k^{-1}$ and $y = h'^{-1}hxkk'^{-1}$. Now, we show that $HxK = HyK$. Let $u \in HxK$ be an arbitrary element. Then, $u = h_1 x k_1$, for some $h_1 \in H$ and $k_1 \in K$. Consequently, we obtain $u = h_1 h^{-1} h' y k' k^{-1} k_1 \in HyK$. Conversely, if $v \in HyK$ is an arbitrary element, then $v = h_2 y k_2$, for some $h_2 \in H$ and $k_2 \in K$. Thus, we have $v = h_2 h'^{-1} hxkk'^{-1} k_2 \in HxK$.   ∎

Let $x, y \in G$ and define $x \sim y$ if and only if $x = hyk$, for some $h \in H$ and $k \in K$. It can be shown easily that $\sim$ is an equivalence relation and the equivalence class of x is HxK.

Lemma 8.38 *If H and K are subgroups of a finite group G and $x \in G$, then*

$$|HxK| = \frac{|H||K|}{|x^{-1}Hx \cap K|}.$$

Proof It is easy to see that the function $f : HxK \to x^{-1}HxK$ defined by $f(hxk) = x^{-1}hxk$, for all $h \in H$ and $k \in K$, is a one to one correspondence. Hence, we deduce that $|HxK| = |x^{-1}HxK| = |x^{-1}Hx||K|/|x^{-1}Hx \cap K|$. Since $|x^{-1}Hx| = |H|$, the result follows.   ∎

Theorem 8.39 *If H and K are subgroups of a finite group G, then*

$$|G| = \sum \frac{|H||K|}{|x^{-1}Hx \cap K|},$$

where summation on right side is taken over elements x chosen from disjoint double cosets HxK.

Proof For each $x \in G$, we have $x \in HxK$. In view of this, we conclude that $G = \bigcup_{x \in G} HxK$. If we write disjoint union, then we get $|G| = \sum |HxK|$. Hence, by Lemma 8.38, the result follows. ∎

Lemma 8.40 *Let H and K be subgroups of a group G and a $\in$ G. Then, the double coset HxK is a union of some right cosets of H in G, and also a union of some left cosets of K in G. Moreover, the number of distinct right cosets of H in HxK is $[K : K \cap x^{-1}Hx]$, and the number of distinct left cosets of K in HxK is $[H : H \cap x^{-1}Kx]$.*

Proof Clearly, we have

$$HxK = \bigcup_{k \in K} Hxk \quad \text{and} \quad HxK = \bigcup_{h \in H} hxK.$$

Now, assume that $k, k' \in K$. Then, we have

$$
\begin{aligned}
Hxk = Hxk' &\Leftrightarrow (xk)(xk')^{-1} \in H \Leftrightarrow xkk'^{-1}x^{-1} \in H \\
&\Leftrightarrow kk'^{-1} \in x^{-1}Hx \Leftrightarrow kk'^{-1} \in K \cap x^{-1}Hx \text{ (since } k, k' \in K) \\
&\Leftrightarrow k(K \cap x^{-1}Hx) = k'(K \cap x^{-1}Hx).
\end{aligned}
$$

This shows that the number of distinct right cosets of H in HxK is $[K : K \cap x^{-1}Hx]$. The proof of the last part is similar. ∎

Theorem 8.41 *Let H and K be subgroups of a group G. If $[G : H]$ is finite, then*

$$[G : H] = \sum_{x \in X} [K : K \cap x^{-1}Hx],$$

where X is the set of representatives of distinct double cosets. Similarly, if $[G : K]$ is finite, then

$$[G : K] = \sum_{x \in X} [H : H \cap x^{-1}Kx],$$

Proof We know that

$$G = \bigcup_{x \in X} Hxk.$$

By Lemma 8.40, for any $x \in X$, the number of distinct right cosets of H in HxK is $[K : K \cap x^{-1}Hx]$. Hence, the number of total distinct right cosets of H in G is

$$[G : H] = \sum_{x \in X} [K : K \cap x^{-1}Hx].$$

The proof of the second part is similar. ∎

Exercises

1. Give an example of a group G and subgroups H and K such that HK is not a subgroup of G.
2. Let H, K, and N be subgroups of a group G such that $H \leq K$, $H \cap N = K \cap N$ and $HN = KN$. Show that $H = K$.
3. If A and B are non-empty finite subsets of a group G, prove that $G = AB$ or $|A| + |B| \leq |G|$.
4. (a) **(Dedekind's Modular Law).** Let H, K, and N be subgroups of a group G and assume that $K \leq N$. Prove that

$$(HK) \cap N = (H \cap N)K.$$

   (b) In particular, if H and K permute, prove that

$$\langle H, K \rangle \cap N = \langle H \cap N, K \rangle.$$

   (c) Can you give an example to show that $(HK) \cap N$ is not equal to $(H \cap N)(K \cap N)$?

* Find the number of left cosets of K which are contained in the double coset HxK.
* Let H, K be two subgroups of G. Prove that every subgroup of G containing both H and K contains the product sets HK and KH.

8.6 Worked-Out Problems

Problem 8.42 Let $G = \langle a \rangle$ be an infinite cyclic group and let $H = \langle a^k \rangle$, where k is a positive integer. Show that

$$H, \ aH, \ a^2H, \ \ldots, \ a^{k-1}H \tag{8.1}$$

is a complete list repetition free of the left (right) cosets of H in G.

Solution The problem is subdivided into proving that

(1) The list (8.1) is complete, i.e., every element of G belongs to one of the left cosets listed;
(2) The list (8.1) is repetition free.

(1) Suppose that x is an arbitrary element of G. Then, there exists positive integer n such that $x = a^n$. Now, by the Division algorithm, we can write $n = qk + r$, for some integers q and r with $0 \leq r < k$. So, we have $x = a^n = a^{qk+r} = a^r(a^k)^q \in a^r H$. Since $1 \leq r \leq k - 1$, it follows that $a^r H$ is one of the left cosets in the list (8.1).

Therefore, every element of G lies in one of the left cosets in the list (8.1).

(2) Suppose (with a view to obtaining a contradiction) that $a^i H = a^j H$ with $0 \leq i < j \leq k - 1$. Then, we get $(a^i)^{-1} a^j \in H$, or equivalently $a^{j-i} \in H$. Hence, we conclude that $a^{j-i} = a^{mk}$, for some integer m. Since a has infinite order, it follows that $j - i = mk$. This is a contradiction since $0 < j - i < k$. From this contradiction it follows that the list (8.1) is repetition free. ∎

Problem 8.43 Let $d(G)$ be the smallest number of elements necessary to generate a finite group G. Prove that $|G| \geq 2^{d(G)}$. Note that by convention $d(G) = 0$ if $|G| = 1$.

Solution We do the proof by mathematical induction. If $d(G) = 1$, then $|G| > 2$, because each non-identity element has order of at least 2. Suppose that if a group is generated by $n - 1$ elements, then its order is at least 2^{n-1}. Now, let $G = \langle a_1, a_2, \ldots, a_n \rangle$, i.e., G is generated by n elements. Then, the subgroup $H = \langle a_1, a_2, \ldots, a_{n-1} \rangle$ is a proper subgroup of G, and hence by assumption we have $|H| \geq 2^{n-1}$. Since $a_n \notin H$, it follows that $a_n H \cap H = \emptyset$. In addition, we have $a_n H \cup H \subseteq G$. Consequently, we obtain

$$|G| \geq |a_n H \cup H| = |a_n H| + |H| \geq 2|H| = 2 \cdot 2^{n-1} = 2^n,$$

and we are done. ∎

Problem 8.44 Prove that a group has exactly three subgroups if and only if it is cyclic of order p^2, for some prime p.

Solution Suppose that G is a cyclic group of order p^2. By Theorem 4.31, G has a unique subgroup H of order p. Therefore, the subgroups of G are $\{e\}$, H and G.

Conversely, assume that G is a group which has exactly three subgroups. Then, we conclude that there exists only one non-trivial proper subgroup H of G. Since H has no non-trivial subgroup, it follows that H is a group of order p, for some prime p. Let $H = \langle a \rangle$. Since $G \neq H$, it follows that there exists $b \in G - H$. Now, $\langle b \rangle$ is a subgroup of G different from H. Hence, we conclude that $\langle b \rangle = G$. This means that G is cyclic, and it has a subgroup of order p. Thus, G is finite and the only prime divisor of $|G|$ is p. Consequently, $|G|$ must be p^2, otherwise G has a subgroup from other divisors. ∎

Problem 8.45 Let H and K be subgroups of a group G, and suppose that L is a left coset of H in G and R_1, R_2 are two right cosets of K in G. If $L \cap R_1 \neq \emptyset$ and $L \cap R_2 \neq \emptyset$, prove that $|L \cap R_1| = |L \cap R_2|$.

Solution Assume that $a \in L \cap R_1$ and $b \in L \cap R_2$. Then, we have $L = aH = bH$, $R_1 = Ka$ and $R_2 = Kb$. So, we obtain

$$L \cap R_1 = aH \cap Ka = (aHa^{-1} \cap K)a,$$
$$L \cap R_2 = bH \cap Kb = (bHb^{-1} \cap K)b.$$

Since $aH = bH$, it follows that $a^{-1}b \in H$ or $b^{-1}a \in H$. This shows that $Ha^{-1} = Hb^{-1}$. Consequently, we can write $L \cap R_1 = (aHa^{-1} \cap K)a$ and $L \cap R_2 = $

$(aHa^{-1} \cap K)b$. Since $aHa^{-1} \cap K$ is a subgroup of G, it follows that $L \cap R_1$ and $L \cap R_2$ are right cosets of $aHa^{-1} \cap K$ in G. This forces $|L \cap R_1| = |L \cap R_2|$. ∎

Problem 8.46 Let G be a group of order 100 that has a subgroup H of order 25. Prove that every element of G of order 5 is in H.

Solution Suppose that a is an element of G of order 5. Then, by Theorem 8.34, we can write

$$|\langle a \rangle H| = \frac{|\langle a \rangle||H|}{|\langle a \rangle \cap H|} = \frac{5 \cdot 25}{|\langle a \rangle \cap H|} = \frac{125}{|\langle a \rangle \cap H|}.$$

Since $\langle a \rangle H \subseteq G$, it follows that $|\langle a \rangle H| \leq 100$. This forces $|\langle a \rangle \cap H| > 1$. Since $|\langle a \rangle| = 5$ and $\langle a \rangle \cap H \subseteq \langle a \rangle$, it follows that $|\langle a \rangle H| = 5$. This yields that $|\langle a \rangle H| = \langle a \rangle$, and hence we conclude that $a \in H$. ∎

Problem 8.47 Prove that if G is a finite group, the index of $Z(G)$ in G cannot be prime.

Solution Suppose that $[G : Z(G)] = p$, where p is a prime. Since $p > 1$, it follows that $G \neq Z(G)$, and so there exists $a \in G - Z(G)$. If $C_G(a)$ is the centralizer of a in G, then $Z(G)$ is a subgroup of $C_G(a)$. Since $a \in C_G(a)$ and $a \notin Z(G)$), it follows that $C_G(a) \neq Z(G)$. Hence, we conclude that

$$[C_G(a) : Z(G)] > 1. \tag{8.2}$$

On the other hand, by Theorem 8.29, we have

$$p = [G : Z(G)] = [G : C_G(a)][C_G(a) : Z(G)]. \tag{8.3}$$

Since p is prime, by (8.2) and (8.3), it follows that $[G : C_G(a)] = 1$ and $[C_G(a) : Z(G)] = p$. This shows that $G = C_G(a)$. It means that every element of G commutes with a, and it is a contradiction as $a \notin Z(G)$. Therefore, we conclude that $[G : Z(G)]$ cannot be prime. ∎

8.7 Supplementary Exercises

1. Let G be a finite group whose order is not divisible by 3. Suppose that $(ab)^3 = a^3 b^3$, for all $a, b \in G$. Prove that G must be abelian.
2. Let $n = n_1 + \cdots + n_r$ be a partition of the positive integer n. Use Lagrange's theorem to show that $n!$ is divisible by

$$\prod_{i=1}^{r} n_i!.$$

3. Let G be a finite abelian group and let m be the least common multiple of the orders of its elements. Prove that G contains an element of order m.

4. Let G be a finite group. Prove that the number of elements $x \in G$ such that $x^7 = e$ is odd.

5. If G is a finite group with precisely 2 conjugacy classes, prove $|G| = 2$.

6. If $\sigma \in S_n$ has order p, p is a prime, and $(n, p) = 1$, show that $r\sigma = r$, for some $1 \le r \le n$.

7. If in a group G of order n, for each positive integer $m|n$, the equation $x^m = e$ has less than $m + \varphi(m)$ solutions, then show G is cyclic.

8. In a cyclic group of order n, show that for each integer m that divides n there exist $\varphi(m)$ elements of order m.

9. Suppose that G is an abelian group with an odd number of elements. Show that the product of all of the elements of G is the identity.

10. Let G be a group such that $|G| < 200$. Suppose that G has subgroups of order 25 and 35. Find the order of G.

11. Let G be a group of order pqr, where $p, q,$ and r are distinct primes. If H and K are subgroups of G with $|H| = pq$ and $|K| = qr$, prove that $|H \cap K| = q$.

12. Let G be the set of all matrices of the form

$$\begin{bmatrix} 2^k & p(x) \\ 0 & 1 \end{bmatrix},$$

where $k \in \mathbb{Z}$ and $p(x)$ is any polynomial with rational coefficients. Show that

(a) G is a group under matrix multiplication;
(b) There exists a subgroup H of G and an element $u \in G$ such that the left coset uH contains an infinite number of right cosets of H in G;
(c) The chain $H \subset uHu^{-1} \subset u^2Hu^{-2} \subset \cdots$ is an infinite proper ascending chain.

13. Let G be an abelian group and suppose that G has elements of orders m and n, respectively. Prove that G has an element whose order is the least common multiple of m and n.

14. In Theorem 8.31, show that equality holds if and only if $G = HK$.

15. Let H be a subgroup of a group G such that $[G : H] = p$, where p is a prime number. If H is a subgroup of $Z(G)$, prove that G is abelian.

16. Let G be an abelian group of order n and $a_1, \ldots, a_n$ be elements of G. Let $x = a_1 a_2 \ldots a_n$. Show that

(a) If G has exactly one element of order 2, then $x = b$;
(b) If G has more than one element of order 2, then $x = e$;
(c) If n is odd, then $x = e$.

17. Let H and K be subgroups of a finite group G. Show that

(a) $[H : H \cap K] \le [H \vee K : K]$, where $H \vee K = \langle H \cup K \rangle$;
(b) If $[G : K] < 2[H : H \cap K]$, then $G = H \vee K$.

18. Let $n > 1$. Show that there exists a proper subgroup H of S_n such that $[S_n : H] \leq n$.

19. Using Lagrange's theorem, for any integers m and n, prove that

$$\frac{(mn)!}{(m!)^n} \quad \text{and} \quad \frac{(mn)!}{(m!)^n n!}$$

are integers.

Hint: Consider the integers 1 to mn in a family of consecutive integers as follows: $\{1, \ldots, m\}, \{m + 1, \ldots, 2m\}, \ldots, \{(n - 1)m + 1, \ldots, nm\}$. The elements of symmetric group S_{mn} that move each set within itself is a subgroup of S_{mn}.

20. Let G be a group and H be a subgroup of G. Prove that

$$\left| \bigcup_{a \in G} a^{-1} H a \right| \leq 1 + |G| - [G : H].$$

21. Suppose that G has exactly m subgroups of order p, where p is a prime number. Show that the total number of elements of order p in G is $m(p - 1)$. Deduce the following results:

(a) A non-cyclic group of order 55 has at least one element of order 5;

(b) A non-cyclic group of order p^2 has altogether $p + 3$ subgroups.

Chapter 9
Normal Subgroups and Factor Groups

The set of cosets of a subgroup H has no group structure. We are now interested in a criterion on H to give the set of its cosets a group structure. In this chapter, we introduce the concept of normal subgroups and we form a group of cosets, say factor group. A factor group is a way of creating a group from another group. This new group often retains some of the properties of the original group.

9.1 Normal Subgroups

There is one kind of subgroup that is especially interesting. If G is a group and H is a subgroup of G, it is not always true that $aH = Ha$ for all $a \in G$. There are certain situations where this does hold, however, and these cases turn out to be of critical importance in the theory of groups. It was Galois, who first recognized that such subgroups were worthy of special attention.

Definition 9.1 Let G be a group and N be a subgroup of G. We say that N is a *normal subgroup* of G, or that N is *normal* in G, if we have $aNa^{-1} \subseteq N$, for all $a \in G$.

Equivalently, a subgroup N of G is normal if $axa^{-1} \in N$, for all $a \in G$ and $x \in N$. Note that, also, we can say $a^{-1}xa^{-1} \in N$, for all $a \in G$ and $x \in N$.

Example 9.2 Clearly, the trivial subgroups G and $\{e\}$ of a group G are normal subgroups.

Example 9.3 If G is an abelian group, then all its subgroups are normal.

Example 9.4 The center of a group G forms a normal subgroup of G.

© The Author(s), under exclusive license to Springer Nature Singapore Pte Ltd. 2021   217
B. Davvaz, *A First Course in Group Theory*,
https://doi.org/10.1007/978-981-16-6365-9_9

Example 9.5 Consider the dihedral group D_n generated by R and S with $R^n = Id$, $S^2 = Id$, and $SRS = R^{-1}$. If N is the subgroup of rotational symmetries, then N is normal in D_n.

We denote it $N \trianglelefteq G$, or $N \lhd G$ when we want to emphasize that N is a proper subgroup of G. The condition for a subgroup to be normal can be stated in many slightly different ways.

Definition 9.6 Let G be a group and X be a non-empty subset of G. The set

$$N_G(X) = \{a \in G \mid aXa^{-1} = X\}$$

is called the *normalizer* of X in G.

Theorem 9.7 *If G is a group and X is a non-empty subset of G, then $N_G(X)$ is a subgroup of G. In particular, if H is a subgroup of G, then $H \trianglelefteq N_G(H)$.*

Proof It is straightforward.   ∎

Theorem 9.8 *Let N be a subgroup of a group G. The following conditions are equivalent:*

(1) N is a normal subgroup of G;
(2) $aNa^{-1} = N$, for all $a \in G$;
(3) $aN = Na$, for all $a \in G$;
(4) Every left coset of N in G is also a right coset of N in G.

Proof $(1 \Rightarrow 2)$: Let N be a normal subgroup of G. Then, for each $a \in G$, we have $aNa^{-1} \subseteq N$. If $a \in G$, then $a^{-1} \in G$, and so $a^{-1}N(a^{-1})^{-1} = a^{-1}Na \subseteq N$. This yields that $N \subseteq aNa^{-1}$. Thus, we conclude that $aNa^{-1} = N$.

$(2 \Rightarrow 3)$: Let $a \in G$ be arbitrary. Since $aNa^{-1} = N$, it follows that $aNa^{-1}a = Na$, or equivalently $aN = Na$.

$(3 \Rightarrow 4)$: It is straightforward.

$(4 \Rightarrow 1)$: Suppose that every left coset of N in G is a right coset of N in G. Thus, for $a \in G$, aN being a left coset, must be a right coset. Let $aN = Nb$, for some $b \in G$. Since $a \in aN$, it follows that $a \in Nb$. Hence, we get $Na \subseteq NNb \subseteq Nb$. This shows that $Na = Nb$, and so $aN = Na$. Consequently, $aNa^{-1} = N$, and this implies that N is a normal subgroup of G.   ∎

Theorem 9.9 *For each positive integer n, A_n is a normal subgroup of S_n.*

Proof Assume that $\alpha \in S_n$ is a product of k transpositions, say $\alpha = \tau_1 \ldots \tau_k$. Then, we have $\alpha^{-1} = \tau_k^{-1} \ldots \tau_1^{-1}$. Now, if $\sigma \in A_n$, then σ is even, say a product of $2m$ transpositions. Consequently, $\alpha\sigma\alpha^{-1}$ is a product of $k + 2m + k = 2(k + m)$ transpositions, and so $\alpha\sigma\alpha^{-1}$ is even.   ∎

Theorem 9.10 *$SL_n(\mathbb{F})$ is a normal subgroup of $GL_n(\mathbb{F})$.*

Proof For every $A \in GL_n(\mathbb{F})$ and $B \in SL_n(\mathbb{F})$, we have

$$\det(ABA^{-1}) = det(A) \det(B) \det(A^{-1}) = \det(B) = 1.$$

This shows that $ABA^{-1} \in SL_2(\mathbb{F})$, and so $SL_n(\mathbb{F}) \trianglelefteq GL_n(\mathbb{F})$. ∎

Lemma 9.11 *Let G be a group and $N \trianglelefteq G$. If $N \leq H \leq G$, then $N \trianglelefteq H$.*

Proof It is clear. ∎

Theorem 9.12 *Let H and N be subgroups of a group G. If $N \trianglelefteq G$, then $HN = NH \leq G$.*

Proof By Theorem 8.32, it is enough to show that $HN = NH$. Let $x \in HN$ be an arbitrary element. Then, there exist $h \in H$ and $n \in N$ such that $x = hn$. Since $N \trianglelefteq G$, it follows that $hnh^{-1} \in N$. Hence, we conclude that $hn \in Nh \subseteq NH$, or equivalently, $x \in NH$. This shows that $HN \subseteq NH$. Analogously, we observe that $NH \subseteq HN$. Therefore, we get $HN = NH$. ∎

Theorem 9.13 *Let H and N be normal subgroups of a group G. Then,*

(1) $H \cap N$ is a normal subgroup of G;
(2) HN is a normal subgroup of G.

Proof (1) Let $a \in G$ and $x \in H \cap N$ be arbitrary. Since $H \trianglelefteq G$, it follows that $axa^{-1} \in H$. Since $N \trianglelefteq G$, it follows that $axa^{-1} \in N$. Consequently, we have $axa^{-1} \in H \cap N$.

(2) By Theorem 9.12, HN is a subgroup of G. Now, let $a \in G$ and $hn \in HN$ be arbitrary. Then, we get $ahna^{-1} = (aha^{-1})(ana^{-1}) \in HN$, using the normality of both H and N. ∎

Theorem 9.14 *Let G be a group, not necessarily finite, and let N be a subgroup of G such that the index $[G : N] = 2$. Then, N is a normal subgroup of G.*

Proof Since $[G : N] = 2$, it follows that N has two left cosets and two right cosets. One of them is always N itself. Take $a \notin N$. Then, aN is the other left coset, Na is the other right coset, and $N \cup aN = N \cup Na = G$. But these are disjoint unions, so $aN = Na$, and therefore $aNa^{-1} = N$. This equation holds for any a in the coset aN. The equation clearly holds for any element of the coset N. Hence, the equation holds for all elements of G, and we conclude that N is normal. ∎

In the following example, we present three groups K, H and G such that $K \trianglelefteq H$ and $H \trianglelefteq G$, but K is not normal in G.

Example 9.15 Let $D_4 = \langle R, S \rangle$ be the dihedral group of order 8. Since $[\langle R^2, S \rangle : \langle S \rangle] = 2$, it follows that $\langle S \rangle \trianglelefteq \langle R^2, S \rangle$. Since $[D_4 : \langle R^2, S \rangle] = 2$, it follows that $\langle R^2, S \rangle \trianglelefteq D_4$. But $\langle S \rangle$ is not a normal subgroup of D_4.

Theorem 9.16 *If p is the smallest prime number that divides the order of a finite group G, then every subgroup of G with the index p is a normal subgroup.*

Proof Let N be a subgroup of a finite group G of order n such that $[G : N] = p$. First we claim that if $a \notin N$, then $a^i \notin N$, for all $1 \leq i \leq p - 1$. Indeed, if this statement is not true, then there is $1 < k \leq p - 1$ such that $a^k \in N$. Suppose that j is the least positive integer such that $a^j \in N$, i.e., for $1 \leq t \leq j - 1$, $a^t \notin N$. Let $o(a) = m$. Since $m | n$, $1 < j < p$ and p is the smallest prime factor of n, we conclude that $j \nmid m$. Hence, there exist integers q and r such that $m = qj + r$ and $0 < r < j$. Now, we can write

$$e = a^m = a^{qj+r} = \left(a^j\right)^q a^r.$$

This implies that $x^r = \left(a^j\right)^{-q}$. Since $a^j \in N$, it follows that $a^r \in N$, and it is a contradiction. Therefore, we proved that our claim holds.

In order to prove the theorem, assume that N is not a normal subgroup of G. So, there exist $a \in G$ and $x \in N$ such that $axa^{-1} \notin N$. Obviously, $a \notin N$, and so by the above discussion, we deduce that $a^i \notin N$, for every $1 \leq i \leq p - 1$. On the other hand, if $a^i N = a^j N$, then there is $y \in N$ such that $a^i = a^j y$, or equivalently $a^{i-j} = y$, and it is a contradiction. Consequently, $N, aN, \ldots, a^{p-1}N$ are disjoint left cosets of N in G. On the other hand, assume that $axa^{-1} = b$. Since $b \notin N$, in a similar way, it follows that $N, bN, \ldots, b^{p-1}N$ are disjoint left cosets of N in G, too. Thus, the following two families of left cosets

$$\{N, aN, \ldots, a^{p-1}N\} \text{ and } \{N, bN, \ldots, b^{p-1}N\}$$

are equal. So, $aN = b^r N$, for some $1 \leq r \leq p - 1$. This means that $a = b^r z$, for some $z \in N$. Now, we obtain

$$a = b^r z = \left(axa^{-1}\right)^r z = ax^r a^{-1} z.$$

Hence, we have $e = x^r a^{-1} z$, or equivalently $a = zx^r$. Since x and z belong to N, it follows that $a \in N$, a contradiction. Therefore, we conclude that N is a normal subgroup of G. ∎

Theorem 9.17 *Let G be a group and H and K be normal subgroups of G. If $|H \cap K| = 1$, then $hk = kh$, for all $h \in H$ and $k \in K$.*

Proof Suppose that $h \in H$ and $k \in K$ are arbitrary. We consider the element $hkh^{-1}k^{-1}$. On the one hand, since $H \trianglelefteq G$, it follows that $h(kh^{-1}k^{-1}) \in H$, and on the other hand, since $K \trianglelefteq G$, it follows that $(hkh^{-1})k^{-1}) \in K$. So, we obtain $hkh^{-1}k^{-1} \in H \cap K = \{e\}$. This implies that $hkh^{-1}k^{-1} = e$, or equivalently, $hk = kh$. ∎

Exercises

1. Let G be a group in which, for some integer $n > 1$, $(ab)^n = a^n b^n$ for all $a, b \in G$. Show that the subset $H = \{x^{n-1} \mid x \in G\}$ is a normal subgroup of G.
2. Find all normal subgroups of S_3.
3. If N is a normal subgroup of G with $|N| = 2$, prove that $N \leq Z(G)$.
4. Let N be a normal subgroup of S_4.

   (a) If N contains a transposition, prove that $N = S_4$;
   (b) If N contains a cycle of length 3, prove that $N = A_4$;
   (c) If N contains a cycle of length 4, prove that $N = S_4$;
   (d) Find all the normal subgroups of S_4.

5. Show that if a finite group G has exactly one subgroup N of a given order, then N is a normal subgroup of G.
6. If G is a group and N is a normal subgroup of G, show that $C_G(H) \trianglelefteq G$.
7. Prove that if N is a normal subgroup of the group G, then $Z(N)$ is a normal subgroup of G. Show by an example that $Z(N)$ need not be contained in $Z(G)$.
8. Let H be a subgroup of order 2 in G. Show that $N_G(H) = C_G(H)$. Deduce that if $N_G(H) = G$, then $H \leq Z(G)$.
9. If A is an abelian group with $A \trianglelefteq G$ and B is any subgroup of G, prove that $A \cap B \trianglelefteq AB$.
10. Show that if H and N are subgroups of a group G, and N is normal in G, then $H \cap N$ is normal in H. Show by an example that $H \cap N$ need not be normal in G.
11. Let N be a normal subgroup of a finite group G. If N is cyclic, prove every subgroup of N is also normal in G.
12. Let G be the set of all triples of the form $(a_1, a_2, 1)$ or $(a_1, a_2, -1)$, where a_1 and a_2 are integers. Define a binary operation on G by the rule

$$(a_1, a_2, 1)(b_1, b_2, c) = (a_1 + b_2, a_1 + b_2, c),$$
$$(a_1, a_2, -1)(b_1, b_2, c) = (a_1 + b_2, a_2 + b_1, -c),$$

where $c = \pm 1$. Prove that

   (a) G is a group;
   (b) $H = \langle (1, 0, 1), (0, 1, 1) \rangle$ is a normal subgroup of G;
   (c) $K = \langle (1, 0, 1) \rangle$ is a normal subgroup of H;
   (d) Is K a normal subgroup of H?

13. If N is a normal subgroup such that $[G : N] = n$, show that $x^n \in N$, for all $x \in G$.
14. Let H be a proper subgroup of G such that for all $x, y \in G \setminus H$, $xy \in H$. Prove that H is a normal subgroup of G.

9.2 Factor Groups

A factor group is a way of creating a group from another group. This new group often retains some of the properties of the original group.

Theorem 9.18 *Let G be a group and N be a normal subgroup of G. The set $G/N = \{aN \mid a \in G\}$ is a group under the binary operation $(aN)(bN) = abN$, for all aN and bN in G/N.*

Proof Our first task is to show that the operation is well defined. In order to do this, suppose that $aN = cN$ and $bN = dN$, for some a, b, c and d in G. Then, we conclude $c \in aN$ and $d \in bN$. Hence, there exist $n_1, n_2 \in N$ such that $c = an_1$ and $d = bn_2$. Now, we have

$$(ab)^{-1}cd = b^{-1}a^{-1}cd = b^{-1}a^{-1}an_1bn_2 = b^{-1}n_1bn_2.$$

Since $N \lhd G$ it follows that $b^{-1}n_1b \in N$. Therefore, we deduce that $(ab)^{-1}cd \in N$. This forces that $abN = cdN$.

   Computing $aN(bNcN) = aN(bc)N = a(bc)N$, and similarly, we have $(aNbN)cN = (ab)NcN = (ab)cN$. So, associativity in G/N follows from associativity in G. Since $aNeN = aeN = aN = eaN = eNaN$, it follows that $eN = N$ is the identity element in G/N. Finally, $a^{-1}NaN = (a^{-1}a)N = N = (aa^{-1})N = aNa^{-1}N$ shows that $a^{-1}N$ is the inverse of aN. This proves that G/N is a group. ∎

   The group G/N which was defined in Theorem 9.18 is called the *factor group* of G by N.

Theorem 9.19 *If G is a finite group and $N \lhd G$, then $|G/N| = |G|/|N|$.*

Proof The proof follows from Lagrange's Theorem.   ∎

   A factor group is also called the *quotient group* of G by N. Further, if the binary operation in G is addition, then each coset of N in G is denoted by $a + N$ and the binary operation in G/N is also denoted additively, i.e., we write $(a + N) + (b + N) = (a + b) + N$.

Example 9.20 Let $G = \mathbb{Z}_{18}$ and $N = \langle 6 \rangle$. Then, $G/N = \{0 + N, \ 1 + N, \ 2 + N, \ 3 + N, \ 4 + N, \ 5 + N\}$.

Example 9.21 Let U_{25} be the group of units in $\mathbb{Z}_{25}$ under multiplication modulo 25. We have $|U_{25}| = \varphi(25) = 20$. If N is the subgroup generated by 7, then $N = \langle 7 \rangle = \{1, \ 7, \ 18, \ 24\}$. Since U_{25} is abelian, it follows that every subgroup including N is normal. Since $[U_{25} : N] = |U_{25}|/N = 5$, it follows that there exist 5 left cosets of N in U_{25}. We write them out

$$N = \{1, 7, 18, 24\},$$
$$2N = \{2, 11, 14, 23\},$$
$$3N = \{3, 4, 21, 22\},$$
$$6N = \{6, 8, 17, 19\},$$
$$9N = \{9, 13, 12, 16\}.$$

There are many ways to name these left cosets, because we can take any number of the left coset to stand in for the whole set. For example, $2N = 11N = 14N = 23N$. Then, the Cayley table for U_{25}/N is:

H	H	$2H$	$3H$	$6H$	$9H$
H	H	$2H$	$3H$	$6H$	$9H$
$2H$	$2H$	$3H$	$6H$	$9H$	H
$3H$	$3H$	$6H$	$9H$	H	$2H$
$6H$	$6H$	$9H$	H	$2H$	$3H$
$9H$	$9H$	H	$2H$	$3H$	$6H$

Example 9.22 Let G be a group such that $(ab)^p = a^p b^p$ for all $a, b \in G$, where p is a prime number. Let

$$N = \{x \in G \mid x^{p^m} = e \text{ for some } m \text{ depending on } x\}.$$

Then, N is a normal subgroup of G. If $\overline{G} = G/N$ and if $\overline{x} \in \overline{G}$ is such that $\overline{x}^p = \overline{e}$, then $\overline{x} = \overline{e}$.

Theorem 9.23 *If G is a group and $N \trianglelefteq G$, then any subgroup of G/N is in the form of H/N such that $N \trianglelefteq H \leq G$.*

Proof Suppose that $\mathcal{S}$ is a subgroup of G/N. Define $H = \{x \in G \mid xN \in \mathcal{S}\}$. First, we show that H is a subgroup of G. Since $\mathcal{S} \leq G/N$, it follows that $N \in \mathcal{S}$. This means that $e \in H$, and so H is non-empty. Now, let $a, b \in H$. Then, $aN, bN \in \mathcal{S}$. Since $\mathcal{S}$ is a subgroup, it follows that $aN(bN)^{-1} = aNb^{-1}N = ab^{-1}N \in \mathcal{S}$. This implies that $ab^{-1} \in H$, and hence H is a subgroup of G. So, by Lemma 9.11, we conclude that $N \trianglelefteq H$. Now, suppose that $A \in H/N$. Then, $A = aN$, for some $a \in H$, and so $A = aN \in \mathcal{S}$. Consequently, we obtain $H/N \subseteq \mathcal{S}$. On the other hand, if $A \in \mathcal{S}$, then $A = aN$, for some $a \in H$, and hence $A \in H/N$. Therefore, we get $\mathcal{S} = H/N$, in which $N \trianglelefteq H \leq G$. ∎

Theorem 9.24 *Let G be a group and $N \trianglelefteq G$. Then, $H/N \trianglelefteq G/N$ if and only if $N \trianglelefteq H \trianglelefteq G$.*

Proof If $N \trianglelefteq H \trianglelefteq G$, then for every $aN \in G/N$ and $hN \in H/N$ we have $a^{-1}NhNaN = a^{-1}ha \in H$, which implies that $(aN)^{-1}hNaN \in H/N$. This shows that $H/N \trianglelefteq G/N$.

Conversely, if $H/N \trianglelefteq G/N$, then by Theorem 9.23, we obtain $N \trianglelefteq H \leq G$. Now, for every $a \in G$ and $h \in H$, we have $(aN)^{-1}hNaN \in H/N$, or equivalently $a^{-1}haN \in H/N$. Hence, there exists $h' \in H$ such that $a^{-1}haN = h'N$, which implies that $h'^{-1}a^{-1}ha \in N$. Since N is a subgroup of H, it follows that $h'^{-1}a^{-1}ha \in H$. Therefore, we conclude that $a^{-1}ha \in H$. This completes the proof.   ∎

Exercises

1. If G is an abelian group and N is a subgroup of G, show that G/N is abelian.
2. If G is a cyclic group and N is a subgroup of G, show that G/N is cyclic.
3. What is the order of the factor group $\mathbb{Z}_{60}/\langle 15\rangle$?
4. Let G be a group with $G/Z(G)$ abelian, and let $m \in \mathbb{N}$ be odd. Prove that $G^m := \{x^m \mid x \in G\}$ is a normal subgroup of G.
5. Prove or disprove: If N is a normal subgroup of G such that N and G/N are abelian, then G is abelian.
6. Prove or disprove: If N and G/N are cyclic, then G is cyclic.
7. Suppose that N is a normal subgroup of a finite group G. If G/N has an element of order n, show that G has an element of order n. Show, by example, that the assumption that G is finite is necessary.
8. Let N be a normal subgroup of a group G and let a belong to G. If the element aN has order 3 in the group G/N and $|N| = 10$, what are the possibilities for the order of a?
9. Suppose that N is a normal subgroup of a group G. If $|N| = 4$ and aN has order 3 in G/N, find a subgroup of order 12 in G.
10. If N is a normal subgroup of a group G such that N and G/N are finitely generated, prove that so is G.
11. Let $UT_2(\mathbb{F})$ be the group of invertible upper triangular matrices with entries in $\mathbb{F}$, that is matrices of the form

$$\begin{bmatrix} a & b \\ 0 & c \end{bmatrix},$$

where $a, b, c \in \mathbb{F}$ and $ac \neq 0$. Let N consists of matrices of the form

$$\begin{bmatrix} 1 & x \\ 0 & 1 \end{bmatrix},$$

where $x \in \mathbb{F}$.

(a) Show that N is an abelian subgroup of $UT_2(\mathbb{F})$;
(b) Prove that N is normal in $UT_2(\mathbb{F})$;
(c) Show that $UT_2(\mathbb{F})/N$ is abelian;
(d) Is $UT_2(\mathbb{F})$ normal in $GL_2(\mathbb{F})$.

12. Let G be a group and let $K \leq H \leq G$ with $K \trianglelefteq G$. Prove that

$$N_G \left(\frac{H}{K} \right) = \frac{N_G(H)}{K}.$$

9.3 Cauchy's Theorem and Class Equation

Cauchy's Theorem gives some converse to Lagrange's Theorem.

Theorem 9.25 (Cauchy's Theorem for Abelian Groups) *Let G be a finite abelian group and p be a prime that divides the order of G. Then, G has an element of order p.*

Proof We establish this theorem by mathematical induction on $|G|$. In other words, we assume that the statement is true for all abelian groups having fewer elements than G, and we use this assumption to show that the statement holds for G as well. To start the induction we note that the statement is vacuously true for group having a single element. So, let $|G| > 1$. If G has no trivial subgroups, then G must be cyclic of prime order, and this prime must be p. Hence, there is $a \in G$ such that $G = \langle a \rangle$. This yields that $|G| = o(a) = p$, and we are finished.

Now, we assume that G has a non-trivial subgroup N. We consider two cases:

Case 1: Let $p \big| |N|$. In this case, by our induction hypothesis, since N is abelian and $|N| < |G|$, it follows that there exists a non-identity element $x \in N \subset G$ such that $o(x) = p$, and we are done.

Case 2: Let $p \nmid |N|$. Since G is abelian, it follows that N is a normal subgroup of G. Hence, we may construct the factor group G/N. Moreover, G/N is abelian. Since $|G/N| = |G|/|N|$ and $p \nmid |N|$, it follows that $p \big| |G/N| < |G|$. Hence, by our induction assumption, there is a left coset $xN \in G/N$ such that $o(xN) = p$. Then, we obtain $(xN)^p = x^p N = N$. This implies that $x^p \in N$ and $x \notin N$, consequently $\left(x^p \right)^{|N|} = x^{p|N|} = e$. Take $y = x^{|N|}$, then $y^p = e$. At the end, we must show that $y \neq e$. Indeed, if $y = e$, then $x^{|N|} = e$, and so $(xN)^{|N|} = N$. On the other hand, since $(xN)^p = N$ and $p \nmid |N|$, we conclude that $xN = N$, or equivalently $x \in N$, it is a contradiction. Consequently, $y \neq e$ and $y^p = e$. Therefore, y is the desired element of order p. ∎

Theorem 9.26 *If G is a finite abelian group and a positive integer m divides $|G|$, then G contains a subgroup of order m.*

Proof We proceed by mathematical induction over $|G|$. If $|G| = 1$, then m must be 1. In this case G is its own subgroup of order m. To apply induction, let $|G| > 1$ and assume that the statement is true for every abelian group of order less than $|G|$. We can suppose that $m > 1$, because for $m = 1$, $\{e\}$ is the subgroup of G of order 1. Let p be a prime number such that $p | m$. We conclude that $p \big| |G|$. Then, by Cauchy's Theorem, there is $a \in G$ such that $o(a) = p$. Let $N = \langle a \rangle$ be the cyclic group generated by a. Then, we have $|N| = p$. Now, G/N is an abelian group such that $|G/N| = |G|/|N| < |G|$. Since $p|m$, it follows that $m = kp$, for some positive integer k. Moreover, we have $k||G/N|$. Hence, by our induction assumption and Theorem 9.23, G/N has a subgroup H/N of order k, where H is a subgroup of G containing N. Consequently, we obtain $|H| = |H/N| \cdot |N| = kp = m$. This completes the proof. ∎

Let $a, b \in G$. We recall that b is a conjugate of a in G if $b = xax^{-1}$, for some $x \in G$, and in this case, we write $a \sim_{Conj} b$. In Theorem 3.53, we showed that the conjugacy relation is an equivalence relation. For each $a \in G$, let $C(a) = \{b \in G \mid a \sim_{Conj} b\}$, the equivalence class of $a \in G$ under $\sim_{Conj}$ relation, and it is usually called the *conjugate class* of $a \in G$.

Our attention now narrow to the case in which G is a finite group. Let G be a finite group, and suppose that $C(a_1), C(a_2), \ldots, C(a_m)$ are the totally of all conjugate classes of G. For each $a \in G$, let $c_a = |C(a)|$. Since conjugacy classes are disjoint and their union is G, we obtain

$$|G| = \sum_{i=1}^{m} c_{a_i}.$$

Theorem 9.27 *If G is a finite group, then*

$$c_a = \frac{|G|}{|C_G(a)|},$$

in other words, the number of elements conjugate to a in G is the index of the centralizer of a in G.

Proof Let $C_G(a)$ has k distinct left cosets, say $x_1 C_G(a), x_2 C_G(a), \ldots, x_k C_G(a)$. Then, we know that $k = [G : C_G(a)]$. We claim that for every $i \neq j$, $x_i a x_i^{-1}$ and $x_j a x_j^{-1}$ are distinct conjugate of a. Indeed, if $x_i a x_i^{-1} = x_j a x_j^{-1}$, for some $i \neq j$, then we conclude that

$$x_j^{-1} x_i a x_i^{-1} x_j = a \Rightarrow (x_j^{-1} x_i) a (x_j^{-1} x_i)^{-1} = a \Rightarrow (x_j^{-1} x_i) a = a (x_j^{-1} x_i)$$
$$\Rightarrow x_j^{-1} x_i \in C_G(a) \Rightarrow x_i C_G(a) = x_j C_G(a) \Rightarrow i = j.$$

Now, we show that $x_1 a x_1^{-1}, \ldots, x_k a x_k^{-1}$ are all distinct conjugate of a. Let $b = xax^{-1}$, for some $x \in G$. Since

$$G = \bigcup_{i=1}^{k} x_i C_G(a),$$

it follows that $x = x_i y$, for some $y \in C_G(a)$ and positive integer i. Then, we can write

$$xax^{-1} = (x_i y)a(x_i y)^{-1} = x_i yay^{-1}x_i^{-1} = x_i ax_i^{-1}.$$

Consequently, any conjugate b of a is equal to one of the $x_i ax_i^{-1}$. Therefore, a has exactly k conjugate. This yields that $C(a)$ contains exactly k elements. ∎

Corollary 9.28 *If G is a finite group, then*

$$|G| = \sum_a \frac{|G|}{|C_G(a)|},$$

where this sum runs over one element a in each conjugate class.

Proof Since $|G| = \sum c_a$, using Theorem 9.27, the result immediately follows. ∎

Lemma 9.29 *Let G be a group. Then, $a \in Z(G)$ if and only if $C_G(a) = G$.*

Proof It is straightforward. ∎

In Lemma 9.29, if G is finite, then $|C_G(a)| = |G|$ is equivalent to $C_G(a) = G$.

Corollary 9.30 *Let G be a finite group, then*

$$|G| = |Z(G)| + \sum_a \frac{|G|}{|C_G(a)|},$$

where the sum runs over elements a, taken one from each of those distinct conjugate classes which contains more than one element.

Proof By Lemma 9.29, $a \in Z(G)$ if and only if $c_a = 1$. Since there are $|Z(G)|$ number of conjugate classes each having only one element, the corollary follows. ∎

The equation in Corollary 9.30 is usually referred to as the class equation of G.

Theorem 9.31 (Cauchy's Theorem) *Let G be a finite group and $p\||G|$. Then, G has an element of order p.*

Proof We do the proof by mathematical induction. The statement is vacuously true for groups of order 1. We assume the statement holds for all groups having fewer elements than G, and then we prove the statement is true for G. We consider the following two cases:

Case 1: There exists a proper subgroup H of G such that $p\||H|$. Then, by our induction assumption there exists an element a of order p in H, and so it is also in G.

Case 2: p is not divisor of the order of any proper subgroup of G. If $G \neq Z(G)$, then there exists $a \notin Z(G)$. Hence, we obtain $C_G(a) \neq G$. Since $C_G(a)$ is a subgroup of G, by our assumption, we conclude that $p \nmid |C_G(a)|$. We write down the class equation:

$$|G| = |Z(G)| + \sum_{C_G(a) \neq G} \frac{|G|}{|C_G(a)|}.$$

Since $p \big| |G|$ and $p \nmid |C_G(a)|$, it follows that

$$p \left| \frac{|G|}{|C_G(a)|} \right.$$

Thus, we obtain

$$p \left| \sum_{C_G(a) \neq G} \frac{|G|}{|C_G(a)|} \right.$$

Since also we have $p \big| |G|$, it follows that

$$p \left| \left(|G| - \sum_{C_G(a) \neq G} \frac{|G|}{|C_G(a)|} \right), \right.$$

and consequently, $p \big| |Z(G)|$. But we assumed that p is not divisor of the order of any proper subgroup of G, so we conclude that $Z(G)$ can not be a proper subgroup of G. This means that $Z(G) = G$, i.e., G is abelian. Now, by Cauchy Theorem for abelian groups, the result follows.   ∎

Exercises

1. In this exercise, we obtain a simple proof of Cauchy's Theorem. Let G be a finite group of order n and p be a prime number such that $p|n$. Consider the set

$$S = \{(a_1, a_2, \ldots, a_p) \mid a_i \in G \text{ and } a_1 a_2 \ldots a_p = e\}.$$

   (a) Compute $|S|$, the number of elements in S;
   (b) Define a relation ρ on S by saying two p-tuples are related if one is a cycle permutation of the other. For example, $(a_1, a_2, \ldots, a_p)\rho(a_2, a_3, \ldots, a_p, a_1)$. Show that ρ is an equivalence relation.
   (c) If all component of a p-tuple are equal, show that its equivalence class contains only one element; otherwise, if two components of a p-tuple are distinct, there are p elements in the equivalence class;
   (d) Let r denote the number of solutions to the equation $x^p = e$. Then, r equals the number of equivalence classes with only one element. Let s denote the number of equivalence classes with p elements. Show that $r + sp = n^{p-1}$;

(e) Deduce Cauchy's Theorem from the above, namely, G has kp solutions to the equation $x^p = e$.

2. If all non-identity elements of a group G have the same order, show that this order is a prime p and $|G|$ is a power of p.
3. Prove that any normal subgroup of order $2p$, where p is a prime number, has a normal subgroup of order p.
4. If G is an abelian group of order $p_1 p_2 \ldots p_k$, where $p_1, p_2, \ldots, p_k$ are distinct primes, prove that G is cyclic.

9.4 Worked-Out Problems

Problem 9.32 Let H be a cyclic subgroup of a group G and let $H \trianglelefteq G$. If N is a subgroup of H, prove that $N \trianglelefteq G$.

Solution Suppose that $H = \langle x \rangle$ is a normal subgroup of G and let $a \in G$ is arbitrary. Since H is normal, it follows that $axa^{-1} \in H$. Hence, we have $axa^{-1} = x^k$, for some integer k. If N is a subgroup of H, then N is cyclic and generated by x^n, for some integer n. Now, we obtain $ax^n a^{-1} = (axa^{-1})^n = x^{kn} = (x^n)^k \in N$. This completes the proof. ∎

Problem 9.33 Let N be a normal subgroup of a finite group G. If $|N|$ and $[G : N]$ are relatively prime, prove that any element $a \in G$ satisfying $a^{|N|} = e$ must belong to N.

Solution Suppose that $a \in G$ such that $a^{|N|} = e$. Since $([G : N], |N|) = 1$, it follows that there exist integers m and n such that $m[G : N] + n|N| = 1$. Then, we deduce that

$$a = a^{m[G:N]+n|N|} = a^{m[G:N]}a^{n|N|} = a^{m[G:N]}\left(a^{|N|}\right)^n = a^{m[G:N]}.$$

Now, we consider aN as an element of G/N. Then, we have $(aN)^{|G/N|} = a^{|G/N|}N = N$, the identity element of G/N. This yields that $a^{|G/N|} \in N$, and consequently, $a = \left(a^{[G:N]}\right)^m \in N$. ∎

Problem 9.34 Let G be a group and $Z(G)$ be the center of G. If $G/Z(G)$ is cyclic, prove that G is abelian.

Solution To start, every elements in the factor group $G/Z(G)$ is a left coset $aZ(G)$ with $a \in G$. Since $G/Z(G)$ is cyclic, it follows that $G/Z(G) = \langle xZ(G) \rangle$, for some $x \in G$. Now, let a and b be elements of G so that $aZ(G)$ and $bZ(G)$ belong to $G/Z(G)$. It follows that there exist integers i and j such that $aZ(G) = \left(xZ(G)\right)^i = x^i Z(G)$ and $bZ(G) = \left(xZ(G)\right)^j = x^j Z(G)$. Hence, $a = x^i z_1$ and $b = x^j z_2$, for some $z_1, z_2 \in Z(G)$. Then, by the definition of center and laws of exponents, we have

$$ab = (x^i z_1)(x^j z_2) = x^i x^j z_1 z_2 = x^j x^i z_2 z_1 = (x^j z_2)(x^i z_1) = ba.$$

Now, we conclude that G is abelian.   ∎

Problem 9.35 If G is a group of order p^n, where p is a prime number, prove that $Z(G) \neq \{e\}$.

Solution Suppose that $|Z(G)| = m$. By the class equation, we have

$$|G| = m + \sum_a \frac{|G|}{|C_G(a)|},$$

where sum runs over element a, taken one from each of conjugate class which has more than one element. Now, for each $a \notin Z((G)$, we have $|C_G(a)|\big||G|$. This implies that $|C_G(a)| = p^{k_a}$, for some integer $1 \leq k_a < n$. Hence, we obtain

$$p\left|\left(p^n - \sum_a \frac{p^n}{p^{k_a}}\right)\right. = m.$$

Since $e \in Z(G)$, it follows that m is non-zero. Hence, m is a positive integer divisible by the prime number p, and so we conclude that $m > 1$. This shows that $Z(G) \neq \{e\}$.   ∎

Problem 9.36 Let G be a group of order p^n, where p is a prime number. Prove that every subgroup of order p^{n-1} of G is normal.

Solution Suppose that H is a subgroup of G and $|H| = p^{n-1}$. We establish the proof by mathematical induction. If $n = 1$, then $|G| = p$ and $H = \{e\} \trianglelefteq G$. Suppose that $n > 1$ and the statement is true for every group of order p^m, where $1 \leq m < n$. We know that $H \trianglelefteq N_G(H)$. We consider the following two cases:

   Case 1: $N_G(H) \neq H$. In this case, $|N_G(H)| = p^n$, and so $N_G(H) = G$. This means that $H \trianglelefteq G$.

   Case 2: $N_G(H) = H$. Since $Z(G) \subseteq N_G(H)$, it follows that $Z(G) \subseteq H$. Since $p\big||Z(G)|$, by Cauchy's Theorem it follows that there exists $a \in Z(G)$ such that $o(a) = p$. Take $K = \langle a \rangle$, then K is a normal subgroup of order p in G. Now, we have $|H/K| = p^{n-2}$, so by our induction assumption, we conclude that $H/K \trianglelefteq G/K$, because $|G/K| = p^{n-1}$. Therefore, H is a normal subgroup of G.   ∎

Problem 9.37 If G is a group of order p^2, prove that G is abelian.

Solution As a group G is abelian if and only if $Z(G) = G$, our aim is to show that $Z(G) = G$. Since $|G| = p^2$, by Lagrange's Theorem, $|Z(G)|\big|p^2$. This implies that $|Z(G)| = 1$, p or p^2. By Problem 9.35, we conclude that $|Z(G)| \neq 1$. Assume that $|Z(G)| = p$. Take $a \in G$ such that $a \notin Z(G)$. Then, $C_G(a)$ is a subgroup of G and $Z(G)$ is a subset of $C_G(a)$. Since $a \notin Z(G)$ and $a \in C_G(a)$, it follows that $Z(G) \neq C_G(a)$. Hence, we have $p = |Z(G)| < |C_G(a)|$. Then, we conclude that

$|C_G(a)| = p^2$ or $C_G(a) = G$. This means that all of elements of G commute with a, and so $a \in Z(G)$, a contradiction. Thus, $|Z(G)| = p$ is not an actual possibility. Therefore, $|Z(G)| = p^2$, and so $G = Z(G)$. ∎

9.5 Supplementary Exercises

1. (a) Let N be a normal subgroup of a group G, with G/N abelian. Does it follow that $G = HN$ for some abelian subgroup H of G? Give a proof or a counterexample.

    (b) Answer the same question, with both occurrences of "abelian" replaced by "cyclic".

2. Suppose that $K \trianglelefteq G$ with $|K| = m$. Let $x \in G$ and n be a positive integer such that $(m, n) = 1$. Prove that

    (a) If $o(x) = n$, then $o(xK) = n$;
    (b) If $o(xK) = n$, then there is an element $y \in G$ such that $o(y) = n$ and $xK = yK$.

3. Let N be a normal subgroup of a group G. If $x, y \in G$ such that $xy \in N$, show that $yx \in N$.

4. Let H be a subgroup of index 2 in a finite group G. If the order of H is odd and every element of $G \setminus H$ is of order 2, prove that H is abelian.

5. Suppose that G is a finite abelian group such that each element of G has order 1 or 3. If H is a subgroup of G, show that each element of G/H has order 1 or 3. Use induction on $|G|$ to show that $|G| = 3^k$, for some positive integer k.

6. For which values of n is it true that the dihedral group D_n has a pair of proper normal subgroups H and K for which $D_n = HK$ and $H \cap K$ is singleton.

7. If $K \leq H \leq G$ and $N \trianglelefteq G$, show that the equations $HN = KN$ and $H \cap N = K \cap N$ imply that $H = K$.

8. Let G be a finite group, H be a subgroup of G, and $N \trianglelefteq G$. If $|H|$ and $[G : N]$ are relatively prime, prove that H is a subgroup of N.

9. Let G be a finite group and $N \trianglelefteq G$. If $|N|$ and $[G : N]$ are relatively prime, prove that N is the unique subgroup of G of order N.

10. Let G be a group and H be the intersection of all subgroups of G that have finite index in G. Show that H is normal in G.

11. Let G be a group and H be a subgroup of G with finite index. Show that there is a normal subgroup N of G for which $N \leq H \leq G$ where N also has finite index in G.

12. Prove that the torsion subgroup T of an abelian group G is a normal subgroup of G, and that G/T is torsion free.

13. Let N be a normal subgroup of G of index n. Show that if $a \in G$, then $a^n \in N$. Give an example to show that this may be false when N is not normal.

14. If $|G| = pq$, where p and q are not necessarily distinct primes, prove $|Z(G)| = 1$ or G is abelian.

15. If H and K are subgroups of finite index in a group G, and $[G : H]$ and $[G : K]$ are relatively prime, prove that $G = HK$.

16. If G is a non-abelian group of order p^3, where p is a prime number, show that $|Z(G)| = p$.
    Hint: Use Problem 9.34.

17. Let G be a finite group and p be the smallest prime number such that $p\,||G|$. If $N \lhd G$ and $|N| = p$, prove that $N \leq Z(G)$.

18. If H is a subgroup of finite index in a group G, prove that H contains a subgroup N which is of finite index and normal in G.

19. The set X is called a *normal subset* of G if G is the normalizer of X. Let X be a finite normal subset of a group G such that for some positive integer n, $x^n = e$, for all $x \in X$. Prove that every element of the group $H = \langle X \rangle$ may be written as a product of not more than $(n - 1)|X|$ elements of X. In particular, H is a finite group.

Chapter 10
Some Special Subgroups

In this section, we introduce the notion of commutator subgroup or derived subgroup of a group. This subgroup is important because it is the smallest normal subgroup such that the quotient group of the original group by this subgroup is abelian. Also, we study other special subgroup of a group called the maximum subgroup.

10.1 Commutators and Derived Subgroups

Let G be a group, and let $a, b \in G$. The *commutator* of a and b is $[a, b] = a^{-1}b^{-1}ab$.

"Commutator" is a good word to use: because $[a, b]$ is a kind of measure as to how near a to b come to commuting, $[a, b]$ being the identity element of G if and only if $ab = ba$. For any $a, b, c \in G$, we denote $[a, b, c] = \big[[a, b], c\big]$ and $b^a = a^{-1}ba$.

Theorem 10.1 *Let G be a group. For every $a, b, c \in G$, the following hold:*

(1) $[ab, c] = [a, c]^b[b, c];$

(2) $[a, bc] = [a, c][a, b]^c;$

(3) $[a, b^{-1}, c]^b[b, c^{-1}, a]^c[c, a^{-1}, b]^a = 1.$

Proof (1) We can write

$$[ab, c] = (ab)^{-1}c^{-1}abc = b^{-1}a^{-1}c^{-1}abc$$
$$= b^{-1}(a^{-1}c^{-1}ac)b(b^{-1}c^{-1}bc) = [a, c]^b[b, c].$$

(2) We have

$$[a, bc]\, a^{-1}(bc)^{-1}abc = a^{-1}c^{-1}b^{-1}abc$$
$$= (a^{-1}c^{-1}ac)c^{-1}(a^{-1}b^{-1}ab)c = [a, c][a, b]^c.$$

© The Author(s), under exclusive license to Springer Nature Singapore Pte Ltd. 2021
B. Davvaz, *A First Course in Group Theory*,
https://doi.org/10.1007/978-981-16-6365-9_10

(3) We can write

$$[a, b^{-1}, c]^b = b^{-1}[[a, b^{-1}], c]b = b^{-1}[a^{-1}bab^{-1}, c]b$$
$$= b^{-1}(ba^{-1}b^{-1}ac^{-1}a^{-1}bab^{-1}c)b = (a^{-1}b^{-1}ac^{-1}a^{-1})(bab^{-1}cb)$$
$$= (aca^{-1}ba)^{-1}(bab^{-1}cb).$$

Hence, we have

$$[a, b^{-1}, c]^b = (aca^{-1}ba)^{-1}(bab^{-1}cb). \tag{10.1}$$

Similarly, we obtain

$$[b, c^{-1}, a]^c = (bab^{-1}cb)^{-1}(cbc^{-1}ac) \tag{10.2}$$

and

$$[c, a^{-1}, b]^a = (cbc^{-1}ac)^{-1}(aca^{-1}ba). \tag{10.3}$$

Now, by (10.1)–(10.3), the results follows. ∎

It is perfectly natural to look at the set of all commutators in a group G. This subset may not form a subgroup of G, so we move to the next best thing.

Definition 10.2 Let G be a group. The subgroup generated by the set $\{[a, b] \mid a, b \in G\}$ is called the *commutator subgroup* or *derived subgroup* of G. It is denoted by G' or $G^{(1)}$ or $[G, G]$.

A group G is called *perfect* if $G = G'$.

Theorem 10.3 *If G is a group, then $G' \trianglelefteq G$ and G/G' is abelian.*

Proof For every $a, b, g \in G$, we have

$$[a^g, b^g] = [g^{-1}ag, g^{-1}bg] = (g^{-1}ag)^{-1}(g^{-1}bg)^{-1}(g^{-1}ag)(g^{-1}bg)$$
$$= (g^{-1}a^{-1}g)(g^{-1}b^{-1}g)(g^{-1}ag)(g^{-1}bg)$$
$$= g^{-1}(a^{-1}b^{-1}ab)g = g^{-1}[a, b]g = [a, b]^g.$$

So, we obtain $[a, b]^g = [a^g, b^g]$. Now let $x \in G'$ be an arbitrary element. Then, we can write $x = \prod_{i=1}^{n}[a_i, b_i]$, where $a_i, b_i \in G$ for all $1 \leq i \leq n$. Therefore, we have

$$g^{-1}xg = \left(\prod_{i=1}^{n}[a_i, b_i]\right)^g = \prod_{i=1}^{n}[a_i, b_i]^g = \prod_{i=1}^{n}[a_i^g, b_i^g] \in G'.$$

This shows that $G' \trianglelefteq G$.

Now, let aG' and bG' be arbitrary elements of G/G'. Since $a^{-1}b^{-1}ab \in G'$, it follows that $abG' = \left(a^{-1}b^{-1}\right)^{-1}G'$, or equivalently $aG'bG' = bG'aG'$, and so G/G' is abelian. ∎

Theorem 10.4 *Let G be a group and N be a normal subgroup of G. Then, G/N is abelian if and only if $G' \leq N$.*

Proof Suppose that G/N is abelian. Then, for every $a, b \in G$, we have $aNbN = bNaN$. This implies that $abN = baN$, or equivalently $a^{-1}b^{-1}ab \in N$. Since N contains every commutator $[a, b]$, it follows that $G' \leq N$.

Conversely, let $G' \leq N$, and assume that aN and bN are arbitrary elements of G/N. Since $a^{-1}b^{-1}ab \in G'$, it follows that $a^{-1}b^{-1}ab \in N$. Hence, we have $abN = baN$. This means that $aNbN = bNaN$, and so we conclude that G/N is abelian. ■

Definition 10.5 A group G is called *metabelian* if there exists a normal subgroup N of G such that both N and G/N are abelian.

Theorem 10.6 *A group G is metabelian if and only if $G'' = \{e\}$.*

Proof Let G be a metabelian group; we show that $G'' = \{e\}$. Since G is metabelian, it has a normal abelian subgroup N, and G/N is abelian. Thus, by Theorem 10.4, $G' \leq N$. Since N is abelian, it follows that G' is abelian, and so $G'' = \{e\}$.

Conversely, suppose that $G'' = \{e\}$. We show that G is metabelian. It is clear that if G is abelian, then G is metabelian. So, we have only to consider the case where G is not abelian, and hence $G' \neq \{e\}$. Thus, assume that G is not abelian. By Theorem 10.3, we know that G' is a normal subgroup of G. Let $a, b \in G'$. We know that $aba^{-1}b^{-1} = e$ since $G'' = \{e\}$ and so we have that $ab = ba$. Thus, arbitrary elements $a, b \in G'$ commute, and so $G' \neq \{e\}$ is a normal abelian subgroup of G. Again, by Theorem 10.3, G/G' is abelian, and so G is metabelian. ■

Theorem 10.7 *Let G be a group such that G' is a subset of the center of G. Then, $[a, bc] = [a, b][a, c]$, for all $a, b, c \in G$.*

Proof We can write

$$[a, bc] = a^{-1}(bc)^{-1}abc = a^{-1}c^{-1}(ac)(c^{-1}a^{-1})b^{-1}abc$$
$$= (a^{-1}c^{-1}ac)c^{-1}(a^{-1}b^{-1}ab)c = [a, c]c^{-1}[a, b]c.$$

Since $[a, b] \in G'$ and $G' \subseteq Z(G)$, it follows that $[a, b] \in Z(G)$. So, we conclude that $[a, b]c = c[a, b]$. Consequently, we get $[a, bc] = [a, c][a, b] = [a, b][a, c]$. This completes the proof. ■

More generally, for subgroups H and K of a group G, we denote $[H, K]$ the subgroup of G generated by the set $\{[h, k] \mid h \in H \text{ and } k \in K\}$.

Theorem 10.8 *If G is a group and H, K are subgroups of G, then*

(1) $[H, K] = [K, H]$;
(2) $[H, K] \trianglelefteq H \vee K$.

Proof (1) Let $[h, k]$ be an arbitrary commutator in $[H, K]$. Then, we have

$$[h, k] = h^{-1}k^{-1}hk = \left(k^{-1}h^{-1}kh\right)^{-1} = [k, h]^{-1}.$$

Since $[K, H] \leq G$ and $[k, h] \in [K, H]$, it follows that $[k, h]^{-1} \in [K, H]$. This implies that $[h, k] \in [K, H]$. So, we have $[H, K] \subseteq [K, H]$. In a similar way, we obtain $[K, H] \subseteq [H, K]$. Therefore, we conclude that $[H, K] = [K, H]$.

(2) Since $H \vee K$ is the subgroup generated by $H \cup K$, it follows that $[h, k] \in H \vee K$, for all $h \in H$ and $k \in K$. This shows that $[H, K]$ is a subgroup of $H \vee K$. Now, let $[h, k]$ be an arbitrary commutator in $[H, K]$ and $x \in H$. Then, by Theorem 10.1 (1), we have $[hx, k] = [h, k]^x[x, k]$, and so $[h, k]^x = [hx, k][x, k]^{-1}$. Since $[hx, k]$ and $[x, k]^{-1}$ belong to $[H, K]$, it follows that $[h, k]^x \in [H, K]$. Also, let $y \in K$, then by Theorem 10.1 (2), we can write $[h, ky] = [h, y][h, k]^y$, and hence $[h, k]^y = [h, y]^{-1}[h, ky]$. Since $[h, y]^{-1}$ and $[h, ky]$ lies in $[H, K]$, it follows that $[h, k]^y \in [H, K]$. Therefore, we conclude that $[H, K] \trianglelefteq H \vee K$. ∎

Theorem 10.9 *If G is a group and H, K are normal subgroups of G, then $[H, K] \leq H \cap K$.*

Proof Suppose that $h \in H$ and $k \in K$ are arbitrary. Since $H \trianglelefteq G$, it follows that $h^{-1}(k^{-1}hk) \in H$. Similarly, since $K \trianglelefteq G$, it follows that $(h^{-1}k^{-1}h)k \in K$. Thus, we obtain $[h, k] \in H \cap K$. Consequently, we have $[H, K] \leq H \cap K$. ∎

Theorem 10.10 *If H, K, and N are normal subgroups of a group G, then $[HK, N] = [H, N][K, N]$.*

Proof Assume that $a \in H$, $b \in K$ and $c \in N$. Then, by Theorem 10.1 (1), we can write

$$[ab, c] = [a, b]^b[b, c] = [a^b, c^b][b, c]. \tag{10.4}$$

Since H and N are normal subgroups of G, it follows that $[a^b, c^b] \in [H, N]$. On the other hand, we have $[b, c] \in [K, N]$. Now, by (10.4), we conclude that $[HK, N] \subseteq [H, N][K, N]$.

Conversely, since H and K are subgroups of HK, it follows that

$$[H, N] \subseteq [HK, N] \text{ and } [K, N] \subseteq [HK, N].$$

Consequently, we get $[H, N][K, N] \subseteq [HK, N]$, and this completes the proof. ∎

Exercises

1. Let G be a group and H be a cyclic subgroup of G. If $G' \leq H$, show that $H \trianglelefteq G$.
2. For any group G, show that G' is the subset of all "*long commutators*":

$$G' = \{a_1 a_2 \ldots a_n a_1^{-1} a_2^{-1} \ldots a_n^{-1} \mid a_i \in G \text{ and } n \geq 2\}.$$

Hint: $\quad (aba^{-1}b^{-1})(cdc^{-1}d^{-1}) = a(ba^{-1})b^{-1}c(dc^{-1})d^{-1}a^{-1}(ab^{-1})bc^{-1}$ $(cd^{-1})d$.

3. Let G be a group and suppose that H and K are subgroups of G. Prove that

   (1) $H \trianglelefteq G$ if and only if $[H, G] \leq H$;
   (2) If K is a subgroup of H and $K \trianglelefteq G$, then $[H, G] \leq K$ if and only if $H/K \leq Z(G/K)$.

4. Let $G = HK$, where H and K are abelian subgroups of G. Prove that G' is abelian.
5. Let N be a normal subgroup of a group G such that $N \cap G' = \{e\}$. Show that $N \leq Z(G)$.
6. If H is a subgroup of a metabelian group G, prove that H is metabelian.

10.2 Derived Subgroups of Some Special Groups

In this section, we investigate the derived subgroups of symmetric groups, alternating groups, quaternion group, general linear groups, special linear groups, and dihedral groups.

Lemma 10.11 *Let $n \geq 5$ and $N \trianglelefteq A_n$. If N contains a cycle of length 3, then $N = A_n$.*

Proof Suppose that $(x\ y\ z) \in N$. By Theorem 5.43, each permutation in A_n is a product of cycles of length 3. Hence, it suffices to prove that every cycle of length 3 lies in N. Assume that $(a\ b\ c) \in S_n$ is an arbitrary cycle. Take $\sigma \in S_n$ such that $x\sigma = a$, $y\sigma = b$ and $z\sigma = c$. Then, we have $\sigma^{-1}(x\ y\ z)\sigma = (a\ b\ c)$.

If $\sigma \in A_n$, then $(a\ b\ c) \in N$, because $N \trianglelefteq A_n$.

If $\sigma \notin A_n$, then σ is an odd permutation. Since $n \geq 5$, we can consider r and s distinct from x, y, and z. So, we have $(r\ s)\sigma \in A_n$. Consequently, we have $\left((r\ s)\sigma\right)^{-1}(x\ y\ z)(r\ s)\sigma \in N$, or equivalently $\sigma^{-1}(r\ s)(x\ y\ z)(r\ s)\sigma \in N$. Since $(r\ s)$ and $(x\ y\ z)$ are disjoint cycles, it follows that $\sigma^{-1}(x\ y\ z)\sigma \in N$. This shows that $(a\ b\ c) \in N$. ∎

Theorem 10.12 *For all $n \geq 2$, $S_n' = A_n$.*

Proof Let $n \geq 2$ and suppose that σ and τ are two arbitrary permutations in S_n. Let $\sigma = \sigma_1 \sigma_2 \ldots \sigma_r$ and $\tau = \tau_1 \tau_2 \ldots \tau_s$, where σ_i's and τ_i's are transpositions. Then, we have

$$[\sigma, \tau] = \sigma^{-1}\tau^{-1}\sigma\tau = \sigma_r \ldots \sigma_2 \sigma_1 \tau_s \ldots \tau_2 \tau_1 \sigma_1 \sigma_2 \ldots \sigma_r \tau_1 \tau_2 \ldots \tau_s.$$

Hence, $[\sigma, \tau]$ is a product of $2(r + s)$ transpositions, and so it is an even permutation. This shows that $[\sigma, \tau] \in A_n$. Therefore, we conclude that $S'_n \leq A_n$.

Now, in order to show that $A_n = S'_n$, we consider the following cases:

Case 1: $n = 2$. In this case, clearly, we have $S'_n = A_2 = \{id\}$.

Case 2: $n = 3$. We have $A_3 = \{id, (1\,2\,3), (1\,3\,2)\}$. Since there are permutations in S_3 that don't commute, it follows that the derived subgroup is not just $\{id\}$. Hence, it must be all of A_3.

Case 3: $n = 4$. For distinct $i, j, k \in \{1, 2, 3, 4\}$, we have

$$[(i\,j), (i\,k)] = (i\,j)(i\,k)(i\,j)(i\,k) = (i\,j\,k)^2 = (i\,k\,j),$$

that is, every cycle of length 3 belongs to derived subgroup. Therefore, we conclude that $S'_4 = A_4$.

Case 4: $n \geq 5$. We know that S'_n is a normal subgroup of S_n. Moreover, we proved that $S'_n \leq A_n$. On the other hand, we have

$$[(1\,2), (2\,3)] = (1\,2)(2\,3)(1\,2)(2\,3) = (1\,2\,3) \in S'_n.$$

This means that S'_n contains a cycle of length 3, and so by Lemma 10.11, we deduce that $S'_n = A_n$. ∎

Theorem 10.13 *We have*

(1) $A'_2 = A'_3 = \{id\}$;
(2) $A'_4 = \{id, (1\,2)(3\,4), (1\,3)(2\,4), (1\,4)(2\,3)\}$;
(3) $A'_n = A_n$, *for all* $n \geq 5$.

Proof (1) It is straightforward.

(2) Assume that $N = \{id, (1\,2)(3\,4), (1\,4)(2\,3), (1\,3)(2\,4)\}$. It can be checked that N is a normal subgroup of A_4. Since $|A_4/N| = 3$, it follows that A_4/N is abelian, and so by Theorem 10.4, we have $A'_4 \leq N$. Since A_4 is non-abelian, it follows that $A'_4 \neq \{id\}$. Hence, there exists a non-identity permutation $(a\,b)(c\,d)$ in A'_4. Since $A'_4 \trianglelefteq A_4$ and every product of disjoint transpositions are conjugate in A_4, we conclude that $A'_4 = N$.

(3) Since $A'_n \trianglelefteq A_n$, by Lemma 10.11, it is enough to prove that A'_n contains a cycle of length 3. Since $n \geq 5$, assume that $\sigma = (a\,c\,b)$ and $\tau = (b\,c)(d\,e)$. Then, we have $[\tau\,\sigma] = \tau^{-1}\sigma^{-1}\tau\sigma = (\tau^{-1}\sigma^{-1}\tau)\sigma$. Also, we have

$$\tau^{-1}\sigma^{-1}\tau = \tau^{-1}(a\,b\,c)\tau = (a\tau\;b\tau\;c\tau) = (a\,c\,b) = \sigma.$$

Consequently, we obtain $[\tau, \sigma] = \sigma^2 = (a\,b\,c)$ as desired. ∎

Theorem 10.14 *If Q_8 is the quaternion group, then $Q_8' = \{1, 1\}$.*

Proof We know that $Q_8 = \{1, -1, I, -I, J, -J, K, -K\}$. The commutator subgroup contains the element

$$[I, J] = I^{-1}J^{-1}IJ = (-I)(-J)IJ = (IJ)(IJ) = K^2 = -1.$$

Similarly, $[J, K] = -1$ and $K, I] = -1$. On the other hand, 1 and -1 commute with all elements of Q_8, hence $[a, -1] = [a, 1] = 1$, for all $a \in Q_8$. Therefore, the commutator subgroup is the subgroup of Q_8 generated by -1 and 1, which is $Q_8' = \{1, 1\}$. ∎

Theorem 10.15 *Let $\mathbb{F}$ be a field and $n \geq 2$ be an integer. Then,*

$$SL_n(\mathbb{F})' = GL_n(\mathbb{F})' = SL_n(\mathbb{F}),$$

except when $n = 2$ and $\mathbb{F}$ consists of 2 or 3 elements. Thus, $SL_n(\mathbb{F})$ is perfect (with the exceptions listed).

Proof Let A and B are two arbitrary matrices in $GL_n(\mathbb{F})$. Then, we have $A^{-1}B^{-1}AB \in GL_n(\mathbb{F})'$. Since

$$\det(A^{-1}B^{-1}AB) = \det(A^{-1})\det(B^{-1})\det(A)\det(B) = 1,$$

it follows that $A^{-1}B^{-1}AB \in SL_n(\mathbb{F})$. This yields that $GL_n(\mathbb{F})' \subseteq SL_n(\mathbb{F})$. On the other hand, since $SL_n(\mathbb{F}) \subseteq GL_n(\mathbb{F})$, it follows that $SL_n(\mathbb{F})' \subseteq GL_n(\mathbb{F})'$. So, we have $SL_n(\mathbb{F})' \subseteq GL_n(\mathbb{F})' \subseteq SL_n(\mathbb{F})$. Consequently, it is enough to show that $SL_n(\mathbb{F})' = SL_n(\mathbb{F})$. To establish this equality, by Theorem 7.41, it is enough to show that each matrix of the form $E_{ij}(\lambda)$ with $i \neq j$ and $\lambda \in \mathbb{F}^*$, is a product of commutators. It is easy to check that for any distinct indices i, j, k,

$$[E_{ik}(\lambda), E_{kj}(1)] = E_{ik}(-\lambda)E_{kj}(-1)B_{ik}(\lambda)E_{kj}(1) = E_{ij}(\lambda).$$

his expresses each $E_{ij}(\lambda)$ as a commutator when $n \geq 3$. For $n = 2$, we have

$$\left[\begin{bmatrix} a^{-1} & 0 \\ 0 & a \end{bmatrix}, \begin{bmatrix} 1 & b \\ 0 & 1 \end{bmatrix}\right] = \begin{bmatrix} a^{-1} & 0 \\ 0 & a \end{bmatrix}^{-1}\begin{bmatrix} 1 & b \\ 0 & 1 \end{bmatrix}^{-1}\begin{bmatrix} a^{-1} & 0 \\ 0 & a \end{bmatrix}\begin{bmatrix} 1 & b \\ 0 & 1 \end{bmatrix}$$

$$= \begin{bmatrix} a & 0 \\ 0 & a^{-1} \end{bmatrix}^{-1}\begin{bmatrix} 1 & -b \\ 0 & 1 \end{bmatrix}^{-1}\begin{bmatrix} a^{-1} & 0 \\ 0 & a \end{bmatrix}\begin{bmatrix} 1 & b \\ 0 & 1 \end{bmatrix}$$

$$= \begin{bmatrix} 1 & b(1 - a^2) \\ 0 & 1 \end{bmatrix}.$$

So, if $\mathbb{F}$ contains an element a such that $a \neq 0$ and $a^2 \neq 1$, then we can express any matrix $E_{12}(\lambda)$ as a commutator by taking $b = (1 - a^2)^{-1}\lambda$, and similarly for $E_{21}(\lambda)$.

This yields that $SL_n(\mathbb{F})' = SL_n(\mathbb{F})$ except when $n = 2$ and $a^3 = a$, for all $a \in \mathbb{F}$, and this happens only when $|\mathbb{F}| = 2$ or 3. ∎

Theorem 10.16 *If D_n is the dihedral group of order $2n$, then by Corollary 7.74, we have $D_n = \langle R, S \rangle$, where S is a reflection and R is a rotation. Then, $D'_{2n} = \langle R^2 \rangle$.*

Proof Since $[S, R]$ is $S^{-1}R^{-1}SR = SSRR = R^2$, it follows that R^2 is a commutator. Moreover, since $[S, R^i] = S^{-1}R^{-i}SR^i = SSR^iR^i = R^{2i}$, it follows that every element of $\langle R^2 \rangle$ is a commutator. Now, we prove that every commutator belong to $\langle R^2 \rangle$. Suppose that A and B are two arbitrary elements of D_{2n}. We consider the following cases:

Case 1: Both A and B are rotations. In this case, we can write $A = R^i$ and $B = R^j$, for some integers i and j. Since R^i and R^j commute, it follows that $[A, B] = I_n$.

Case 2: A is a rotation and B is a reflection. Then, we have $A = R^i$ and $B = R^j S$, for some integers i and j. Since $B = B^{-1}$, it follows that

$$[A, B] = A^{-1}B^{-1}AB = A^{-1}BAB = R^{-i}R^jSR^iR^jS = R^{j-i}SR^{i+j}S.$$

By Theorem 7.75, since $RS = SR^{-1}$, it follows that $[A, B] = R^{-2i} \in \langle R^2 \rangle$.

Case 3: A is a reflection and B is a rotation. In this case, we have $[A, B]^{-1} = B^{-1}A^{-1}BA$. By the case (2), $[A, B]^{-1} \in \langle R^2 \rangle$, and this implies that $[A, B] \in \langle R^2 \rangle$.

Case 4: Both A and B are reflections. Then, we have $A = R^i S$ and $B = R^j S$. Hence, we obtain $A = A^{-1}$ and $B = B^{-1}$. So, we conclude that

$$[A, B] = A^{-1}B^{-1}AB = ABAB = (AB)^2 = (R^i SR^j S)^2$$
$$= (R^i R^{-j}SS)^2 = R^{2(i-j)} \in \langle R^2 \rangle.$$

This completes the proof. ∎

Exercises

1. Compute $[GL_n(\mathbb{F}) : SL_n(\mathbb{F})]$, where $\mathbb{F}$ is finite.
2. Suppose that $O_n(\mathbb{F}) = \{A \in GL_n(\mathbb{F}) \mid A^t A = I_n\}$ and $SO_n(\mathbb{F}) = \{A \in O_n(\mathbb{F}) \mid \det(A) = 1\}$. Prove that $SO_n(\mathbb{F}) \trianglelefteq O_n(\mathbb{F})$ and compute $[O_n(\mathbb{F}) : SO_n(\mathbb{F})]$.
3. Determine the derived subgroup of $SO_n(\mathbb{F})$.

10.3 Maximal Subgroups

A maximal subgroup of a given group G is a proper subgroup of G, that is, a subgroup which is neither the identity nor G itself, and in order to be maximal, it can not be contained in a larger proper subgroup of G. More precisely:

Definition 10.17 A subgroup M of a group G is called a *maximal subgroup* if $M \neq G$ and the only subgroups containing M are M and G.

Example 10.18 We know that the subgroups of $\mathbb{Z}$ are of the form $n\mathbb{Z}$. Note that $n\mathbb{Z} \subseteq m\mathbb{Z}$ if and only if $m|n$, and so the maximal subgroups of $\mathbb{Z}$ are precisely $p\mathbb{Z}$ for p prime.

Example 10.19 In each Hasse diagram, the lower subgroup is a maximal subgroup in the upper one.

Theorem 10.20 *If H is a proper subgroup of a finite group G, then there is a maximal subgroup of G containing H.*

Proof If H is a maximal subgroup of G, then we have nothing to do. If H is not a maximal subgroup, we pick a proper subgroup H_1 of G such that $H \subset H_1$. Again, if H_1 is maximal in G, then we are done; otherwise, we pick a proper subgroup H_2 of G such that $H \subset H_1 \subset H_2$. Continuing in this way, we construct a chain of proper subgroups of G:

$$H \subset H_1 \subset H_2 \subset \ldots,$$

and this gives a strictly increasing chain of integers

$$|H| < |H_1| < |H_2| < \ldots,$$

which is bounded above by $|G|$, and so must terminate. Consequently, we obtain a maximal subgroup containing H. ∎

We conclude that in non-trivial finite groups, maximal subgroups will always exist. However, not all groups will have maximal subgroups.

Example 10.21 The rational numbers $\mathbb{Q}$ under addition have no maximal subgroups. To see the reason, suppose that M is a maximal subgroup of $\mathbb{Q}$. There exists a rational number r/s ($r, s \in \mathbb{Z}$ and $s \neq 0$) such that $r/s \notin M$. Since $\mathbb{Z}$ is a proper subgroup of $\mathbb{Q}$, it follows that $M \neq \{0\}$. Let $m/n \in M$ with $m, n \in \mathbb{Z}$ and $mn \neq 0$. Since $m/n \in M$, it follows that $nm/n \in M$, and so $m \in M$. Since M is a maximal subgroup, it follows that

$$M + \langle r/s \rangle = \mathbb{Q}. \tag{10.5}$$

Consider $r/sms \in \mathbb{Q}$. Then, by (10.5), there exist $h \in M$ and $t \in \mathbb{Z}$ such that $r/sms = h + tr/s$. Hence, we have $r/s = msh + tmr = m(sh + tr)$. Since $m \in M$, it follows that $r/s \in M$. This is a contradiction.

Theorem 10.22 *The center of a group G is properly contained in every maximal subgroup of G having a composite index.*

Proof Let M be a maximal subgroup of G with composite index. Suppose that $Z(G)$ is not contained in M. Then, there exists $a \in Z(G)$ such that $a \notin M$. We consider

$\langle M, a \rangle$, the subgroup generated by $M \cup \{a\}$. Since M is a maximal subgroup and $M \subset \langle M, a \rangle$, it follows that $\langle M, a \rangle = G$. Assume that $x \in G$ is an arbitrary element. Since $a \in Z(G)$, it follows that $x = a^k b$, for some $b \in M$ and $k \in \mathbb{Z}$. Now, for every $y \in M$, we have

$$xyx^{-1} = (a^k b) y (a^k b)^{-1} = a^k byb^{-1} a^{-k} = byb^{-1},$$

because $a \in Z(G)$. Since $b, y \in M$, it follows that $byb^{-1} \in M$. This means that $xyx^{-1} \in M$, and so M is a normal subgroup of G. Since G/M has composite order, we conclude that G/M has a non-trivial subgroup, say H/M. Hence, $\{M\} \subset H/M \subset G/M$. This yields that $M \subset H \subset G$, and this is a contradiction with the maximality of M.

Note that if $Z(G) = M$, then M is a normal subgroup of G and by the same arguments, we obtain a contradiction.   ∎

Lemma 10.23 *If an element of a group is not contained in any maximal subgroup, then the group is cyclic.*

Proof Suppose that G is a group and $x \in G$ with x is not in any maximal subgroup of G. Since x is not in any maximal subgroup, it follows that $\langle x \rangle$ is not maximal. Moreover, since x is not in any maximal subgroup, it follows that $\langle x \rangle$ is not a subset of any maximal subgroup. This yields that $\langle x \rangle = G$.   ∎

Theorem 10.24 *Let G be a finite group. If G has exactly one maximal subgroup, then the order of G is a power of a prime.*

Proof First, we show that G is cyclic. Assume that M is the unique maximal subgroup of G. Let $x \in G - M$. Then, the subgroup generated by x is contained in no maximal subgroup and hence is equal to G. Consequently, G is a cyclic group.

Clearly, we have $|G| > 1$. Assume that $|G|$ has more than one prime factor. Then, it may be written $|G| = p^k m$, where p is prime, k, m are positive integers and $(p, m) = 1$. Since G is cyclic, if k is a divisor of $|G|$, then G has a subgroup of order k. In particular, let $m = qr$ with q prime. So, G has a subgroup of order $p^k r$ and $p^{k-1} m$. Since $(p^k r, p^{k-1} m) = 1$, neither of these is a subgroup of the other. Since the order of each of these is less than that of order G, it follows that each of them is a proper subgroup of G. Since there are no proper divisors of $|G|$ greater than the order of either of these, no proper subgroup contains either of them. Therefore, we conclude that both of these are maximal subgroups. Since G is defined to have exactly one maximal subgroup, the assumption that $|G|$ has more than one prime factor must be invalid.   ∎

Theorem 10.25 *If a finite group has exactly two maximal subgroups, then it is a cyclic group.*

Proof Suppose that G is a finite group. Let M_1 and M_2 be the only distinct maximal subgroups of G. By Lagrange's Theorem, if a group is finite, the order of any of its proper subgroups can be no greater than half the order of the group. So, we have $|M_1| \leq |G|/2$ and $M_2| \leq |G|/2$. Therefore, we can write

$$|M_1 \cup M_2| \leq |M_1| + |M_2| - 1 \leq |G|/2 + |G|/2 - 1 = |G| - 1 < |G|.$$

Consequently, there exists $x \in G$ such that $x \notin M_1 \cup M_2$. Hence, by Lemma 10.23, we conclude that G is cyclic. $\blacksquare$

If M is a maximal subgroup of a group G, then also every conjugate aMa^{-1} of M is maximal in G. Indeed, $aMa^{-1} \subset H < G$ implies that $M < a^{-1}Ha < G$. For this reason the maximal subgroups are studied up to conjugation.

Exercises

1. Let M be a maximal subgroup of a group G. Prove that if $M \trianglelefteq G$, then $[G : M]$ is finite and equal to a prime number.
2. Let G be the additive group consisting of all rational numbers. Prove that G contains no maximal subgroup.
3. Let G be a group in which each proper subgroup is contained in maximal subgroup of finite index in G. If every two maximal subgroups on G are conjugate in G, prove that G is a cyclic group.
4. If G is a perfect group and M is a maximal subgroup of G, prove that

   (1) M contains the center of G;
   (2) $M/Z(G)$ is maximal in $G/Z(G)$.

10.4 Worked-Out Problems

Problem 10.26 If $n \geq 5$, prove that A_n is the unique proper non-trivial normal subgroup of the symmetric group S_n.

Solution Suppose that N is a proper non-trivial normal subgroup of S_n and let σ be a non-identity permutation in N. Then, there exists i such that $i\sigma \neq i$. We choose $j \neq i, i\sigma$. Now, if $\tau = (i\ j)$, then $\alpha = \sigma\tau\sigma^{-1}\tau^{-1}$ is non-identity and lies in N. Moreover, α is a product of the transpositions $\sigma\tau\sigma^{-1}$ and τ. Consequently, it is either a cycle of length 3 or a permutation of the form $(a\ b)(c\ d)$. Since N is normal, it contains either all cycles of length 3 or all permutations of type $(a\ b)(c\ d)$. Therefore, by Theorem 5.43, we deduce that $N = A_n$. $\blacksquare$

Problem 10.27 The fact that the set of all commutators in a group need not be a subgroup is an old result; the following example is due to P.J. Cassidy (1979).

(a) Let $\mathbb{F}[x, y]$ denote the ring of all polynomials in two variables over a field $\mathbb{F}$, and let $\mathbb{F}[x]$ and $\mathbb{F}[y]$ denote the subrings of all polynomials in x and in y, respectively. Define G to be the set of all matrices of the form

$$A = \begin{bmatrix} 1 & f(x) & h(x,y) \\ 0 & 1 & g(y) \\ 0 & 0 & 1 \end{bmatrix},$$

where $f(x) \in \mathbb{F}[x]$ and $g(y) \in \mathbb{F}[y]$ and $h(x,y) \in \mathbb{F}[x,y]$. Prove that G is a multiplicative group and that G' consists of all those matrices for which $f(x) = 0 = g(y)$.

(b) If $(0,0,h)$ is a commutator, show that there are polynomials $f(x), f'(x) \in \mathbb{F}[x]$ and $g(x), g'(x) \in \mathbb{F}[y]$ with $h(x,y) = f(x)g'(y) - f'(x)g(y)$.

(c) Show that $h(x,y) = x^2 + xy + y^2$ does not posses a decomposition as in part (b), and conclude that $(0,0,h) \in G'$ is not a commutator.

solution (a) If A denoted by triple (f,g,h), then

$$(f,g,h)(f',g',h') = (f + f', g + g', h + h' + fg'). \tag{10.6}$$

Since G is a subset of a matrix group, we need only to check the subgroup axioms, i.e., it is not necessary to check the associativity. The operation in (10.6) shows that the multiplication of two elements of G is also in G. It is clear that $I_3 \in G$ with $f(x) = h(x,y) = g(y) = 0$. Moreover, if $f + f' = 0$, $g + g' = 0$ and $h + h' + fg' = 0$, then we get $f' = f$, $g' = -g$ and $h' = -h + fg$. This yields that the element $(-f, -g, -h + fg)$ is the inverse of (f,g,h). Hence, we conclude that G is a group.

If $h = h(x,y) = \sum a_{ij} x^i y^j$, then

$$(0,0,h) = \prod_{i,j} [(a_{ij}x^i, 0, 0), (0, y^j, 0)].$$

This shows that G' consists of all those matrices for which $f(x) = 0 = g(y)$.

(b) Let $(0,0,h)$ be a commutator. Then we can write

$$\begin{aligned}
(0,0,h) &= (f,g,h_1)(f',g',h_2)(f,g,h_1)^{-1}(f',g',h_2)^{-1} \\
&= (f,g,h_1)(f',g',h_2)(-f,-g,-h_1+fg)(-f',-g',-h_2+f'g') \\
&= (f+f', g+g', h_1+h_2+fg')(-f-f', -g-g', -h_1-h_2+fg+f'g'+fg') \\
&= (0,0, fg'+fg'+fg+f'g'+(f+f')(-g-g')) \\
&= (0,0, fg'-f'g).
\end{aligned}$$

(c) If $f(x) = \sum b_i x^i$ and $f'(x) = \sum c_i x^i$, then there are equations

$$h(0,y) = f(0)g'(y) - f'(0)g(y),$$

$$\frac{\partial}{\partial x} h(x,y)\Big|_{x=0} = \frac{\partial}{\partial x} f(x)\Big|_{x=0} g'(y) - g(y) \frac{\partial}{\partial x} f'(x)\Big|_{x=0},$$

$$\frac{\partial^2}{\partial x^2} h(x,y)\Big|_{x=0} = \frac{\partial^2}{\partial x^2} f(x)\Big|_{x=0} g'(y) - g(y) \frac{\partial^2}{\partial x^2} f'(x)\Big|_{x=0}.$$

Hence, we obtain the following three equations

$$b_0 g'(y) - c_0 g(y) = y^2,$$
$$b_1 g'(y) - c_1 g(y) = y,$$
$$b_2 g'(y) - c_2 g(y) = 1.$$

Considering $\mathbb{F}[x, y]$ as a vector space over $\mathbb{F}$, one obtains the contradiction that the independent set $\{1, y, y^2\}$ is in the subspace spanned by $\{g, g'\}$.

10.5 Supplementary Exercises

1. If N is a normal subgroup of the group G, prove that $(G/N)' = G'N/N$.
2. Let a and b be two elements of order m and n, respectively, in group G. Prove that if a and b both commute with $[a, b]$, and d is the greatest common divisor of m and n, then $[a, b]^d = e$.
3. Suppose that G is a group and $|G| = p^n$, where p is a prime number. If $[G : C_G(x)] \leq p$, for all $x \in G$, prove that

   (a) $C_G(x) \trianglelefteq G$, for all $x \in G$;
   (b) $G' \leq Z(G)$;
   (c) $|G'| \leq p$.

4. Let G be a subgroup of a group G. If $[H, G'] = e$, prove that $[H', G] = e$.
5. Let p be prime and let G be a non-abelian group of order p^3. Show that $Z(G) = [G, G]$ and this is a subgroup of order p.
6. Let $K \leq M < G$, with $K \trianglelefteq G$. Prove that M/K is a maximal subgroup of G/K if and only if M is a maximal subgroup of G.

Chapter 11
Group Homomorphisms

A homomorphism is a function between groups satisfying a few natural properties. A homomorphism that is both one to one and onto is an isomorphism. This chapter presents different isomorphism theorems which are important tools for proving further results. The first isomorphism theorem, that will be the second theorem to be proven after the factor theorem, is easier to motivate, since it will help us in computing quotient groups. Cayley's Theorem states that a permutation group of a group is isomorphic to the given group.

11.1 Homomorphisms and Their Properties

Here are the Cayley tables for a cyclic group of order 4 and U_{10}:

$\cdot$	e	a	a^2	a^3
e	e	a	a^2	a^3
a	a	a^2	a^3	e
a^2	a^2	a^3	e	a
a^3	a^3	e	a	a^2

$\cdot$	1	3	7	9
1	1	3	7	9
3	3	9	1	7
7	7	1	9	3
9	9	7	3	1

There is obvious sense in which these two groups are "the same". Indeed, we can obtain the right table from the left table by replacing e, a, a^2 and a^3 with 1, 7, 9 and 3, respectively. Then, although the two groups look different, they are essentially the same. We may think of saying two groups are the same if it is possible to obtain one of them from the other by substitution as above. One way to implement a substitution is to use a function. In a sense, a function is a thing which substitutes its output for its input. We will define what it means for two groups to be the same by using certain kinds of functions between groups. These functions are called group homomorphisms; a special kind of homomorphism, called an isomorphism, will be

© The Author(s), under exclusive license to Springer Nature Singapore Pte Ltd. 2021
B. Davvaz, *A First Course in Group Theory*,
https://doi.org/10.1007/978-981-16-6365-9_11

used to define sameness for groups. The term homomorphism comes from the Greek words homo, "like", and morphe, "form".

Definition 11.1 Let G and H be groups. A function $f : G \rightarrow H$ is called a *homomorphism* if

$$f(ab) = f(a)f(b),$$

for all $a, b \in G$. Here, the multiplication in ab is in G and the multiplication in $f(a)f(b)$ is in H.

This definition can be visualized as shown in Fig. 11.1. The pairs of dashed arrows represent the group operations.

A short description of a homomorphism is that it preserve the operation of G. The set of all homomorphism from G to H is denoted by $Hom(G, H)$. This set is always non-empty because it contains the *zero homomorphism*, the homomorphism which sends every elements of G to the identity element of H.

In the definition of homomorphism, we assumed multiplicative notation for the operations in both G and H. If the operation in one or both is something else, we must adjust the definition accordingly. For instance, see Table 11.1.

Before working out some facts about homomorphisms, we present some examples.

Example 11.2 For any pair of groups G and H, one can always define the trivial homomorphism. This is the rather uninteresting function that maps every element of the domain to the identity element in the range.

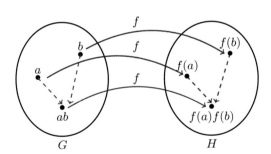

Fig. 11.1 Homomorphism between two groups

Table 11.1 Operations of groups in a homomorphism

Operation in G	Operation in H	Homomorphism definition
$+$	$+$	$f(x + y) = f(x) + f(y)$
$+$	$\cdot$	$f(x + y) = f(x) \cdot f(y)$
$\cdot$	$+$	$f(x \cdot y) = f(x) + f(y)$
$\star$	$\times$	$f(x \star y) = f(x) \times f(y)$

Example 11.3 Let $\mathbb{Z}$ be the group of integers under addition and $\mathbb{Z}_n$ be the group of integers under addition modulo n. If we define $f : \mathbb{Z} \to \mathbb{Z}_n$ by $f(x)$ equals to remainder of x on division by n, then n is a homomorphism.

Example 11.4 If $G = \langle a \rangle$, then $f : \mathbb{Z} \to G$ defined by $f(m) = a^m$ is a homomorphism.

Example 11.5 Let H denote the group $\{1, -1\}$ under multiplication (the 1 and -1 here are just the ordinary numbers, H is a group of order 2). We may define a function f from the group $\mathbb{Z}$ of integers under addition to H by

$$f(n) = \begin{cases} 1 & \text{if } n \text{ is even} \\ -1 & \text{if } n \text{ is odd.} \end{cases}$$

Then f is a homomorphism.

Example 11.6 The logarithm function is homomorphism from the group of positive real numbers under multiplication, to the group of all real numbers under addition.

Example 11.7 Let the circle group H consists of all complex numbers z such that $|z| = 1$. We can define a homomorphism f from the additive group of real numbers $\mathbb{R}$ to H by $f(\theta) = \cos\theta + i\sin\theta$. Geometrically, we are simply wrapping the real line around the circle in a group-theoretic fashion.

Example 11.8 Let $\mathbb{R}[x]$ denote the group of all polynomials with real coefficients under addition. For any $p \in \mathbb{R}[x]$, let p' denote the derivative of p. Then, the function $p \mapsto p'$ is a homomorphism from $\mathbb{R}[x]$ to itself.

Example 11.9 Let $G = C([0, 1])$ be the additive group of all continuous real-valued functions. Then, the integration function from G to $\mathbb{R}$ given by $f \mapsto \int_0^1 f(x)dx$ is a homomorphism.

Theorem 11.10 *If G and H are groups and $f : G \to H$ is a homomorphism, then*

(1) $f(e) = e$, *where e on the left is the identity in G and e on the right is the identity in H;*
(2) $f(a^{-1}) = f(a)^{-1}$, *for all $a \in G$.*

Proof (1) Since $ee = e$, it follows that $f(e)f(e) = f(e) = f(ee) = f(e) = f(e)e$. So, by cancellation property in H, we obtain $f(e) = e$.

(2) Let $a \in G$ be an arbitrary element. Since $f(aa^{-1}) = f(e) = e$, it follows that $f(a)f(a^{-1}) = e$, and so by the definition of inverse, we conclude that $f(a^{-1}) = f(a)^{-1}$. ∎

Informally, we can speak about a homomorphism as a function that respects structure. A homomorphism of groups "respects" the property of the identity element, multiplication, and inversion.

Theorem 11.11 *If G and H are groups and $f : G \to H$ is a homomorphism, then*

(1) For any integer n and a ∈ G, $f(a^n) = f(a)^n$;
(2) For any a ∈ G, if the order of a is finite, then $o(f(a))|o(a)$.

Proof (1) It follows from the first part of Theorem 11.10 trivially when $n = 0$, and by mathematical induction for $n > 0$. If $n < 0$, then put $m = -n$. Hence, we obtain

$$f(a^n) = f(a^{-m}) = f((a^m)^{-1}) = f(a^m)^{-1} = f(a)^{-m} = f(a)^n.$$

(2) If $o(a) = n$, then $a^n = e$, and so we obtain $f(a^n) = f(e) = e$. This implies that $f(a)^n = e$. Thus, we conclude that $o(f(a)|n$. ∎

Remark 11.12 If the operation of H is addition, then the property (1) in Theorem 11.11 becomes $f(a^n) = nf(a)$. If both the operation of G and H are addition, then the property (1) becomes $f(na) = nf(a)$.

Example 11.13 We want to determine all homomorphisms from $\mathbb{Z}_{24}$ to $\mathbb{Z}_{18}$. Suppose that $f : \mathbb{Z}_{24} \to \mathbb{Z}_{18}$ be a homomorphism. For every non-negative integer m, we have

$$f(\overline{m}) = f(\underbrace{\overline{1} + \overline{1} + \cdots + \overline{1}}_{m \text{ times}}) = \underbrace{f(\overline{1}) + f(\overline{1}) + \cdots + f(\overline{1})}_{m \text{ times}} = mf(\overline{1}),$$

for all $0 \le m \le 23$. This means that a homomorphism f is completely determined by the value $f(\overline{1})$. Let $f(\overline{1}) = \overline{n}$, where $0 \le n \le 17$. In additive group $\mathbb{Z}_{18}$, we can write

$$\overline{24n} = \overline{24}\overline{n} = \overline{24}f(\overline{1}) = f(\overline{24}) = f(\overline{0}).$$

Since any homomorphism maps the identity to identity, it follows that $f(\overline{0}) = \overline{0}$. Hence, $\overline{24n} = \overline{0}$. This yields that $18|24n$ or $3|4n$. Since $(3, 4) = 1$, it follows that $3|n$, and so we conclude that $n \in \{0, 3, 6, 9, 12, 15\}$. Therefore, any homomorphism must send $\overline{1}$ to $\overline{3k}$, for $k = 0, 1, \ldots, 5$. This shows that there exist at most six possible homomorphisms from $\mathbb{Z}_{24}$ to $\mathbb{Z}_{18}$, namely $f_0, f_1, \ldots, f_5$, where $f_k(\overline{m}) = \overline{3km}$, for every $k = 0, 1, \ldots, 5$.

Now, we investigate that for each $0 \le k \le 5$, f_k is well defined. Assume that $m, m' \in \mathbb{Z}$ such that $\overline{m} = \overline{m'}$ in $\mathbb{Z}_{24}$. Then, we have $24|m - m'$, and so $3km - 3km'$ is a multiple of 72. It follows that $3km - 3km'$ is a multiple of 18. Thus, $\overline{3km} = \overline{3km'}$ or $f_k(\overline{m}) = f_k(\overline{m'})$.

In the rest, we show that f_k is a homomorphism. If $m, m' \in \mathbb{Z}$, then we have

$$f_k(\overline{m} + \overline{m'}) = f_k(\overline{m + m'}) = \overline{3k(m + m')}$$
$$= \overline{3km} + \overline{3km'} = f(\overline{m}) + f(\overline{m'}).$$

Therefore, each f_k is a homomorphism from $\mathbb{Z}_{24}$ to $\mathbb{Z}_{18}$.

Since homomorphisms preserve the group operation, it should not be a surprise that they preserve many group properties.

Theorem 11.14 *Let G and H be groups and $f : G \to H$ be a homomorphism.*

(1) The image of a subgroup of G is a subgroup of H. In particular, $f(G)$ (or another notation, Imf) is a subgroup of H;
(2) The inverse image of a subgroup of H is a subgroup of G.

Proof (1) Let A be a subgroup of G. Since $e = f(e) \in f(A)$, it follows that $f(A)$ is non-empty. Let $f(a)$ and $f(b)$ be two arbitrary elements of $f(A)$. Then, we have $f(a)f(b)^{-1} = f(ab^{-1})$. Since A is a subgroup, it follows that $ab^{-1} \in A$. Therefore, we conclude that $f(a)f(b)^{-1} \in f(A)$, and so $f(A)$ is a subgroup of H.

(2) Let B be a subgroup of H. Again, since $f(e) = e \in B$, it follows that $e \in f^{-1}(B)$. Hence, $f^{-1}(B)$ is non-empty. Assume that x and y are two arbitrary elements of $f^{-1}(B)$. Then, we have $f(x) \in B$ and $f(y) \in B$. Since B is a subgroup, it follows that $f(x)f(y)^{-1} \in B$, and so $f(xy^{-1}) \in B$. Therefore, we conclude that $xy^{-1} \in f^{-1}(B)$. This completes the proof. ∎

Theorem 11.15 *Let G and H be groups and $f : G \to H$ be a homomorphism. If $A \trianglelefteq G$ and $B \trianglelefteq H$, then $f(A) \trianglelefteq f(G)$ and $f^{-1}(B) \trianglelefteq G$.*

Proof By Theorem 11.14, we know that $f(A) \leq f(G)$ and $f^{-1}(B) \leq G$.

Now, let $f(a) \in f(A)$ and $f(x) \in f(G)$ be arbitrary. Then, we can write $f(x)f(a)f(x)^{-1} = f(xax^{-1})$. Since $A \trianglelefteq G$, it follows that $xax^{-1} \in A$. So, we conclude that $f(x)f(a)f(x)^{-1} \in f(A)$. This shows that $f(A) \trianglelefteq f(G)$.

Finally, suppose that $x \in G$ and $y \in f^{-1}(B)$ are arbitrary. Then, we have $f(x) \in H$ and $f(y) \in B$. Since $B \trianglelefteq H$, it follows that $f(x)f(y)f(x)^{-1} \in B$, or equivalently $f(xyx^{-1}) \in B$. This implies that $xyx^{-1} \in f^{-1}(B)$, and so $f^{-1}(B) \trianglelefteq G$. ∎

The following is an important concept for homomorphisms.

Definition 11.16 The *kernel* of a homomorphism f from a group G to a group H is the set $\{x \in G \mid f(x) = e\}$. The kernel of f is denoted by $Kerf$.

Example 11.17 Let $D(\mathbb{R})$ be the additive group of all differentiable functions, $f : \mathbb{R} \to \mathbb{R}$, with continuous derivative. Let $C(\mathbb{R})$ be the additive group of all continuous functions $f : \mathbb{R} \to \mathbb{R}$. Suppose that $F : D(\mathbb{R}) \to C(\mathbb{R})$ be defined by $F(f) = df/dx$. Then

(1) F is a homomorphism;
(2) $KerF = \{f \in D(\mathbb{R}) \mid df/dx = 0\}$, which is the set of all constant functions.

The following is a fundamental theorem about the kernel of a homomorphism.

Theorem 11.18 *If G and H are groups and $f : G \to H$ is a homomorphism, then $Kerf$ is a normal subgroup of G.*

Proof Since $f(e) = e$, it follows that $Kerf$ is non-empty. Now, we suppose that a and b are two arbitrary elements of $Kerf$. Then, we have $f(a) = e$ and $f(b) = e$. Consequently, we get $f(ab^{-1}) = f(a)f(b)^{-1} = ee^{-1} = e$. This shows that $ab^{-1} \in Kerf$ and so $Kerf$ is a subgroup of G. On the other hand, for any $x \in G$, we have $f(xax^{-1}) = f(x)f(a)f(x)^{-1} = f(x)ef(x)^{-1} = e$. This yields that $xax^{-1} \in Kerf$. Therefore, we deduce that $Kerf$ is also a normal subgroup of G. ∎

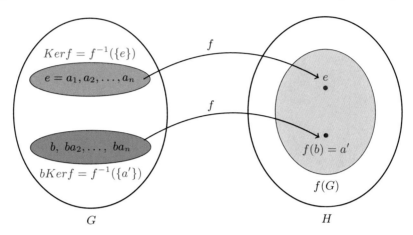

Fig. 11.2 Illustration for the kernel

Theorem 11.19 *If G and H are groups and $f : G \to H$ is a homomorphism, then*

(1) $f(a) = f(b)$ if and only if $aKerf = bKerf$;
(2) If $f(b) = a'$, then $f^{-1}(\{a'\}) = \{x \in G \mid f(x) = a'\} = bKerf$. Look at the illustration in Fig. 11.2.

Proof (1) We have

$$f(a) = f(b) \Leftrightarrow f(b)^{-1}f(a) = e \Leftrightarrow f(b^{-1}a) = e$$
$$\Leftrightarrow b^{-1}a \in Kerf \Leftrightarrow aKerf = bKerf.$$

(2) We must show that $f^{-1}(\{a'\}) \subseteq bKerf$ and $bKerf \subseteq f^{-1}(\{a'\})$. To demonstrate the first inclusion, assume that $x \in f^{-1}(\{a'\})$ is an arbitrary element. Then, we have $f(x) = a'$, and so $f(x) = f(b)$. Now by part (1), we conclude that $xKerf = bKerf$, and hence $x \in bKerf$.

To prove the second inclusion, suppose that $x \in Kerf$ is an arbitrary element. Then, we have $f(bx) = f(b)f(x) = a'e = a'$ or $bx \in f^{-1}(\{a'\})$. This completes the proof. ∎

Theorem 11.20 *Let G and H be groups and $f : G \to H$ be a homomorphism. Then, f is one to one if and only if $Kerf = \{e\}$.*

Proof Suppose that f is one to one. If $x \in Kerf$, then $f(x) = e$. On the other hand, we always have $f(e) = e$. Now, since $f(x) = f(e)$ and f is one to one, it follows that $x = e$. This proves that $Kerf = \{e\}$.

Conversely, let $Kerf = \{e\}$ and assume that a and b are elements of G such that $f(a) = f(b)$. Then, $f(a)f(b)^{-1} = e$, or equivalently $f(ab^{-1}) = e$. This means that $ab^{-1} \in Kerf = \{e\}$, and so $a = b$. Hence, f is a one to one function. ∎

Exercises

1. Let X be a non-empty set of generators of G. Let $f : G \to H$ and $g : G \to H$ be homomorphisms. If for every $x \in X$, $f(x) = g(x)$, prove that $f = g$.

2. Let G and H be groups and $f : G \to H$ be a homomorphism. Show that $f([x, y]) = [f(x), f(y)]$, for all $x, y \in G$.

3. Describe all the homomorphisms from $\mathbb{Z}_{12}$ to itself.

4. Let $f : \mathbb{Z}_8 \to \mathbb{Z}_{32}$ be a homomorphism such that $f(1) = 4$. Compute $f(5)$. What is the kernel of f?

5. Show that there is a homomorphism from S_n to $\mathbb{Z}_2$ whose kernel is A_n.

6. Determine all homomorphisms from $\mathbb{Z}$ onto S_3. Determine all homomorphisms from $\mathbb{Z}$ to S_3.

7. Find all group homomorphisms from $\mathbb{Z}_4$ into $\mathbb{Z}_{10}$.

8. How many homomorphisms are there from $\mathbb{Z}_{20}$ onto $\mathbb{Z}_8$?

9. If G is a finitely generated group by a set with n elements and H is a finite group, prove that $|Hom(G, H)| \leq |H|^n$.

10. If f is a homomorphism from $\mathbb{Z}_{30}$ onto a group of order 5, determine the kernel of f.

11. Let G and H be groups and $f : G \to H$ be an onto homomorphism. If X is a subset of G that generates G, show that $f(X)$ generates H.

12. Find a homomorphism f from U_{30} to U_{30} with kernel $\{\overline{1}, \overline{11}\}$ and $f(\overline{7}) = \overline{7}$.

13. Let $\mathbb{Z}[x]$ be the group of polynomials in x with integer coefficients under addition. Prove that the function from $\mathbb{Z}[x]$ into $\mathbb{Z}$ given by $f(x) \mapsto f(3)$ is a homomorphism. Give a geometrical description of the kernel of this homomorphism.

14. Prove that the function from $\mathbb{R}$ under addition to $GL_2(\mathbb{R})$ that takes x to

$$\begin{bmatrix} \cos x & \sin x \\ -\sin x & \cos x \end{bmatrix}$$

is a homomorphism. What is the kernel of this homomorphism?

15. Let G be a subgroup of some dihedral group. For each $x \in G$, we define

$$f(x) = \begin{cases} 1 & \text{if } x \text{ is a rotation} \\ -1 & \text{if } x \text{ is a reflection.} \end{cases}$$

Prove that f is a homomorphism from G to the multiplicative group $\{1, -1\}$. What is the kernel?

16. (a) Let $n \geq 5$ and $m \geq 3$. Suppose that m is not divisible by 3. Let $f : S_n \to D_m$ be a homomorphism. Prove that A_n is contained in the kernel of f. How many homomorphisms $S_n \to D_m$ are there?

   (b) Let $n \geq 5$ and $m \geq 3$ be arbitrary, so m may be divisible by 3. How many homomorphisms $S_n \to D_m$ are there?

11.2 Isomorphism Theorems

We wish to have a way to determine if two groups have similar properties. The advantage of this is that if we could tell that two groups G and H have similar properties and we already know all the properties of G, then we would immediately know all the properties of H. The tool which will allow us to do this is called an isomorphism, from the Greek words "isos" which mean "same" and "morphe" which means "form".

Definition 11.21 Let G and H be groups. A function $f : G \to H$ is said to be an *isomorphism* if the following conditions are satisfied:

(1) f is a homomorphism;
(2) f is one to one;
(3) f is onto.

If there is an isomorphism from G onto H, we say that G and H are *isomorphic* and write $G \cong H$.

Example 11.22 If G is a group, then the identity function $id : G \to G$ defined by $id(x) = x$, for all $x \in G$, is an isomorphism.

Example 11.23 The group of real numbers $\mathbb{R}$ under addition and positive real numbers $\mathbb{R}^+$ under multiplication are isomorphic. Indeed, if we consider $f : \mathbb{R} \to \mathbb{R}^+$ by $f(x) = e^x$, then for every real numbers a and b we have $f(a + b) = e^{a+b} = e^a e^b = f(a) f(b)$. So, f is a homomorphism. Moreover, by the well-known results of calculus, f is one to one and onto.

An injective (or one to one) homomorphism is called a *monomorphism* and a surjective (or onto) homomorphism is called an *epimorphism*. Of course a bijective homomorphism is what we have been calling an isomorphism. A group H is said to be a *homomorphic image* of a group G, if there exists an epimorphism from G onto H.

Let G and H be groups and $f : G \to H$ be a homomorphism. Clearly, we have

(1) f is a monomorphism if and only if $Ker f = \{e\}$;
(2) f is an epimorphism if and only if $Im f = H$;
(3) f is an isomorphism if and only if $Ker f = \{e\}$ and $Im f = H$.

How do we demonstrate that two groups G and H are not isomorphic, if this is the case? A structural property of a group is one that must be shared by any isomorphic group. It is not concerned with names or some other non-structural characteristics of the elements. In order to prove that two groups G and H are not isomorphic, one needs to demonstrate that there is no isomorphism from G onto H. Usually, in practice, this is much easier than it sounds in general, and is accomplished by finding some structural property that holds in one group, but not in the other.

Example 11.24 $\mathbb{Z}_6 \ncong S_3$ because $\mathbb{Z}_6$ is abelian and S_3 is not abelian.

Example 11.25 The dihedral group D_{12} is not isomorphic to S_4 because D_{12} has 13 elements of order 2 (12 reflections and the rotation for π), while S_4 has only 9 such elements (transpositions and product of disjoint transpositions).

Lemma 11.26 *Let G and H be groups and $f : G \to H$ be an isomorphism. Then, $f^{-1} : H \to G$ is also an isomorphism.*

Proof Since f is a bijective function, by Theorem 5.5, it follows that f^{-1} exists and it is a bijective function from H onto G. So, it remains to be seen that f^{-1} is a homomorphism. Assume that x and y are arbitrary elements of H. Then, there exist $a, b \in G$ such that $f(a) = x$ and $f(b) = y$. This yields that $a = f^{-1}(x)$ and $b = f^{-1}(y)$. Hence, we can write $xy = f(a)f(b) = f(ab)$, and so $f^{-1}(xy) = ab$. This shows that $f^{-1}(xy) = f^{-1}(x)f^{-1}(y)$. Therefore, f^{-1} is a homomorphism. ■

Lemma 11.27 *If $f : G \to H$ and $g : H \to K$ are isomorphisms between groups, then so is the function $g \circ f$.*

Proof By Theorem 5.6, gof is a bijective function from G onto K. To check that $g \circ f$ is a homomorphism, let a and b be arbitrary elements of G. Then, we obtain

$$(g \circ f)(ab) = g\big(f(ab)\big) = g\big(f(a)f(b)\big)$$
$$= g\big(f(a)\big)g\big(f(b)\big) = (g \circ f)(a)(g \circ f)(b),$$

where we used in turn the facts that f and g are homomorphisms. Therefore, $g \circ f$ is an isomorphism. ■

Theorem 11.28 *The relation of isomorphism between groups is an equivalence relation on the family of all groups.*

Proof The result follows by Lemmas 11.26 and 11.27. ■

Lemma 11.29 *Let G and H be two cyclic groups of the same order. Then, G and H are isomorphic.*

Proof Suppose that $G = \langle a \rangle$ and $H = \langle b \rangle$. If $x \in G$, then $x = a^i$, for some integer i. Define $f : G \to H$ by $f(x) = b^i$.

We first have to check that f is well defined. If G is infinite, then so is H and every element of G may be uniquely represented in the form a^i. Thus, f is automatically well defined in this case. Now, assume that G has order n, and suppose that $x = a^j$, too. We have to check that $b^i = b^j$.

Since $a^i = a^j$, it follows that $a^{i-j} = e$, and so $m | i - j$. Since $|H| = m$, it follows that $i - j = k|H|$, for some integer k. Consequently, we can write $b^{i-j} = b^{k|H|} = \left(b^{|H|}\right)^k = e$. This shows that $b^i = b^j$. Therefore, we conclude that f is well defined. Now, let a^i and a^j be two elements of G. Then, we have

$$f(a^i a^j) = f(a^{i+j}) = b^{i+j} = b^i b^j = f(a^i)f(a^j).$$

Fig. 11.3 Vertices of the
cube with the same number
are endpoints of the four
diagonals of the cube

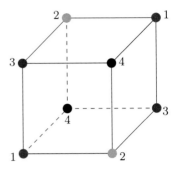

This means that f is a homomorphism. Moreover, the function $g : H \to G$ defined by $g(b^i) = a^i$ is the inverse of f, and so f is a one-to-one correspondence. This completes the proof. ∎

Corollary 11.30 *Any infinite cyclic group is isomorphic to the additive group $\mathbb{Z}$. Any finite cyclic group of order n is isomorphic to $\mathbb{Z}_n$, the additive group of modulo n.*

Proof The result follows directly from Lemma 11.29. ∎

Example 11.31 (*The Rotation Group of a Cube*) A cube has 8 vertices, 12 edges, and 6 faces (see Fig. 11.3), each of which is a square. The cube has exactly 24 rotational symmetries, which comprise:

(1) 1 trivial rotation or the identity symmetry;
(2) 9 non-trivial rotations (by $\pi/2$, π and $3\pi/2$) about 3 axes joining the centers of opposite faces;
(3) 8 non-trivial rotations (by $2\pi/3$ and $4\pi/3$) about the 4 great diagonals;
(4) 4 non-trivial rotations (by π) about the 4 axes joining the midpoints of opposite edges.

Since the group of rotations of a cube has the same order as S_4, we need only prove that the group of rotations is isomorphic to a subgroup of S_4. We number the vertices of the cube from 1 to 4, and where opposite vertices are given the same number. Now a cube has 4 diagonals and any rotation induces a permutation of these diagonals. But we cannot just assume that different rotations correspond to different rotations. Observe that vertices with the same number are endpoints of the four diagonals of the cube. We can number the diagonals according to their endpoints. The permutations of the numbers of vertices in the front face of the cube correspond to permutations of the diagonals. Therefore, there is a one to one correspondence between the rotations of the cube and the permutations of the diagonals of the cube. The composition of rotations coincides with the product of permutations and so the group of rotations of a cube is isomorphic to the symmetric group S_4. We see two perpendicular axes where $\pi/2$ rotations give the permutations $\sigma = (1\,2\,3\,4)$ and $\tau = (1\,4\,3\,2)$. These induce the subgroup $\{id, \sigma, \sigma^2, \sigma^3, \tau^2, \tau^2\sigma, \tau^2\sigma^2, \tau^2\sigma^3\}$ and the subgroup $\{id, \sigma\tau, (\sigma\tau)^2\}$. Consequently, the rotations induce all 24 permutations since $24 = (8, 3)$.

There are three theorems, formulated by E. Noether, describing the relationship between factor groups, normal subgroups and homomorphisms. These theorems are based on a basic result on homomorphisms presented in the following lemma.

Lemma 11.32 *If N is a normal subgroup of a group G, then the canonical map $\pi : G \to G/N$, given by $\pi(a) = aN$, is an onto homomorphism with kernel N.*

Proof Clearly, we have $\pi(ab) = abN = aNbN = \pi(a)\pi(b)$, for all $a, b \in G$. Moreover, since every element of G/N is of the form aN, for some $a \in G$, it follows that π is onto. Finally, for each $a \in G$, we have $a \in Ker\pi$ if and only if $\pi(a) = aN = N$ if and only if $a \in N$. ∎

Theorem 11.33 *Let G and H be groups and $f : G \to H$ be a homomorphism and suppose that N is a normal subgroup of G satisfies $N \leq Kerf$. Then, there exists a unique homomorphism $g : G/N \to H$ which satisfies $g \circ \pi = f$.*

Proof Let aN be a left coset of N in G. Choose any $x \in aN$, then $x = ay$, for some $y \in N$. Moreover, we have $f(x) = f(ay) = f(a)f(y) = f(a)e = f(a)$, because $N \leq Kerf$. Therefore, f has the same effect on every element of the coset aN. So, if we define $g : G/N \to H$ by $g(aN) = f(a)$, for all $a \in G$, then g is well defined. Now, assume that aN and bN are arbitrary elements of G/N. Then, we have

$$g(aNbN) = g(abN) = f(ab) = f(a)f(b) = g(aN)g(bN).$$

Hence, g is a homomorphism such that $g \circ \pi = f$. At the end, we prove the uniqueness. Suppose that $g, g' : G/N \to H$ are homomorphisms which satisfy $g \circ \pi = f$ and $g' \circ \pi = f$. Then, we have $g(\pi(a)) = g'(\pi(a))$, for all $a \in G$. Since π is onto, any element of G/N has the form $\pi(a)$, for some $a \in G$. Thus, we conclude that $g = g'$. This completes the proof. ∎

Theorem 11.34 (First Isomorphism Theorem) *Let G and H be groups and $f : G \to H$ be a homomorphism. Then, there exists an isomorphism $g : G/Kerf \to Imf$ such that $g \circ \pi = f$. In particular, if f is onto, then $G/Kerf \cong H$. In this case, we say the diagram presented in Fig. 11.4 is commutative.*

Proof Since Imf is a subgroup of H, without loss of generality we may assume that $H = Imf$. With $N = Kerf$ we conclude that by Theorem 11.33 that there is an onto homomorphism $g : G/N \to Imf$ given by $g(aN) = f(a)$, for all $a \in G$. Since $Kerg = \{aN \mid f(a) = e\} = \{aN \mid a \in N\} = \{N\}$, it follows that g is one to one. This completes our proof. ∎

Theorem 11.34 also is called the *fundamental theorem of homomorphism*, and it is one of the most basic theorems in group theory.

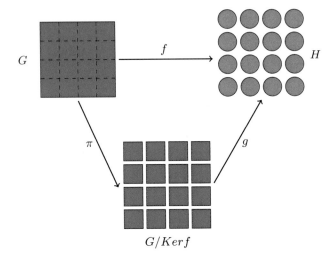

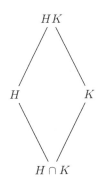

Fig. 11.4 Commutative diagram for the first isomorphism theorem

Fig. 11.5 The second
isomorphism theorem

Example 11.35 Let $f : \mathbb{Z} \to \mathbb{Z}_n$ be the homomorphism defined in Example 11.3. We obtain $Imf = \mathbb{Z}_n$ and $Kerf = n\mathbb{Z}$, and so we conclude that $\mathbb{Z}/n\mathbb{Z} \cong \mathbb{Z}_n$.

Example 11.36 The function $f : \mathbb{R} \to \mathbb{C}^*$ given by $f(x) = e^{2\pi xi}$, for all $x \in \mathbb{R}$, is a homomorphism. We obtain $Imf = \{z \in \mathbb{C} \mid |z| = 1\} := S^1$, the unit complex numbers and $Kerf = \mathbb{Z}$. Therefore, we can write $\mathbb{R}/\mathbb{Z} \cong S^1$.

Theorem 11.37 (Second Isomorphism Theorem) *Let G be a group and H, N be subgroups of G. If $N \trianglelefteq G$, then $H \cap N$ is normal in H and $H/(H \cap N) \cong HN/N$.*

The second isomorphism theorem can be represented pictorially as in Fig. 11.5.

Proof Since N is a normal subgroup of G, it follows that $NH = HN$, and so HN is a subgroup of G and also we have $N \leq HN$. Moreover, for every $x \in HN$, there exist $h \in H$ and $n \in N$ such that $xNx^{-1} = hnNn^{-1}h^{-1} = hNh^{-1} = N$. This yields that $N \trianglelefteq HN$. Now, we define a function $f : H \to HN/N$ by $f(h) = hN$, for all $h \in H$. Clearly, f is a homomorphism. Consider $xN \in HN/N$, where $x \in HN$. Then, $x = hn$, for some $h \in H$ and $n \in N$. Hence, we have $xN = hnN = hN = f(h)$. This shows that f is onto. Now, by the first isomorphism theorem, we obtain $H/Kerf \cong HN/N$. If we can establish that $Kerf = H \cap N$, we shall obtain that $H \cap N$ is a normal subgroup of H and $H/(H \cap N) \cong HN/N$. Indeed, we have

$$Kerf = \{h \in H \mid f(h) = N\} = \{h \in H \mid hN = N\}$$
$$= \{h \in H \mid h \in N\} = H \cap N.$$

This completes the proof. ∎

Theorem 11.38 (Third Isomorphism Theorem) *Let H and N be normal subgroups of a group G such that $N \leq H$. Then, H/N is a normal subgroup of G/N and $(G/N)/(H/N) \cong G/H$.*

Proof We define $f : G/N \to G/H$ by $f(aN) = aH$, for every $a \in G$. Since f is defined on cosets, we should check that f is well defined. To begin with, if $aN = bN$, then $a^{-1}b \in N$. Since $N \leq H$, it follows that $a^{-1}b \in H$ or $aH = bH$. This shows that f is well defined. For every aN and bN in G/N, we have $f(aNbN) = f(abN) = abH = aHbH = f(aN)f(bN)$. Hence, f is a homomorphism. Clearly, f is onto, for if $aH \in G/H$, then $f(aN) = aH$. Furthermore, we have

$$Kerf = \{aN \in G/N \mid f(aN) = H\} = \{aN \in G/N \mid aH = H\}$$
$$= \{aN \in G/N \mid a \in H\} = H/N,$$

as required. The result now follows by the first isomorphism theorem. ∎

We may picture the third isomorphism theorem as illustration in Fig. 11.6.

Exercises

1. Find a group which is isomorphic to one of its proper subgroups.
2. Let $G = \mathbb{R} \setminus \{-1\}$ and define a binary operation on G by $a \star b = a + b + ab$. Prove that G is a group under this operation. Show that $(G, \star)$ is isomorphic to the multiplicative group of non-zero real numbers.
3. Prove that

    (a) The multiplication groups $\mathbb{R}^*$ and $\mathbb{C}^*$ are not isomorphic;
    (b) The additive groups $\mathbb{R}$ and $\mathbb{Q}$ are not isomorphic;
    (c) The additive groups $\mathbb{Z}$ and $\mathbb{Q}$ are not isomorphic;

4. Prove that every cyclic group of finite order n is isomorphic to the multiplicative group of all complex nth roots of 1.

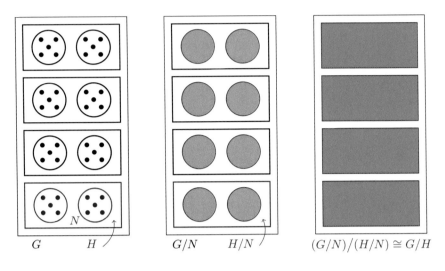

Fig. 11.6 Illustration for the third isomorphism theorem

5. Let $1 \leq n \leq 3$. Prove that any two groups containing exactly n elements are isomorphic.
6. If G is a non-abelian group of order 6, prove that $G \cong S_3$.
7. Let X_1 and X_2 be two sets. Suppose that there exists a one to one correspondence between X_1 and X_2. Show that there exists an isomorphism of S_{X_1} onto S_{X_2}.
8. Prove that any group of order 4 is isomorphic to the group $\mathbb{Z}_4$ or to the group $K = \{id, (1\,2)(3\,4), (1\,3)(2\,4), (1\,4)(2\,3)\}$.
9. Is $\mathbb{Q}/\mathbb{Z} \cong \mathbb{Q}$?
10. Show that the groups $\mathbb{Q}/\mathbb{Z}$ and $\mathbb{R}/\mathbb{Q}$ can not be isomorphic.
11. Let $f, g : \mathbb{R} \to \mathbb{R}$ be real valued functions defined by $f(x) = 1/x$ and $g(x) = (x-1)/x$. Then, f and g generate a group G with the operation given by function composition. Prove that $G \cong S_3$.
12. Let X be a non-empty set and $G = \{f \mid f : X \to \mathbb{Z}_2\}$. Show that

    (a) G is a group addition of functions;
    (b) $\mathcal{P}(X)$ is a group under the binary operation

    $$A \triangle B = (A \cup B) - (A \cap B);$$

    (c) $(\mathcal{P}(X), \triangle) \cong (G, +)$.

13. Show that the nth roots of unity are isomorphic to $\mathbb{Z}_n$.
14. Show that to each positive integer n there exist only a finite number of pairwise non-isomorphic groups of order n.
15. Let G be a non-abelian group. Prove that the group of automorphisms of G is not cyclic.

16. Prove that D_4 and Q_8 are not isomorphic.
17. Show that U_{17} is isomorphic to Z_{16}.
18. Show that the three groups $\mathbb{Z}_6$, U_9 and U_{18} are isomorphic to each other.
19. How many pairwise non-isomorphic groups can you find which are homomorphic images of S_3?
20. Suppose that for each prime p, $\mathbb{Z}_p$ is the homomorphic image of a group G. What can we say about $|G|$?
21. Prove that $GL_2(\mathbb{Z}_2) \cong S_3$.
22. Show that D_4 is isomorphic to the subgroup of all lower triangular matrices in $GL_3(\mathbb{Z}_2)$.
23. Show that $SL_2(\mathbb{Z}_3)$ is not isomorphic to S_4.
24. Explain why S_n ($n \geq 3$) contains a subgroup isomorphic to D_n.
25. For real numbers a and b with $a \neq 0$, define $f_{ab} : \mathbb{R} \to \mathbb{R}$ by $f_{ab}(x) = ax + b$, for all $x \in \mathbb{R}$. Let $G = \{f_{ab} \mid a, b \in \mathbb{R} \text{ and } a \neq 0\}$ and $N = \{f_{1b} \mid b \in \mathbb{R}\}$. Prove that N is a normal subgroup of G and G/N is isomorphic to the group of non-zero real numbers under multiplication.
26. Let $G = \{z \in \mathbb{C} \mid z^n = 1, \text{ for some positive integer } n\}$. Prove that for any fixed integer $k > 1$, the function f from G to itself defined by $f(z) = z^k$, for all $z \in G$, is an onto homomorphism but is not an isomorphism.
27. Let a group G contain two normal subgroups K and N. Let H be a subgroup of G. Prove that $HK/K \cong HN/N$ if $H \cap K = H \cap N$.
28. In the group $GL_2(\mathbb{Z}_3)$, let

$$H = \langle \begin{bmatrix} 0 & -1 \\ 1 & 0 \end{bmatrix}, \begin{bmatrix} -1 & 1 \\ 1 & 1 \end{bmatrix} \rangle.$$

Prove that $H \cong Q_8$ and $H \trianglelefteq GL_2(\mathbb{Z}_3)$.

29. Let G be a group, and suppose that S be any set for which there exists a bijective function $f : G \to S$. Define a binary operation on S by setting $a \cdot b = f\left(f^{-1}(a)f^{-1}(b)\right)$, for all $a, b \in S$. Prove that S is a group under this binary operation, and that f is actually a group isomorphism.
30. Let G be defined as all formal symbols $x^i y^j$, $i = 0, 1$ and $j = 0, 1, \ldots, n-1$, where we assume

$$x^i y^j = x^{i'} y^{j'} \Leftrightarrow i = i', j = j',$$
$$x^2 = y^n = e, n > 2,$$
$$xy = y^{-1}x.$$

(a) Find the form of the product $(x^i y^j)(x^k y^l)$ as $x^r y^s$;
(b) Prove that G is non-abelian of order $2n$;
(c) If n is odd, prove that the center of G is $\{e\}$, while if n is even the center of G is larger than $\{e\}$;
(d) Can you interpret this group as the dihedral group of order $2n$.

11.3 Cayley's Theorem

Our next subject is a classic theorem of Cayley. The proof of this theorem is not difficult, and it is a good exercise in group theory as it uses many concepts previously studied. Then, an important generalization of it is given.

Theorem 11.39 (Cayley's Theorem) *Let G be a given group. Then, there exists a set X such that G is isomorphic to a permutation group on X.*

Proof We choose the set X consisting of all the elements of G. For each element $a \in G$, let ρ_a be a function on X defined by the formula

$$\rho_a(x) = xa,$$

for all $x \in X$. It follows that ρ_a is a permutation on X. Furthermore, the associative law proves

$$\rho_{ab} = \rho_a \circ \rho_b,$$

for all $a, b \in G$. Thus, the function ρ is a homomorphism from G into the symmetric group S_X. Clearly, ρ_a is the identity function on X if and only if $a = e$. This means that ρ is one to one. Hence, G is isomorphic to the image $\rho(G)$ which is a permutation group on X. ∎

If a group G is isomorphic to a subgroup of a group H, we say G *embeds* in H. In this case there is an embedding (another word for one to one homomorphism) $G \hookrightarrow H$ which identifies G with its image in H. So, Cayley's Theorem says that every finite group embeds in a symmetric group.

Theorem 11.40 (Generalized Cayley's Theorem) *Let H be a subgroup of a group G, and X be the set of all left cosets of H in G. Then, there exists a homomorphism from G into the permutation group S_X whose kernel is the largest normal subgroup of G that is contained in H.*

Proof Suppose that $H \leq G$ and $X = \{xH \mid x \in G\}$. Note that X need not be a group itself. If a is a fixed element of G, we define $\theta_a : X \to X$ by $\theta_a(xH) = axH$, for all $x \in G$. Simulating the proof of Theorem 11.39, we find that $\theta_a \in S_X$ and $\theta_{ab} = \theta_a \theta_b$. This shows that the function $f : G \to S_X$ defined by $f(a) = \theta_a$ is a homomorphism of G into S_X. Next, we identify the kernel of f. We have

$$
\begin{aligned}
Ker f &= \{a \in G \mid f(a) = \theta_e\} \\
&= \{a \in G \mid \theta_a(xH) = \theta_e(xH), \text{ for all } x \in G\} \\
&= \{a \in G \mid axH = xH, \text{ for all } x \in G\} \\
&= \{a \in G \mid x^{-1}axH = H, \text{ for all } x \in G\} \\
&= \{a \in G \mid x^{-1}ax \in H, \text{ for all } x \in G\} \\
&= \{a \in G \mid a \in xHx^{-1}, \text{ for all } x \in G\} \\
&= \bigcap_{x \in G} xHx^{-1} \leq H.
\end{aligned}
$$

Now, we claim that from this characterization of $Kerf$, $Kerf$ must be the largest normal subgroup of G which is contained in H. To prove this, suppose that N is a normal subgroup of G contained in H. Using $N \leq H$, it is immediate that $N = xNx^{-1} \subseteq xHx^{-1}$, for all $x \in G$. Thus, we conclude that

$$N \leq \bigcap_{x \in G} xHx^{-1} = Kerf.$$

This completes the proof.   ∎

Note that Theorem 11.40 reduces to Cayley's Theorem in the special case of $H = \{e\}$.

In Theorem 3.56, we proved that a Cayley table is a Latin square, i.e., the rows and columns are permutation of one another. However, the associative law is not easy to discern by the naked eye. In Theorem 3.57, we studied a method for the verification of the associative law. Now, by using a method similar the proof of Cayley's Theorem, we give another procedure for verification of the associative law.

Theorem 11.41 *A Latin square is a Cayley table if composite of two rows is some row in the table.*

Proof We use $1, 2, \ldots, n$ to denote the entries in a Latin square and a_{ij} denote the entry at the ith row and the jth column. In order to prove the statement, it remains to check if associative law holds in the set $G = \{1, 2, \ldots, n\}$ with the binary operation $\star$ defined by the given Latin square. We assume 1 is the identity element.

Simulating the proof of Theorem 11.39, we define $\sigma : G \to S_n$ by $\sigma(j) = \sigma_j$ such that $i\sigma_j = a_{ij} = i \star j$, for all $i, j \in G$. The right cancellation law guarantees that σ_j is indeed in S_n. Although G is a set for all we know, S_G is a group. Associative law in G amounts to the relation $\sigma_i \sigma_j = \sigma_{i \star j}$, for all $i, j \in G$. If that is true, then $\sigma(G)$ is a subgroup of S_G. Conversely, if $\sigma(G)$ is a subgroup of S_G, then the associative law holds, for we claim that $\sigma_i \sigma_j$, which is σ_k, for some $k \in G$, is indeed $\sigma_{i \star j}$. To see this, we show that k must be $i \star j$. Toward this end, consider $1\sigma_i \sigma_j$ and $1\sigma_k$. The former is equal to $(1 \star i)\sigma_j = i\sigma_j = i \star j$, while the later is equal to $1 \star k = k$. Hence, it reduces to check if $\sigma(G)$ is a subgroup of S_G. Since S_G is a finite group, we need only to check that if the multiplication of any two permutations in $\sigma(G)$ is still a permutation in $\sigma(G)$. This can be achieved by checking the row of the Latin square.   ∎

Example 11.42 As a quick test, examine the following two Latin squares of order 4. One of these turns out to be a Latin square that we can get from a group, while the other does not. Which is which?

x	a	b	c
c	x	a	b
b	c	x	a
a	b	c	x

x	a	b	c	d
a	d	c	x	b
b	x	a	d	c
c	b	d	a	x
d	c	x	b	a

It is easy to check that the first square can be the Cayley table for the group $\mathbb{Z}_4$, if we consider $x = \overline{0}, a = \overline{1}, b = \overline{2}$ and $c = \overline{3}$.

The second table seems so different. Assume that it is a Cayley table of a group. Then, for some ordering $[\sigma_1, \sigma_2, \sigma_3, \sigma_4, \sigma_5]$ of our group elements and another ordering $[\tau_1, \tau_2, \tau_3, \tau_4, \tau_5]$, we have

$\cdot$	τ_1	τ_2	τ_3	τ_4	τ_5
σ_1	x	a	b	c	d
σ_2	a	d	c	x	b
σ_3	b	x	a	d	c
σ_4	c	b	d	a	x
σ_5	d	c	x	b	a

One of the $\sigma_1, \sigma_2, \sigma_3, \sigma_4, \sigma_5$ is the identity element. One of the five rows of the table must be $[\tau_1, \tau_2, \tau_3, \tau_4, \tau_5]$. In particular, we can actually assume that our top row is formed by taking σ_1^{-1} and acting to each element of $[x, a, b, c, d]$, because the row $[x, a, b, c, d]$ is just multiplying $[\tau_1, \tau_2, \tau_3, \tau_4, \tau_5]$ to σ_1, i.e., $[\tau_1\sigma_1, \tau_2\sigma_1, \tau_3\sigma_1, \tau_4\sigma_1, \tau_5\sigma_1] = [x, a, b, c, d]$, or equivalently, $[\tau_1, \tau_2, \tau_3, \tau_4, \tau_5] = [x\sigma_1^{-1}, a\sigma_1^{-1}, b\sigma_1^{-1}, c\sigma_1^{-1}, d\sigma_1^{-1}]$. We can think of these as permutations of the row $[x, a, b, c, d]$. For example, the row corresponding to σ_2, can be considered as the permutation

$$\sigma_2 = \begin{pmatrix} x & a & b & c & d \\ a & d & c & x & b \end{pmatrix}.$$

Similarly, the row corresponding to σ_3, gives us the following permutation:

$$\sigma_3 = \begin{pmatrix} x & a & b & c & d \\ b & x & a & d & c \end{pmatrix}.$$

If we consider the product of σ_2 and σ_3, then we obtain

$$\sigma_2\sigma_3 = \begin{pmatrix} x & a & b & c & d \\ a & d & c & x & b \end{pmatrix}\begin{pmatrix} x & a & b & c & d \\ b & x & a & d & c \end{pmatrix} = \begin{pmatrix} x & a & b & c & d \\ x & c & d & b & a \end{pmatrix}.$$

Exercises

1. How many ways are there to embed $\mathbb{Z}_4$ in S_4?
2. How many ways are there to embed D_4 in S_4?
3. Let G be a group and H be a subgroup of G. Let $X = \{Ha \mid a \in G\}$ be the set of all right cosets of H in G. Define, for $b \in G, T_b : X \to X$ by $T_b(Ha) = Hab^{-1}$.

(a) Prove that the function $f : G \to S_X$ defined by $f(b) = T_b$, for all $b \in G$, is a homomorphism;
(b) Describe $Kerf$;
(c) Show that $Kerf$ is the largest normal subgroup of G lying in H.

4. Show that the function $f : S_n \to A_{n+2}$ given by

$$f(\sigma) = \begin{cases} \sigma & \text{if } \sigma \text{ is even} \\ \sigma (n+1 \; n+2) & \text{if } \sigma \text{ is odd,} \end{cases}$$

is an isomorphism between S_n and a subgroup of A_{n+2}. Deduce that every finite group is isomorphic to a subgroup of A_{n+2}.
5. Show that, for $n \leq 4$, any Latin square of order n can be obtained from the Cayley table of a group by permuting rows, columns, and symbols; but this is not true for $n = 5$.

11.4 Automorphisms

A homomorphism of a group G into itself is called an *endomorphism*. The identity function is clearly an endomorphism. The set of all endomorphism from G to itself is denoted by $End(G)$.

Definition 11.43 By an *automorphism* of a group G, we shall mean an isomorphism of G onto itself.

Lemma 11.44 *Let G be a group and $a \in G$ is a fixed element of G. If a function $\phi_a : G \to G$ is defined by $\phi_a(x) = axa^{-1}$, for every $x \in G$, then ϕ_a is an automorphism of G.*

Proof Assume that x and y are arbitrary elements of G. Then, we have $\phi_a(x)\phi_a(y) = (axa^{-1})(aya^{-1} = a(xy)a^{-1} = \phi_a(xy)$. So, ϕ_a is a homomorphism. Now, we investigate that ϕ_a is a bijective function. In fact, if $\phi_a(x) = \phi_a(y)$, then $axa^{-1} = aya^{-1}$, and so $x = y$. This shows that ϕ_a is one to one. On the other hand, for each $y \in G$, we can write $\phi_a(a^{-1}ya) = aa^{-1}yaa^{-1} = y$. This means that ϕ_a is onto. ∎

The function ϕ_a is called the *inner automorphism* by a.
Let $Aut(G)$ denote the set of all automorphisms of G. For the product of elements of $Aut(G)$, we can use the composition of functions.

Theorem 11.45 *If G is a group, then $Aut(G)$ is also a group.*

Proof Note that the identity function belongs to $Aut(G)$. Since the resultant composition for functions is in general associative, it follows that the composition in $Aut(G)$ is also associative. Now, the result follows immediately from Lemmas 11.26 and 11.27. ∎

Aut (*G*) is called the *group of automorphism* of *G*.

Theorem 11.46 *The set of all inner automorphisms is a normal subgroup of Aut* (*G*), *which is written Inn* (*G*).

Proof Clearly, the identity function lies in $Inn(G)$ as $id(x) = x = exe^{-1} = \phi_e(x)$, for all $x \in G$. Now, let a, b and x be arbitrary elements of G. We can write $\phi_a \circ \phi_b(x) = \phi_a(bxb^{-1}) = abxb^{-1}a^{-1} = (ab)x(ab)^{-1} = \phi_{ab}(x)$. This means that

$$\phi_a \circ \phi_b = \phi_{ab}. \tag{11.1}$$

Thus, $\phi_a \circ \phi_b \in Inn(G)$. Moreover, using (11.1), we can write $\phi_a \circ \phi_{a^{-1}} = \phi_{a^{-1}} \circ \phi_a = \phi_e$. This implies that $(\phi_a)^{-1} = \phi_{a^{-1}} \in Inn(G)$. Hence, we conclude that $Inn(G)$ is a subgroup of $Aut(G)$. To prove that $Inn(G)$ is a normal subgroup of $Aut(G)$, let $\phi_a \in Inn(G)$ and $f \in Aut(G)$ be arbitrary elements. Then, for any $x \in G$, we have

$$(f \circ \phi_a \circ f^{-1})(x) = f \circ \phi_a\big(f^{-1}(x)\big) = f\big(af^{-1}(x)a^{-1}\big)$$
$$= f(a)f\big(f^{-1}(x)\big)f(a^{-1}) = f(a)xf(a)^{-1} = \phi_{f(a)}(x).$$

This shows that $f \circ \phi_a \circ f^{-1} \in Inn(G)$, and so $Inn(G) \trianglelefteq Aut(G)$. ∎

Definition 11.47 $Inn(G)$ is called the *group of inner automorphisms* of *G*. If *G* is abelian, then $Inn(G) = \{e\}$. An automorphism of *G* which is not inner is called *outer*. The factor group $Out(G) = Aut(G)/Inn(G)$ is called the *group of outer automorphisms* of *G* even although its elements are not automorphism.

Theorem 11.48 *The group of inner automorphisms of G is isomorphic to the quotient group* $G/Z(G)$, *where* $Z(G)$ *is the center of G*.

Proof We define $f : G \to Inn(G)$ by $f(a) = \phi_a$, for all $a \in G$. Then, we can write $f(ab) = \phi_{ab} = \phi_a \circ \phi_b = f(a)f(b)$, for all $a, b \in G$. This means that f is a homomorphism. Since each element of $Inn(G)$ is of the form ϕ_a, it follows that f is onto. Hence, by applying the first isomorphism theorem, we obtain $G/Kerf \cong Inn(G)$. Now, the result follows if we show that $Kerf = Z(G)$. An element a of G lies in the kernel K if and only if $\phi_a = id$, i.e., $a^{-1}xa = x$, for all $x \in G$. This yields that $Kerf = Z(G)$. ∎

Theorem 11.49 *Let* $G = \langle a \rangle$ *be a cyclic group, and suppose that* f *is an endomorphism of G. Then,* f *is an automorphism if and only if* $f(a)$ *is a generator of G*.

Proof Suppose that f is an automorphism of G. Let x be an arbitrary element of G. Since f is onto, it follows that there is an element $y \in G$ such that $f(y) = x$. On the other hand, since $y \in \langle a \rangle$, it follows that $y = a^n$, for some integer n. So, we have $x = f(y) = f(a^n) = f(a)^n$. This shows that $G = \langle f(a) \rangle$.

Conversely, assume that $G = \langle f(a) \rangle$. We must show that f is bijective. Let $y \in G$

be an arbitrary element. Then, $y = f(a)^k$, for some integer k. Hence, we obtain $y = f(a^k)$, and so we conclude that f is onto. To prove that f is one to one, we consider the following two cases:

   Case 1: Let G be finite. Then, f is clearly one to one.

   Case 2: Let G be infinite, and suppose that $f(x) = f(y)$. Then, $f(a^i) = f(a^j)$, where $a^i = x$ and $a^j = y$, for some integers i and j. So, we can write $f(a)^i = f(a)^j$, or equivalently $f(a)^{i-j} = e$. Since G is infinite, it follows that $i = j$, and so $x = y$. Consequently, f is one to one. ∎

Theorem 11.50 *If G is an infinite group, then $Aut(G)$ is of order 2.*

Proof The result follows from Theorem 11.49 and the fact that the number of generators of an infinite cyclic group is 2. ∎

Theorem 11.51 *If $G = \langle a \rangle$ is a finite cyclic group of order n, then $Aut(G) \cong U_n$.*

Proof Since G is finite, by Corollary 4.33, the number of generators of G is $\varphi(n)$. This yields that $|Aut(G)| = \varphi(n)$. Suppose that $f : G \to G$ be an automorphism of G. Then, we have $f(a) = a^k$, for some integer k. Since $f(a)$ is a generator of G, it follows that $k < n$ and $(k, n) = 1$. Now, we define $\theta : Aut(G) \to U_n$ by $\theta(f) = \bar{k}$, for all $f \in Aut(G)$. Let f and g be two elements of $Aut(G)$ such that $f(a) = a^k$ and $g(a) = a^l$, for some integers k and l. So, we have $(f \circ g)(a) = f(a^l) = a^{kl}$. Therefore, we obtain $\theta(f \circ g) = \overline{kl} = \bar{k}\,\bar{l} = \theta(f)\theta(g)$. This shows that θ is a homomorphism. Now, if $\theta(f) = \theta(g)$, then $\bar{k} = \bar{l}$. This implies that $n|k - l$. Since $k < n, l < n, (k, n) = 1$ and $(l, n) = 1$, we conclude that $k = l$, and so $f = g$. This proves that f is one to one. Finally, since $|Aut(G)| = |U_n| = \varphi(n)$, it follows that θ is onto. Therefore, θ is an isomorphism. ∎

Theorem 11.52 *If G is a group of order p^n, then $|Aut(G)| = p^n(p - 1)$.*

Proof It follows directly from Theorem 11.51. ∎

Exercises

1. Find all of the automorphisms of $\mathbb{Z}_8$.
2. Prove that $a + ib \mapsto a - ib$ is an automorphism of $\mathbb{C}^*$.
3. Prove that $Aut(S_3) \cong S_3$.
4. What are the inner automorphisms of the quaternion group Q_8? Is $Inn(G) = Aut(G)$ in this case?
5. Prove that $Aut(D_4) \cong D_4$ and yet D_4 has outer automorphism.
6. Prove that $Aut(A_5) \cong S_5$.
7. Let G be any group. Prove that the function f from G to itself defined by $f(a) = a^{-1}$, for all $a \in G$, is an automorphism if and only if G is abelian.

8. If G is a group such that $Aut(G)$ is the trivial group, prove that G is abelian and every element of G is of order 2.
9. If $G \cong H$, prove that $Aut(G) \cong Aut(H)$ and $Inn(G) \cong Inn(H)$.
10. Find two non-isomorphic groups G and H such that $Aut(G) \cong Aut(H)$.
11. Prove that every finite group having more than two elements has a non-trivial automorphism.
12. If an automorphism fixes more than half of the elements of a finite group, prove that it is the identity automorphism.
13. Let G be a group and $f : G \rightarrow G$ defined by $f(x) = x^n$, for all $x \in G$, be an automorphism. Prove that for each $a \in G$, we have $a^{n-1} \in Z(G)$.
14. Prove that if G is a group in which every non-identity element is of order 2, then G has a non-trivial automorphism.
15. Let G be the group of order 9 generated by elements a and b, where $a^3 = b^3 = e$. Find all the automorphisms of G.
16. Prove that the group of inner automorphisms of the symmetric group S_n ($n \geq 3$) is isomorphic to S_n.
17. Prove that every automorphism of S_4 is inner. What is the order of the group of automorphisms of S_4?
18. Let G be the additive group of numbers of the form mp^n, where m and n are integers and p is a fixed prime. Describe $End(G)$ and $Aut(G)$.
19. Prove that if G is a finite non-cyclic abelian group, then $Aut(G)$ is not abelian.
20. Let f be an automorphism of a finite group G with the property that $f(x) = x$ if and only if $x = e$. Suppose further that $f^2 = id$, the identity element of $Aut(G)$. Show that G is abelian.
    Hint: First prove that every $a \in G$ can be represented as $x^{-1}f(x)$, for some $x \in G$.

11.5   Characteristic Subgroups

Recall that a subgroup of a group G is normal if it is invariant under conjugation. Now, conjugation is just a special case of an automorphism of G.

Definition 11.53 Let G be a group and H be a subgroup of G. We say that H is a *characteristic subgroup* of G, if for every automorphism f of G, $f(H) = H$.

Example 11.54 Let G be a group. Then, G itself and the identity subgroup $\{e\}$ are characteristic subgroups of G.

Example 11.55 The center of any group is a characteristic subgroup. Indeed, let $f \in Aut(G)$ be arbitrary. Pick $z \in Z(G)$, then z commutes with every element of G. Pick an element $a \in G$. Since f is onto, it follows that $a = f(b)$, for some $b \in G$. Hence, we can write

$$af(z) = f(b)f(z) = f(bz) = f(zb) = f(z)f(b) = f(z)a.$$

Since a is arbitrary, it follows that $f(z)$ commutes with every element of G, and so $f(z) \in Z(G)$. Consequently, we get $f(Z(G)) \subseteq Z(G)$. Applying the same result to the inverse of f, we obtain $f^{-1}(Z(G)) \subseteq Z(G)$. This implies that $Z(G) \subseteq f(Z(G))$. Therefore, we have $f(Z(G)) = Z(G)$, and this shows that $Z(G)$ is a characteristic subgroup of G.

Example 11.56 Derived subgroup G' of a group G is a characteristic subgroup of G, since for every $f \in AUt(G)$ and $a, b \in G$, we have $f([a, b]) = [f(a), f(b)]$.

Example 11.57 If H is the only subgroup of G of order m, then it must be characteristic, because $f(H)$, for all $f \in Aut(G)$, is again a subgroup of G of order m.

Note that f restricted to H a characteristic subgroup (denoted by $f|_H$) is an automorphism of H (it is an endomorphism by definition of H being characteristic).

Here are a few immediate properties of characteristic subgroups.

Theorem 11.58 *Let G be a group, and let H, K be subgroups of G.*

(1) If H is characteristic in K and K is characteristic in G, then H is characteristic in G (being characteristic is transitive);

(2) If H is characteristic in K, and K is normal in G, then H is normal in G.

Proof (1) Note that by assumption $H \leq K \leq G$. Let f be an automorphism of G. Since K is characteristic in G, it follows that $f(K) = K$ by definition, and so $f|_K$ is an automorphism of K. Now, since H is characteristic in K, it follows that $f|_K(H) = H$. But $f|_K$ is just the restriction of f (recall $H \leq K$), and hence $f(H) = H$.

(2) Consider the automorphism of K given by $k \mapsto aka^{-1}, a \in G$, which is well defined since K is normal in G. For any choice of a, we get a different automorphism of K, which will always preserve H, because H is characteristic in K. Consequently, $aHa^{-1} \subseteq H$. This shows that H is normal in G. ∎

Theorem 11.59 *Let G be a finite group and N be a normal subgroup of G. If $(|N|, [G : N]) = 1$, then N is a characteristic subgroup of G.*

Proof Suppose that $|N| = m$, $[G : N] = n$ and $(m, n) = 1$. Then, we conclude that $|G| = mn$. Let $f \in Aut(G)$ be an arbitrary automorphism. Since N is a normal subgroup of G, it follows that $Nf(N)$ is a subgroup of G. Moreover, we have

$$|Nf(N)| = \frac{|N| \, |f(N)|}{|N \cap f(N)|}.$$

Assume that $k = |N \cap f(N)|$. Since $|N| = |f(N)| = m$, it follows that $|Nf(N)| = m^2/k$. Since $Nf(N) \leq G$, it follows that $|Nf(N)| \big| |G|$, or equivalently $m^2/k | mn$. Since $(m, n) = 1$, it follows that $m = k$. Hence, we obtain $|N \cap f(N)| = |N|$. This yields that $N \cap f(N) = N$, and so $N \subseteq f(N)$. Finally, we conclude that $N = f(N)$, and this completes the proof. ∎

Exercises

1. Prove that every subgroup of a cyclic group is characteristic.
2. Give an example of a group G containing a normal subgroup that is not a characteristic subgroup.
   Hint: Let G be abelian.
3. If H is a characteristic subgroup of a group G, prove that $C_G(H)$ is also a characteristic subgroup of G.
4. Let G be a finitely generated group and let H be a subgroup of G with finite index. Show that there is a subgroup K of H that is characteristic in G and has finite index in G.
5. Let G be a group and H, K be subgroups of G. If H is a characteristic subgroup of G and $H \subseteq K \subseteq G$, prove that K/H is a characteristic subgroup of G/H implies that K is a characteristic subgroup of G.

11.6   Another View of Linear Groups

Let V be a finite dimensional vector space over a field $\mathbb{F}$. The set of all invertible transformations of V to V is denoted by $GL(V, \mathbb{F})$. This set has a group structure under composition of transformations. On the other hand, in Sect. 7.1, we applied the notation $GL_n(\mathbb{F})$ for the general linear group, the group of $n \times n$ invertible matrices.

Theorem 11.60 *Let V be a finite vector space of dimension n over a field $\mathbb{F}$ with ordered basis $\{v_1, \ldots, v_n\}$. Then the map $\theta : GL(V, \mathbb{F}) \to GL_n(\mathbb{F})$ corresponding to this basis is an isomorphism of groups.*

Proof The result immediately follows by the properties of linear transformation. ∎

Hence, since $GL(V, \mathbb{F}) \cong GL_n(\mathbb{F})$, we can consider the elements of general linear group as invertible matrices as well as invertible linear operators on V. The special linear group $SL_n(\mathbb{F})$ is the set of all matrices $A \in GL_n(\mathbb{F})$ such that $\det(A) = 1$. We note that if V is a finite vector space of dimension n over a field $\mathbb{F}$ with ordered basis $\{v_1, \ldots, v_n\}$, and if $\theta : GL(V, \mathbb{F}) \to GL_n(\mathbb{F})$ is as above, then the set

$$\{T \in GL(V, \mathbb{F}) \mid \det\big(\theta(T)\big) = 1\}$$

is independent of the choice of the basis. This subset is the subgroup $\theta^{-1}\big(SL(n, \mathbb{F})\big)$, and we denote it by $SL(V, \mathbb{F})$. It is clear that $SL(V, \mathbb{F}) \cong SL_n(\mathbb{F})$.

Theorem 11.61 *If $\mathbb{F}$ is a finite field with q elements, then*

$$|SL_n(\mathbb{F})| = q^{\binom{n}{2}} \prod_{i=2}^{n} (q^i - 1).$$

Proof Let $\mathbb{F}^*$ be the multiplicative group of $\mathbb{F}$. We define $f : GL_n(\mathbb{F}) \to \mathbb{F}^*$ by $f(A) = \det(A)$, for all $A \in GL_n(\mathbb{F})$. It is easy to check that f is an onto homomorphism. Moreover, we see that $Kerf = SL_n(\mathbb{F})$. So, by the first isomorphism theorem, we conclude that

$$\frac{GL_n(\mathbb{F})}{SL_n(\mathbb{F})} \cong \mathbb{F}^*.$$

This implies that $|SL_n(\mathbb{F})| = |GL_n(\mathbb{F})| \cdot |\mathbb{F}^*|$. Now, by Theorem 7.32, the result follows.   ■

Definition 11.62 (*Projective Linear Groups*) Let $\mathbb{F}$ be a field and n a positive integer. The set $Z = \{aI_n \mid a \in \mathbb{F}^*\}$ of scalar matrices is a normal subgroup of $GL_n(\mathbb{F})$. The *projective general linear group* over $\mathbb{F}$ is defined by

$$PGL_n(\mathbb{F}) = \frac{GL_n(\mathbb{F})}{Z},$$

and the *projective special linear group* over $\mathbb{F}$ is defined by

$$PSL_n(\mathbb{F}) = \frac{SL_n(\mathbb{F})}{SL_n(\mathbb{F}) \cap Z} \cong \frac{SL_n(\mathbb{F})Z}{Z}.$$

The projective groups are obtained from the corresponding ordinary linear groups by identifying matrices that are scalar multiples of each other.

An *automorphism of a field* $\mathbb{F}$ is a one to one function σ from $\mathbb{F}$ onto itself such that $\sigma(a + b) = \sigma(a) + \sigma(b)$ and $\sigma(ab) = \sigma(a)\sigma(b)$, for all $a, b \in \mathbb{F}$. Let $Aut(\mathbb{F})$ be the set of all automorphism of $\mathbb{F}$. We denote $\sigma(a) := a^\sigma$.

Lemma 11.63 *If $\mathbb{F}$ is a field, then $Aut(\mathbb{F})$ is a group under composition of functions.*

Proof It is straightforward.   ■

Definition 11.64 Let V be a finite dimensional vector space on a field $\mathbb{F}$. A function $T : V \to V$ is called a *semi-linear transformation* on V with associated field automorphism $\sigma \in Aut(\mathbb{F})$, if $T(v_1 + v_2) = T(v_1) + T(v_2)$ and $T(cv_1) = c^\sigma T(v_1)$, for all $v_1, v_2 \in V$ and $c \in \mathbb{F}$.

Example 11.65 Let $V = \mathbb{C}^n$ be the vector space of dimension n over $\mathbb{C}$, and σ be an automorphism of $\mathbb{C}$ defined by $(a + bi)^\sigma = a - bi$, for all $a + bi \in \mathbb{C}$. Then, $T : V \to V$ with $T(a_1 + b_1 i, \ldots, a_n + b_n i) = (a_1 - b_1 i, \ldots, a_n - b_n i)$, for all $a_1 + b_1 i, \ldots a_n + b_n i \in \mathbb{C}$, is a semi-linear transformation.

Let V be a vector space on a field $\mathbb{F}$. We denote $\Gamma L(V, \mathbb{F})$ the set of all invertible semi-linear transformations.

Theorem 11.66 $\Gamma L(V, \mathbb{F})$ *is a group under composition of semi-linear transformations.*

Proof It is straightforward.   ∎

The group $\Gamma L(V, \mathbb{F})$ is called the *semi-linear group*. Also, the group

$$P\Gamma L(V, \mathbb{F}) = \frac{\Gamma L(V, \mathbb{F})}{Z(\Gamma L(V, \mathbb{F}))}$$

is called the *projective semi-linear group*. If $dim V = n$, then we write $\Gamma L_n(\mathbb{F})$ and $P\Gamma L_n(\mathbb{F})$, respectively.

Corollary 11.67 $GL(V, \mathbb{F})$ *is a normal subgroup of* $\Gamma L(V, \mathbb{F})$ *and we have*

$$\frac{\Gamma L(V, \mathbb{F})}{GL(V, \mathbb{F})} \cong Aut(\mathbb{F}).$$

Proof We define $\theta : \Gamma L(V, \mathbb{F}) \to Aut(\mathbb{F})$ by $\theta(T) = \sigma$, where σ is the associated field automorphism of T. It is clear that θ is an onto homomorphism with $Ker\theta = GL(V, \mathbb{F})$. This completes the proof.   ∎

Exercises

1. Prove that

    (a) $PSL_n(\mathbb{F}) \trianglelefteq PGL_n(\mathbb{F}) \trianglelefteq P\Gamma L_n(\mathbb{F})$;

    (b) $\dfrac{P\Gamma L_n(\mathbb{F})}{PGL_n(\mathbb{F})} \cong Aut(\mathbb{F})$.

2. Let $\mathbb{F}_q$ be a finite field with q elements. Prove that

    (a) $PSL_2(\mathbb{F}_2) \cong SL_2(\mathbb{F}_2) \cong GL_2(\mathbb{F}_2) \cong S_3$;
    (b) $PSL_2(\mathbb{F}_3) \cong A_4$.

3. Show that the set of all matrices of the form

$$\begin{bmatrix} \pm 1 & k \\ 0 & 1 \end{bmatrix},$$

    is a group isomorphic to D_n, where all entries in the matrix are in $\mathbb{Z}_n$.

4. Let $\mathbb{C}^\infty$ be the complex plane augmented by an extra point ∞. Let $a, b, c, d \in \mathbb{C}$ be such that $ad - bc \neq 0$. Define $f : \mathbb{C}^\infty \to \mathbb{C}^\infty$ by

    if $c \neq 0$, then $\begin{cases} f(z) = \dfrac{az + b}{cz + d} \text{ if } z \neq \infty, z \neq -\dfrac{d}{c} \\ f\left(-\dfrac{d}{c}\right) = \infty \\ f(\infty) = \dfrac{a}{c}, \end{cases}$

and by

if $c = 0$, then $\begin{cases} f(z) = \dfrac{az + b}{a^d} & \text{if } z \neq \infty, \\ f(\infty) = \dfrac{a}{c}. \end{cases}$

Show that under composition these functions form a group G and G is a homomorphic image of $GL_2(\mathbb{C})$.

5. Give an example of an outer automorphism of a group G which maps each conjugate class of G onto itself.

   Hint: Take

$$G = \left\{ \begin{bmatrix} 1 & a \\ 0 & b \end{bmatrix} \mid a, b \in \mathbb{Z}_8 \text{ and } (b, 8) = 1 \right\}.$$

Show that G is a group under matrix multiplication. Define $f : G \to G$ by

$$f\left(\begin{bmatrix} 1 & a \\ 0 & b \end{bmatrix} \right) = \begin{bmatrix} 1 & a + \dfrac{b^2 - 1}{2} \\ 0 & b \end{bmatrix}.$$

11.7   Worked-Out Problems

Problem 11.68 Let G and H be finite groups such that $(|G|, |H|) = 1$, prove that the only homomorphism from G to H is the zero homomorphism.

Solution Suppose that f is a homomorphism from G to H. By the first isomorphism theorem, we can write $G/Kerf \cong Imf$. Since $Kerf \leq G$, it follows that $|Kerf| \big| |G|$. Consequently, we get $(|G/Kerf|, |H|) = 1$. Analogously, since $Imf \leq H$, it follows that $|Imf| \big| |H|$. Hence, we conclude that $|G/Kerf| = |Imf| = 1$. This yields that $G = Kerf$. Therefore, f sends every element of G to the identity element of H, and this means that f is the zero homomorphism. ∎

Problem 11.69 If G is a perfect group, prove that the center of $G/Z(G)$ is trivial.

Solution For convenience we set $H = Z(G)$. Assume that the center of G/H is non-trivial. This means that there is an element $aH \in Z(G/H)$ such that $a \notin H$. Since $aH \in Z(G/H)$, it follows that $aHxH = xHaH$, for all $xH \in G/H$. This implies that $axH = xaH$, or equivalently $a^{-1}x^{-1}ax \in H$. Now, we can define a function $f : G \to H$ by $f(x) = a^{-1}x^{-1}ax$, for all $x \in G$. We show that f is a homomorphism. Let x and y be elements of G. Then, we find that

$$\begin{aligned} f(xy) &= a^{-1}(xy)^{-1}a(xy) = a^{-1}y^{-1}x^{-1}axy \\ &= (a^{-1}y^{-1}ay)y^{-1}(a^{-1}x^{-1}ax)y \\ &= (a^{-1}x^{-1}ax)(a^{-1}y^{-1}ay) = f(x)f(y). \end{aligned}$$

Thus, f is a homomorphism. By the first isomorphism theorem, we have $G/Kerf \cong Imf \leq H$. As H is abelian, we conclude that $G/Kerf$ is abelian. Consequently, by Theorem 10.4, we obtain $G' \subseteq Kerf$. Since G is perfect, we have $G = G'$, which implies that $G = Kerf$. Now, if $G = Kerf$, then $f(x) = e$, for all $x \in G$. So, we obtain $a^{-1}x^{-1}ax = e$ or $ax = xa$. This shows that $a \in H$, which is a contradiction. Therefore, we conclude that $aH = H$ or $a \in H$. This proves that $Z(G/H) = \{H\}$.   ∎

Problem 11.70 Suppose that

(1) G and H are groups and G is finite;
(2) $f : G \to H$ is a homomorphism;
(3) A and B are subgroups of G such that $G = AB$ and $(|A|, |B|) = 1$.

Prove that for each normal subgroup N of G, $N = (N \cap A)(N \cap B)$.

Solution Let $\pi : G \to G/N$ be the canonical map. Since π is onto, it follows that $\pi(G) = G/N$. Then, we have

$$\pi(G) = \pi(AB) = \pi(A)\pi(B) = G/N,$$
$$\pi(A) = \{\pi(a) \mid a \in A\} = \{aN \mid a \in A\} = (AN)/N,$$
$$\pi(B) = \{\pi(b) \mid b \in B\} = \{bN \mid b \in B\} = (BN)/N.$$

Thus, we conclude that

$$\frac{G}{N} = \left(\frac{AN}{N}\right)\left(\frac{BN}{N}\right).$$

On the other hand, by the second isomorphism theorem, we have

$$\frac{A}{A \cap N} \cong \frac{AN}{N} \text{ and } \frac{B}{B \cap N} \cong \frac{BN}{N}.$$

So, we deduce that

$$\left|\frac{AN}{N}\right| \Big| |A| \text{ and } \left|\frac{BN}{N}\right| \Big| |B|.$$

Since $(|A|, |B|) = 1$, it follows that $(|(AN)/N|, |(BN)/N|) = 1$. Hence, we obtain

$$\frac{AN}{N} \cap \frac{BN}{N} = \{N\} \text{ and } A \cap B = \{e\}.$$

Now, we can write $|G| = |AB| = |A|\,|B|$ and

$$\left|\frac{G}{N}\right| = \left|\frac{AN}{N}\right|\left|\frac{BN}{N}\right| = \frac{|A|\,|B|}{|A \cap N|\,|B \cap N|} = \frac{|G|}{|A \cap N|\,|B \cap N|}.$$

This implies that

$$|N| = |A \cap N|\,|B \cap N|. \tag{11.2}$$

Moreover, since $A \cap N \le N$ and $B \cap N \le N$, we have

$$(A \cap N)(B \cap N) \leq N. \qquad (11.3)$$

Now, by (11.2) and (11.3), we conclude that $N = (A \cap N)(B \cap N)$. ∎

Problem 11.71 Prove that there are only two groups of order 6, one is cyclic and another is isomorphic to S_3.

Solution Suppose that G is a group of order 6. Since there exist 5 non-identity elements and $x \leftrightarrow x^{-1}$ is a one to one correspondence between the elements of G and their inverses, it follows that there exists a non-identity element $a \in G$ such that $o(a) = 2$.

Let G be abelian and $H = \langle a \rangle$. Then, we have $|G/H| = 3$, which implies that G/H is cyclic. Assume that $G/H = \langle bH \rangle$, for some $b \in G$. Since $o(bH)|o(b)$, it follows that $o(b) = 3$ or 6. If $o(b) = 6$, then G is cyclic. If $o(b) = 3$, then since $ab = ba$ and $(o(a), o(b)) = 1$, it follows that $o(ab) = 6$. Hence, again in this case we deduce that G is a cyclic group generated by ab.

Now, assume that G is non-abelian. Then, all non-identity elements of G can not be of order 2. Thus, there exists $c \in G$ such that $o(c) = 3$ and $ac \neq ca$. Take $N = \langle c \rangle$. Since $[G : N] = 2$, it follows that N is a normal subgroup of G. Hence, we get $aca^{-1} \in N$. This yields that $aca^{-1} = c$ or $aca^{-1} = c^2$. Since $ac \neq ca$, we conclude that $aca^{-1} = c^2$, or $ca = ac^3$. On the other hand, we know that a can not be in N, so we can write $G = N \cup aN = \{e, c, c^2, a, ac, ac^2\}$. Now, it is a routine verification to check that the function $f : G \to S_3$ given by $f(e) = id$, $f(a) = (1\ 2)$, $f(c) = (1\ 2\ 3)$, $f(c^2) = (1\ 3\ 2)$, $f(ac) = (2\ 3)$ and $f(ac^2) = (1\ 3)$ is an isomorphism. This completes the proof. ∎

Problem 11.72 Prove that any non-abelian group of order 8 is either D_4 or Q_8.

Solution Suppose that G is a non-abelian group of order 8. Since G is non-abelian, it follows that G is not cyclic. So, G don't have any element of order 8. If every non-identity element of G is of order 2, then again G must be abelian. Hence, there is a non-identity element $a \in G$ such that the order of a is neither 8 nor 2. Since $o(a)\big||G|$, it follows that $o(a) = 4$. If $N = \langle a \rangle$, then $[G : N] = 2$, and so $N \trianglelefteq G$. Take $b \in G$ such that $b \notin N$. Then, we obtain

$$G = N \cup bN = \{e, a, a^2, a^3, b, ba, ba^2, ba^3\}.$$

This yields that $G = \langle a, b \rangle$. Since G is non-abelian, it follows that $ab \neq ba$. Since $N \trianglelefteq G$, it follows that $bab^{-1} \in N$. Moreover, $o(bab^{-1}) = o(a) = 4$. In the subgroup N, the only element of order 4 other than a is a^3, so we conclude that $bab^{-1} = a^3 = a^{-1}$. Now, G/N is of order 2, so $(bN)^2 = N$, or equivalently $b^2 \in \{e, a, a^2, a^3\}$. If $b^2 = a$ or $b^2 = a^3$, then $o(b) = 8$. This is impossible, because G is non-abelian. Thus, we conclude that $b^2 = e$ or $b^2 = a^2$. If $b^2 = e$, then G is D_4 (see Exercise 30 in Sect. 11.2). If $b^2 = a$, then G is the quaternion group. ∎

Problem 11.73 Let G be a group and H be a subgroup of G. Prove that $N_G(H)/C_G(H)$ is isomorphic to a subgroup of $Aut(H)$.

Solution Define $\theta : N_G(H) \to Aut(H)$ by $\theta(a) = \theta_a$, for all $a \in N_G(H)$, in which $\theta_a(h) = aha^{-1}$, for all $h \in H$. Since $a \in N_G(H)$, it follows that $aHa^{-1} = H$. This shows that $\theta_a(h) \in H$. Moreover, we observe that θ_a is a bijective homomorphism from H onto itself, and so $\theta_a \in Aut(H)$. Now, it is easy to check that θ is a homomorphism with $Ker\theta = C_G(H)$. At the end, the result follows by the first isomorphism theorem.   ∎

Problem 11.74 Let G be a finite group and suppose that automorphism f sends more than three-quatres of the elements of G onto their inverses. Prove that $f(x) = x^{-1}$, for all $x \in G$ and that G is abelian.

Solution Set $A = \{a \in G \mid f(a) = a^{-1}\}$. If $x \in A$, then $|A \cup xA| = |A| + |xA| - |A \cap xA|$. Since $A \cup xA \subseteq G$, it follows that $|A \cup xA| \leq |G|$. Since the function $h : A \to xA$ defined by $f(a) = xa$, for all $a \in A$, is a one to one correspondence, it follows that $|G| \geq |A \cup xA| = |A| + |A| - |A \cap xA|$. This implies that $|G| > 3|G|/4 + 3|G|/4 - |A \cap xA|$ or $|A \cap xA| > |G|/2$. Now, assume that xa_1 and xa_2 are two elements of $A \cap xA$, then $f(xa_i) = f(x)f(a_i) = x^{-1}a_i^{-1}$ and $f(xa_i) = (xa_i)^{-1} = a_i^{-1}x^{-1}$. Hence, we conclude that $a_i^{-1}x^{-1} = x^{-1}a_i^{-1}$, which implies that x and a_i commute. Since there are more than $|G|/2$ elements in $A \cap xA$, it follows that $|C_G(x)| > |G|/2$. Now, by Lagrange's theorem, we must have $G = C_G(x)$, and this yields that $x \in Z(G)$. Since this is true for all $x \in A$, it follows that $A \subseteq Z(G)$. Thus, we obtain $3|G|/4 < |A| \leq |Z(G)|$. This shows that $G = Z(G)$, because $Z(G)$ is a subgroup of G. Consequently, G is abelian. So, A is a subgroup of G such that $3|G|/4 < |A|$. By Lagrange's theorem we deduce that $G = A$.   ∎

Problem 11.75 Prove that an automorphism of S_n which sends transpositions to transpositions is an inner automorphism.

Solution Suppose that $f : S_n \to S_n$ be an automorphism mapping transpositions to transpositions. For two distinct transpositions we can consider two cases:

(1) $(a\ b)(c\ d)$, which is of order 2;
(2) $(a\ b)(a\ c) = (a\ b\ c)$, which is of order 3.

Therefore, we can say that f maps any pair of disjoint transpositions to a pair of disjoint transpositions. Let $f\big((1\ 2)\big) = (r\ s)$ and suppose that $3 \leq x \leq n$. Since $(1\ 2)(1\ x)$ is a cycle of length 3, it follows that

$$f\big((1\ 2)(1\ x)\big) = f\big((1\ 2)\big)f\big((1\ x)\big) = (r\ s)f\big((1\ x)\big).$$

Hence, $f\big((1\ x)\big)$ moves either r or s. Without loss of generality, we may suppose that it moves r. Consequently, we can write $f\big((1\ x)\big) = (r\ t)$, for some t different from r and s.

Now, we claim that for each $y \neq 1$, we have $f\big((1\ y)\big) = (r\ u)$, for some $1 \leq u \leq n$ and different from r.

If $y = 2$ or $y = x$, the claim is true.

If $y \neq 2$ and $y \neq x$, then both permutations $(1\ y)(1\ 2)$ and $(1\ y)(1\ x)$ are cycles of length 3, so are their images under f. As $f\big((1\ 2)\big) = (r\ s)$ and $f\big((1\ x)\big) = (r\ t)$, if $f\big((1\ y)\big)$ is not moving r, then it must move both s and t. Since $f\big((1\ y)\big)$ is a transposition, it follows that $f\big((1\ y)\big) = (s\ t)$. It is easy to see that $(r\ s)(r\ t)(s\ t) = (r\ t)$, and so $f^{-1}\big((r\ s)\big) f^{-1}\big((r\ t)\big) f^{-1}\big((s\ t)\big) = f^{-1}\big((r\ t)\big)$. Consequently, we get $(1\ 2)(1\ x)(1\ y) = (1\ x)$. Since $y \notin \{1,\ 2,\ x\}$ and $x \neq y$, we obtain a contradiction. Therefore, we conclude that $f\big((1\ y)\big)$ must move r, and hence $f\big((1\ y)\big) = (r\ u)$, for some $1 \leq u \leq n$ and different from r.

Now, we define $\sigma \in S_n$ as follows:

$$y\sigma = \begin{cases} r & \text{if } y = 1 \\ u & \text{if } y \neq 1, \end{cases}$$

where u is the unique element for which $f\big((1\ y)\big) = (r\ u)$. Also, assume that $g : S_n \to S_n$ denote the conjugation by σ. For every y we obtain

$$g^{-1}\big(f\big((1\ y)\big)\big) = g^{-1}\big(r\ u\big) = \sigma(r\ u)\sigma^{-1} = (1\ y).$$

This means that $g^{-1}f$ fixes all permutations of the form $(1\ y)$, for all y. Since $(a\ b) = (1\ b)(1\ a)(1\ b)$, for all a and b, it follows that $g^{-1}f$ fixes every permutation of S_n. Thus, we obtain $f = g$. This completes the proof. ∎

Problem 11.76 A group G is *complete* if it is centerless and every automorphism of G is inner. If $n \neq 2$ and $n \neq 6$, prove that S_n is complete.

Solution Let $C(k)$ denote the the conjugate class in S_n consisting of all products of k disjoint transpositions. We know that a permutation in S_n is of order 2 if and only if it belongs in some $C(k)$. So, if $f \in Aut(S_n)$, then $f\big(C(1)\big) = \big(C(k)\big)$, for some positive integer k. We claim that if $n \neq 6$, then $|C(1)| \neq |C(k)|$, for $k \neq 1$. Assuming this, then $f\big(C(1)\big) = C(1)$, and Problem 11.76 completes the proof.

Clearly, we have $C(1) = n(n - 1)/2$. In order to count the elements of $C(k)$, we observe that there exist

$$\frac{n(n - 1)}{2} \cdot \frac{(n - 2)(n - 3)}{2} \cdot \ldots \cdot \frac{(n - 2k - 2)(n - 2k + 1)}{2}.$$

k-tuples of disjoint transpositions. Since disjoint transpositions commute and there are $k!$ orderings obtained from any k-tuples, we obtain

$$|C(k)| = \frac{n(n - 1)(n - 2) \ldots (n - 2k + 1)}{k!2^k}.$$

Now, the question whether $|C(1)| = |C(k)|$ leads to the question whether there is some $k > 1$ such that

$$(n - 2)(n - 3) \ldots (n - 2k + 1) = k!2^k. \tag{11.4}$$

Since the right side of (11.4) is positive, it follows that $n \geq 2k$. Hence, we have

$$(n-2)(n-3)\ldots(n-2k+1) \geq (2k-2)(2k-3)\ldots(2k-2k+1)$$
$$= (2k-2)!.$$

Moreover, by mathematical induction, we find that if $k \leq 4$, then $(2k-2)! > k!2^{k-1}$. Consequently, (11.4) is true if $k = 2$ or $k = 3$. If $k = 2$, then the right side of (11.4) is 4 and it is easy to see that equality never holds. So, we suppose that $k = 3$. Since $n \geq 2k$, it follows that $n \geq 6$. If $n > 6$, then the left side of (11.4) is greater than or equal to $5 \cdot 4 \cdot 3 \cdot 2 = 120$, while the right side of (11.4) is 24. Therefore, we have proved that if $n \neq 6$, then $|C(1)| \neq |C(k)|$, for all $k > 1$, as desired. ∎

Remark 11.77 $S_2 \cong \mathbb{Z}_2$ is not complete because it has a center. We shall see in future that S_6 is not complete too.

Problem 11.78 Let H be a subgroup of a group $G \leq GL_n(\mathbb{C})$. If the set

$$\{[x, y] \mid x \in G \text{ and } y \in H\}$$

is a set of order m, prove that $[G : C_G(H)] \leq m^{n^2}$.

Solution If $y \in H$, then y has at most m conjugate $x^{-1}yx = y[x, y]^{-1}$ in G (where $x \in G$). Therefore, we have
$$[G : C_G(y)] \leq m.$$

Suppose that $L(H) = \{c_1h_1 + \cdots + c_mh_m \mid c_i \in \mathbb{C} \text{ and } h_i \in H\}$. If we define addition and multiplication by scalars in an obvious manner for elements in $L(H)$, then $L(H)$ is a vector space over $\mathbb{C}$. The set of all $n \times n$ matrices with entries in $\mathbb{C}$ is a vector space over $\mathbb{C}$ containing $L(H)$ as a subspace. Since the former space has dimension n^2, it follows that the dimension $L(H)$ is at most n^2. Assume that $dim L(H) = k \leq n^2$. Then, we can find elements $y_1, \ldots, y_k$ in H which form a basis for $L(H)$. Consequently, we have

$$C_G(H) = \bigcap_{i=1}^{k} C_G(y_i).$$

So, we have

$$[G : C_G(H)] = [G : \bigcap_{i=1}^{k} C_G(y_i)] \leq \prod_{i=1}^{k} [G : C_G(y_i)] \leq m^{n^2}.$$

This completes the proof. ∎

11.8 Supplementary Exercises

1. Let G be a group, and let S be any set for which there exists a one to one and onto function $f : G \to S$. We define a binary operation on S by setting $xy = f\left(f^{-1}(x)f^{-1}(y)\right)$, for all $x, y \in S$. Prove that S is a group under this operation, and that f is actually a group isomorphism.

2. Let G be a group of order 12. Show that either G has a normal subgroup of order 3 or G is isomorphic to A_4.

3. Let $f : G_1 \to G_2$ be an onto homomorphism. Let H_1 be a normal subgroup of G_1 and suppose that $f(H_1) = H_2$. Prove or disprove that $G_1/H_1 \cong G_2/H_2$.

4. Let N_1 and N_2 be two normal subgroups of G. Prove or disprove that

   (a) $N_1 \cong N_2$ implies $G/N_1 \cong G/N_2$;
   (b) $G/N_1 \cong G/N_2$ implies $N_1 \cong N_2$.

5. Let H be the subgroup of $GL_2(\mathbb{Z}_3)$ as follows:

$$H = \left\{ \begin{bmatrix} a & b \\ 0 & 1 \end{bmatrix} \in GL_2(\mathbb{Z}_3) \mid a, b \in \mathbb{Z}_3, \ a \neq 0 \right\}.$$

   Show that H is isomorphic to the symmetric group S_3.

6. Show that the subgroup $\{id, \ (1\ 2)(3\ 4), \ (1\ 3)(2\ 4), \ (1\ 4)(2\ 3)\}$ of A_4 is isomorphic to the group of plane symmetries of a chessboard.

7. Show that to each positive integer n there exist only a finite number of pairwise non-isomorphic groups of order n.

8. Prove that distinct automorphisms of the symmetric group S_4 induce distinct automorphisms of the alternating subgroup A_4. What is the order of the group of automorphisms of A_4?

9. Let G and H be groups, $f : G \to H$ be a homomorphism with kernel N, and K be a subgroup of G. Prove that $f^{-1}\left(f(K)\right) = KN$. Hence, $f^{-1}\left(f(K)\right) = K$ if and only if $N \leq K$.

10. If G is metabelian and $f : G \to H$ is a homomorphism, prove that $f(G)$ is metabelian.

11. Prove that

    (a) If G is a non-abelian group, $Aut(G)$ can not be cyclic;
    (b) There is no finite group G for which $Aut(G)$ is cyclic of odd order greater than 1.

12. Show that if G is a group with trivial center ($Z(G) = \{e\}$), then its group of automorphisms, $Aut(G)$, is also a group with trivial center.
    Hint: Let $f \in Z(Aut(G))$. For any $a \in G$, let $\phi_a \in Inn(G)$. Then $f \circ \phi_a = \phi_a \circ f$ (Why?). Use this to show that for any $x \in G$, $a^{-1}f(a) \in C_G(f(x))$. Infer the result from this.

13. Suppose that $Z(G) = \{e\}$ and $Inn(G)$ is a characteristic subgroup of $Aut(G)$. Show that any automorphism of A is inner.

14. Let G be a group and H a normal subgroup of G with the property that $[G : H] = 4$. Let K be an arbitrary subgroup of G. Show that if the factor group $K/K \cap H$ is non-trivial, then it is isomorphic to either $\mathbb{Z}_2$, or $\mathbb{Z}_4$, or K_4.

15. Let H be a subgroup of G. Prove that the factor group $N_G(H)/C_G(H)$ is isomorphic to a subgroup of $Aut(H)$.

16. Prove that if $G \cong H$, then $G' \cong H'$ and $Z(G) \cong Z(H)$.

17. Suppose that G_1 and G_2 are finite perfect groups such that $G_1/Z(G_1) \cong G_2/Z(G_2)$. Prove that there exists a finite perfect group G and subgroups $H_1, H_2 \leq Z(G)$ with $G/Z(G) \cong G_i/Z(G_i)$ and $G/H_i \cong G_i$, for $i = 1, 2$.

18. Show that no group can have its automorphism group cyclic of odd order.

19. Let G be a finite group. If $k(G)$ is the number of conjugate classes of G, show that

   (a) $k(G) = 2$ if and only if $G \cong \mathbb{Z}_2$;
   (b) $k(G) = 3$ if and only if $G \cong \mathbb{Z}_3$ or $G \cong S_3$;
   (c) What can you say about finite groups with 4 or 5 conjugate classes?

20. Let p be a prime and $Z(p^\infty)$ be the following subset of the group $\mathbb{Q}/\mathbb{Z}$:

$$Z(p^\infty) = \{\overline{a/b} \in \mathbb{Q}/\mathbb{Z} \mid a, b \in \mathbb{Z} \text{ and } b = p^i \text{ for some } i \geq 0\},$$

where $\overline{a/b} = a/b + \mathbb{Z}$. Show that

   (a) $Z(p^\infty)$ is an infinite group under the addition operation of $\mathbb{Q}/\mathbb{Z}$;
   (b) Every element of $Z(p^\infty)$ has finite order p^n for some $n \geq 0$;
   (c) If at least one element of a subgroup H of $Z(p^\infty)$ has order p^k, and no element of H has order greater than p^k, then H is a cyclic subgroup generated by $\overline{1/p^k}$, whence $H \cong \mathbb{Z}_{p^k}$;
   (d) The only proper subgroups of $Z(p^\infty)$ are the cyclic groups $\langle \overline{1/p^n} \rangle$ ($n = 1, 2, \ldots$);
   (e) If $x_1, x_2, \ldots$ are elements of an abelian group G such that $o(x_1) = p$, $px_2 = x_1$, $px_3 = x_2, \ldots, px_{n+1} = x_n, \ldots$, then the subgroup generated by x_i's ($i \geq 1$) is isomorphic to $Z(p^\infty)$;
   (f) If H is a proper subgroup of $Z(p^\infty)$, then $Z(p^\infty)/H \cong Z(p^\infty)$.

21. If G has order $n > 1$, prove that

$$|Aut(G)| \leq \prod_{i=0}^{k}(n - 2^i),$$

   where $k = [\log_2(n - 1)]$.

22. Find a finite group G with a normal subgroup N such that $|Aut(N)| > |Aut(G)|$.

23. Let G be a group of order pq, where p and q are distinct prime numbers, and if G has a normal subgroup of order p and a normal subgroup of order q, prove that G is cyclic.

24. Let G be a group of order pq, where $p > q$ are primes. Prove that

(a) G has a subgroup of order p and a subgroup of order q;
(b) If $q \nmid p - 1$, then G is cyclic;
(c) Given two primes p and q with $q \mid p - 1$, there exists a non-abelian group of order pq;
(d) Any two non-abelian groups of order pq are isomorphic.

References

1. S.S. Abhyankar, C. Christensen, Semidirect products: $x \mapsto ax + b$ as a first example. Math. Mag. **75**(4), 284–289 (2002)
2. M.A. Armstrong, *Groups and Symmetry*. Undergraduate Texts in Mathematics (Springer, New York, 1988)
3. R.B.J.T. Allenby, *Rings, Fields and Groups: An Introduction to Abstract Algebra*, 2nd edn. (Edward Arnold, London, 1991)
4. T.M. Apostol, *Introduction to Analytic Number Theory*. Undergraduate Texts in Mathematics (Springer, New York, 1976)
5. H. Bacry, *Group Theory and Constellations* (Publibook, Paris, 2004)
6. T. Barnard, H. Neill, *Discovering Group Theory*. A transition to advanced mathematics. Textbooks in Mathematics (CRC Press, Boca Raton, 2017)
7. M.V. Bell, *Polya's Enumeration Theorem and its Applications*. Master's thesis, Primary Supervisor: A.-M. Ernvall-Hytönen, Secondary Supervisor: E. Elfving, University of Helsinki, 2015
8. O. Bogopolski, Introduction to Group Theory. Translated, revised and expanded from the 2002 Russian original. EMS Textbooks in Mathematics. European Mathematical Society (EMS), Zürich, 2008
9. C. Boyer, U. Merzbach, *A History of Mathematics*, 2nd edn. (Wiley, 1989)
10. P.M. Cohn, *Further Algebra and Applications* (Springer, London, 2003)
11. M.J. Collins, *Representations and Characters of Finite Groups*. Cambridge Studies in Advanced Mathematics, 22 (Cambridge University Press, Cambridge, 1990)
12. H.S.M. Coxeter, *Regular Polytopes* (Dover Publication Inc., New York, 1973)
13. H.S.M. Coxeter, *Introduction to Geometry* (Wiley, 1969)
14. H.S.M. Coxeter, W.O.J. Moser, *Generators and Relations for Discrete Groups*, 4th edn. Ergebnisse der Mathematik und ihrer Grenzgebiete [Results in Mathematics and Related Areas], 14 (Springer, New York, 1980)
15. M.R. Darafsheh, *Algebra I*, (in Persian), 3rd edn. (University of Tehran Press, Iran, 2010)
16. B. Davvaz, *Polygroup Theory and Related Systems* (World Scientific Publishing Co. Pte. Ltd., Hackensack, 2013)
17. B. Davvaz, M.A. Iranmanesh, *Fundamentals of Group Theory*, (in Persian) (Yazd University Press, Iran, 2005)
18. D.C. DeHovitz, *The Platonic Solids-An Exploration of the Five Regular Polyhedra and the Symmetries of Three-Dimensional Space* (Whitman College, Walla Walla, 2016)
19. I.M.S. Dey, J. Wiegold, Generators for alternating and symmetric groups. J. Aust. Math. Soc. **12**, 63–68 (1971)
20. J.D. Dixon, *Problems in Group Theory* (Blaisdell Publishing Co. Ginn and Co., Waltham, Mass.-Toronto, Ont.-London 1967)

© The Editor(s) (if applicable) and The Author(s), under exclusive license to Springer
Nature Singapore Pte Ltd. 2021
B. Davvaz, *A First Course in Group Theory*,
https://doi.org/10.1007/978-981-16-6365-9

21. J.D. Dixon, B. Mortimer, *Permutation Groups*. Graduate Texts in Mathematics, vol. 163 (Springer, New York, 1996)
22. N. Do, *Topics in Geometry* (McGill University, 2010)
23. A. Doerr, K. Levasseur, *Applied Discrete Structures* (University of Massachusetts Lowell, Department Of Mathematical Sciences, 2013)
24. L. Dornhoff, *Group Representation Theory. Part A: Ordinary Representation Theory*. Pure and Applied Mathematics, 7 (Marcel Dekker Inc., New York, 1971)
25. C. Drutu, M. Kapovich, *Geometric Group Theory*. With an appendix by Bogdan Nica. American Mathematical Society Colloquium Publications, 63 (American Mathematical Society, Providence, 2018)
26. H.B. Enderton, *Elements of Set Theory* [Harcourt Brace Jovanovich, Publishers] (Academic Press, New York, 1977)
27. J.B. Fraleigh, *A First Course in Abstract Algebra*, 7th edn. (Pearson Education Limited, USA, 2014)
28. J.A. Gallian, *Contemporary Abstract Algebra*, 9th edn. (Cengage Learning, 2016)
29. J.A. Gallian, Classroom notes: group theory and the design of a letter facing machine. Am. Math. Mon. **84**(4), 285–287 (1977)
30. D. Guichard, *An Introduction to Combinatorics and Graph Theory* (Whitman College, 2020)
31. H.W. Guggenheimer, *Plane Geometry and Its Groups* (Holden-Daay, 1967)
32. A. Hammel, Verifying the associative property for finite groups. Math. Teach. **61**(2), 136–139 (1968)
33. B. Hernandez, *Burnside's Lemma*. Undergraduate honors thesis, University of Redlands, 2013
34. R. Hartshorne, *Geometry: Euclid and Beyond* (Springer Science + Business Media, New York, 2000)
35. I.N. Herstein, *Topics in Algebra*, 2nd edn. (Xerox College Publishing, Lexington, Mass.-Toronto, Ont., 1975)
36. I.N. Herstein, *Abstract Algebra*, 3rd edn. With a preface by Barbara Cortzen and David J. Winter (Prentice Hall Inc., Upper Saddle River, 1996)
37. V.E. Hill, *Groups, Representations and Characters* (Hafner Press, New York, 1975). Binder
38. K. Hoffman, R. Kunze, *Linear Algebra*, 2nd edn. (Prentice-Hall Inc., Englewood Cliffs, 1971)
39. J.F. Humphreys, *A Course in Group Theory* (Oxford Science Publications, The Clarendon Press, Oxford University Press, New York, 1996)
40. T.W. Hungerford, *Algebra*. Reprint of the 1974 original. Graduate Texts in Mathematics, vol. 73 (Springer, New York, 1980)
41. I.M. Isaacs, *Character Theory of Finite Groups* (AMS Chelsea Publishing, Providence, 2006)
42. K. Iwasawa, Über die Einfachheit der speziellen projektiven Gruppen, (German). Proc. Imp. Acad. Tokyo **17**, 57–59 (1941)
43. G. James, M. Liebeck, *Representations and Characters of Groups*. Cambridge Mathematical Textbooks (Cambridge University Press, Cambridge, 1993)
44. D.L. Johnson, *Symmetries*. Springer Undergraduate Mathematics Series (Springer, London, 2001)
45. H. Kurzweil, B. Stellmacher, *The Theory of Finite Groups- An Introduction*. Translated from the 1998 German Original. Universitext (Springer, New York, 2004)
46. T. Leinster, *Basic Category Theory*. Cambridge Studies in Advanced Mathematics, vol. 143 (Cambridge University Press, Cambridge, 2014)
47. S.-Y.T. Lin, Y.-F. Lin, *Set Theory with Applications*, 2nd edn. (Marine Publishing Company Inc., 1981)
48. E.S. Lyapin, A.Y. Aizenshtat, M.M. Lesokhin, *Exercises in Group Theory*. Translated from the Russian by David E. Zitarelli (Plenum Press, New York; Wolters-Noordhoff Publishing, Groningen, 1972)
49. R.C. Lyndon, *Groups and Geometry* (Cambridge University Press, Cambridge, 1985)
50. K. Mcgerty, *The Classical Groups* (University of Shicago, 2006)
51. J.H. McKay, Another proof of Cauchy's group theorem. Am. Math. Mon. **66**, 119 (1959)

52. P.M. Neumann, G.A. Stoy, E.C. Thompson, *Groups and Geometry* (Oxford Science Publications, The Clarendon Press, Oxford University Press, New York, 1994)
53. G. Polya, R.C. Read, *Combinatorial Enumeration of Groups, Graphs, and Chemical Compounds* (Springer, New York, 1987)
54. D.J.S. Robinson, *A Course in the Theory of Groups*, vol. 80, 2nd edn. Graduate Texts in Mathematics (Springer, New York, 1996)
55. S. Roman, *Fundamentals of Group Theory- An Advanced Approach* (Birkhäuser/Springer, New York, 2012)
56. R.N. Umble, Z. Han, *Transformational Plane Geometry*. With a foreword by Maria Cristina Bucur and Lindsay Eisenhut. Textbooks in Mathematics (CRC Press, Boca Raton, 2015)
57. H.E. Rose, *A Course on Finite Groups* (Springer, London, 2009)
58. S. Ross, *A First Course in Probability*, 2nd edn. (Macmillan Co., New York; Collier Macmillan Ltd., London, 1984)
59. J.J. Rotman, *An Introduction to the Theory of Groups*, vol. 148, 4th edn. Graduate Texts in Mathematics (Springer, New York, 1995)
60. D. Schattschneider, The plane symmetry groups: their recognition and notation. Am. Math. Mon. **85**(6), 439–450 (1978)
61. D. Schattschneider, The 17 plane symmetry groups. Math. Gaz. **58**, 123–131 (1974)
62. W.R. Scott, *Group Theory*, 2nd edn. (Dover Publications Inc., New York, 1987)
63. S. Siksek, *Introduction to Abstract Algebra*, Mathematics Institute, University of Warwick, 2013
64. S. Singh, *Modern Algebra* (Vikas Publishing House PVT LTD, India, 1990)
65. M.K. Siu, Which Latin squares are Cayley tables? Am. Math. Mon. **98**(7), 625–627 (1991)
66. D.A. Suprunenko, *Matrix Groups*. Translated from the Russian. Translation ed. by K.A. Hirsch. Translations of Mathematical Monographs, vol. 45 (American Mathematical Society, Providence, 1976)
67. M. Suzuki, *Group Theory I*. Translated from the Japanese by the author. Grundlehren der Mathematischen Wissenschaften [Fundamental Principles of Mathematical Sciences], vol. 247 (Springer, Berlin, 1982)
68. T. Tsuzuku, *Finite Groups and Finite Geometries*. Translated from the Japanese by A. Sevenster, T. Okuyama [Tetsuro Okuyama]. Cambridge Tracts in Mathematics, vol. 70 (Cambridge University Press, Cambridge-New York, 1982)
69. R.N. Umble, *Transformational Plane Geometry* (Millersville University of Pennsylvania, USA, 2012)
70. O. Viro, *Isometries*, Lecture Notes (Stony Brook University, 2014)

Index

A

Abelian group, 61
Addition modulo n, 88
Adjoining element, 57
Affine geometry, 129
Algebra of sets, 2
Alternating group, 120
Antisymmetric relation, 11
Arithmetical function, 131
Ascending chain condition, 86
Associative binary operation, 51
Associativity law, 59
Automorphism, 265
Automorphism of a field, 271
Axial symmetry, 32
Axis of the symmetry, 32

B

Basic principle of counting, 14
Bezout's Lemma, 20
Bicycle semigroup/monoid, 82
Bijective function, 9
Binary operation, 49
Binomial Theorem, 15

C

Cancellative, 58
Canonical map, 257
Cardinality, 9
Cartesian product, 2
Cauchy's Theorem, 227
Cauchy's Theorem for abelian groups, 225

Cayley's Theorem, 262
Cayley table, 51
Center of a group, 79
Centralizer, 79
Chain, 11
Characteristic subgroup, 268
Closed under an operation, 49
Collineation, 130
Column equivalent, 153
Combinatorial analysis, 14
Common divisor, 20
Commutative binary operation, 51
Commutative diagram, 257
Commutator of two elements, 233
Commutator subgroup, 234
Compatible elements, 11
Complement, 2
Complete group, 277
Completely multiplicative function, 140
Complex number, 3
Composite number, 22
Composition of functions, 8
Congruence modulo n, 23
Congruence transformation, 41
Congruent of plane figures, 41
Conjugate, 70
Conjugate class, 70, 226
Constant function, 8
Cycle, 109
Cycle decomposition, 112
Cycle structure, 112
Cyclic group, 92
Cyclic subgroup, 92

B. Davvaz, *A First Course in Group Theory*,
https://doi.org/10.1007/978-981-16-6365-9

Printed in the United States
by Baker & Taylor Publisher Services